A SURVEY OF EMBRYOLOGY

A
SURVEY
OF
EMBRYOLOGY

Francis G. Gilchrist

PROFESSOR OF BIOLOGY, EMERITUS
LEWIS AND CLARK COLLEGE

McGRAW-HILL BOOK COMPANY

New York, St. Louis, San Francisco
Toronto, London, Sydney

A SURVEY OF EMBRYOLOGY

Library of Congress Catalog Card Number 67–28083

ISBN 07-023208-3

56789-MAMM-765

PREFACE

We hear today of an "old" biology and a "new." The former is mainly descriptive and deals with organs and organisms; the latter is analytical and explanatory and is concerned with cells and the molecules which compose them. This book combines the two viewpoints. Its core is traditional, descriptive vertebrate embryology, but it departs from the usual presentation of the subject in that it incorporates, in as natural a way as possible, many of the results of modern experiments on embryos. It seeks to present the new against the broad background of the "old."

The book is addressed to undergraduates. Some who use it as a text will be biology majors, some will be preprofessional students intent on medicine, dentistry, and allied vocations, but it is hoped that all who use it will do so because they are interested in how living things originate. After all, we do not fully understand anything until we know how it came to be. In biology this is a dual quest: It includes knowledge of the evolutionary origin of species and also an understanding of the development of the individual. Those who are interested in the long history of life on earth will surely be moved by the "evolution" which takes place in a matter of days or weeks every time a speck of living matter grows into a fully functioning being.

Since most of those who take undergraduate courses in embryology do not continue into advanced work in the subject, it has seemed important to present the material in a balanced manner. The author does not share the view that a biology student should take three successive courses in embryology: one descriptive, one experimental, and one biochemical. A broad perspective should be sought from the start. Specialization may come later. Certainly a student needs descriptive embryology if he is to understand what experimental embryology and molecular embryology are all about. Also, descriptive embryology is important to an understanding of anatomy. But to fail to introduce the beginning student to the discoveries and concepts of the experimentalists is to rob him of some of the most rewarding and exciting insights of recent research. The author is quite aware that in pursuing the goal of breadth he has had to forego depth. However, excellent texts are available for those who wish to delve more deeply. (See Sources of Embryological Information listed at the end of this book.)

The book is entitled *A Survey of Embryology,* yet it has been necessary to omit large areas of the science. Invertebrate embryology has been touched on lightly. The important field of insect development has been omitted. Metamorphosis, regeneration, and asexual reproduction have been left out almost entirely. The new insights of chemical embryology and developmental genetics are only lightly covered. However, in spite of all that has been omitted, the book covers more ground than is usual in a basic undergraduate course in embryology.

This is not a book to be memorized by rote. To attempt to do so would be self-defeating. Rather, the student's goal should be to gain an overview which will enable him to choose where he wishes to place his emphasis. In doing this, he should give free rein to his imagination and curiosity. Questions will arise which will drive him to search the literature or, better yet, to go to the embryo itself for an answer.

An introduction to embryology should be accompanied by laboratory work. To this end, the book is organized in such a way that it makes correlation between classwork and laboratory assignments rather easy. Many of the figures in the text are drawn from materials which are generally available for laboratory study. In the laboratory the student gains a mental picture of an embryo as an object developing in time and space. Some use of living material will transform what would otherwise be a series of dead anatomies into an unfolding picture. Some experimental work will bring the student into closer contact with life itself. However, nothing can take the place of patiently and thoughtfully examining complete series of prepared slides (whole mounts and sections) of embryos, while trying always to visualize each embryo as an object in three dimensions.

I began the project of writing a textbook of embryology for undergraduates in 1960-1961 while on leave of absence from Lewis and Clark College. Several months were spent at Chapel Hill, where I profited from the use of the excellent library of the Department of Zoology of the University of North Carolina. However, I did not find time to complete the project until I retired from teaching.

The figures are mostly my own work, although many of them are adaptations from the work of others. The influence of the texts by L. B. Arey and B. M. Patten is acknowledged. These books have been mainstays in my classwork for many years.

I acknowledge with gratitude the generosity of Dr. Donald P. Costello of the University of North Carolina who has read the entire manuscript and has made many valuable suggestions. Dr. George W. Nace of the University of Michigan has reviewed most of the chapters and has offered important constructive comment. Dr. Malcolm S. Steinberg has read the manuscript and has given many significant criticisms. The two chapters on the nervous system were read by Dr. Olaf Larsell of the University of Oregon Medical School before he passed away. His comments and encouragement were extremely helpful. Dr. Robert Bacon of the same school and Dr. Laurens Ruben of Reed College have helpfully commented on portions of the book. My wife, Pearl, has been of immeasurable aid with my syntax and orthography. For all the assistance which I have received, I am indeed grateful.

Francis G. Gilchrist

CONTENTS

3

**THE
CYTOPLASM
IN
DEVELOPMENT**

4

**GENERAL
FEATURES
OF
DEVELOPMENT**

5

THE EARLY DEVELOPMENT OF AMPHIBIANS

6

7

THE EARLY DEVELOPMENT OF THE CHICK

8

**THE
YOUNG
CHICK
EMBRYO**

12

13

14

18

THE
UROGENITAL
SYSTEM

chapter **1**

INTRODUCTION

One characteristic of living things is that they reproduce. This fact is so familiar that we are apt to lose sight of the wonder of it. But it is one of the astounding realities of nature; and it is the responsibility of embryologists to describe it and to try to explain how it is accomplished.

In a broad sense, embryology includes more than the study of the development of eggs. It belongs to the larger field of *developmental biology*. Some organisms grow from buds, or from fragments of parent organisms. Moreover, the regeneration of lost parts and the healing of wounds reawaken processes related to embryonic development and are therefore of concern to embryologists. Aging and rejuvenescence (if such there be) and cancer are aspects of development. In short, *embryology* is concerned with the origin of structure, in contrast to *physiology,* which studies the processes by which structure is maintained. The structure of living things, however, is not to be thought of as something static like that of a stick or stone. On the contrary, the structure of cells and tissues is sensitive and changing. It cannot be separated from function any more than function can exist apart from structure. It is maintained by a constant expenditure of energy. Hence the study of the origin of structure and function (morphogenesis) and the study of the maintenance of structure and function (homeostasis—an important aspect of physiology) cannot be wholly separated. When we study embryology, we see developmental processes in action which are never absent wherever and whenever life exists.

For many years embryologists were concerned with describing the stages of development: first as they saw it with their naked eyes; then with the help of magnifying lenses; and then with the aid of compound microscopes which magnify up to 1,000 times. Techniques were devised by which tissues could be hardened, sectioned, and stained for detailed study. As a result, today we have for some animals a rather full descriptive account of what happens at the tissue level and even at the cellular level when an organism develops. Other animals remain to be studied and many details need to be filled in, but the outlines are clear.

Mere description, however, does not explain. It does not tell us of the processes which are at work in time and space by which a new individual comes into being. To answer such questions, experiments have been undertaken on eggs and embryos. The eggs of certain marine invertebrates, as well as those of frogs, salamanders, and chicks, have proven especially useful. This work is progressing in many laboratories throughout the world at the present time. Early leads have sometimes proved false, but gradually the principles and processes which underlie development are coming to light.

In recent years, especially since the Second World War, electron microscopes have been devised which magnify as many as 200,000 times. These have provided new and exciting insights into cellular structure and activities. The observations have confirmed, for the most part, what the light microscope had already revealed, while at the same time extending our knowledge. Unfortunately, living material cannot be studied with the electron microscope. It must first be killed and usually sectioned before being photographed under the electron beam. By piecing together the information obtained with the light microscope about materials living and dead with what the electron microscope reveals, and by using all sorts of materials, a picture is being gained of what goes on within cells when development takes place.

Methods have been devised by which embryos have been broken down into separated cells and the behavior of the dissociated cells studied *in vitro*. Cells also have been broken down into their structural components (organelles), and the chemical performances of the organelles have been observed in suitable solutions. Thus, by one means or another, organisms have been "taken apart" and analyzed.

Many fundamental answers to biological problems have come from biochemical studies. The molecular changes which take place during development have been followed in various ways. For example, the cells of an embryo may be "homogenized" by grinding them with crushed glass, or stirring them with a blender, or subjecting them to high-frequency vibrations. Next the constituent chemicals of the protoplasm can be separated by taking advantage of their physical differences such as size (dialysis and filtration), density (centrifuging), adsorptive and solubility properties (chromatography), or electric charge (electrophoresis). Finally the fractions thus obtained can be analyzed chemically. In other studies the molecules are first "labeled" with radioactive isotopes so that their history within the cell or embryo can be traced. Again proteins and other antigens have been traced by their immunological properties. In this method sensitized warm-blooded animals such as rabbits and guinea pigs are used in the assay. Molecular biology, however, is still in its infancy, and, although numerous insights have been gained, the relation of the findings to morphogenesis is not always clear. Often what is learned seems to have more to do with the physiological processes of living than with the morphogenetic processes of developing.

Much is being learned today from the study of simple organisms, such as bacteria and viruses. It now appears that all living things, whether one-celled or many-celled, whether plant or animal, are composed of similar chemicals and carry on their life processes in basically the same physical and chemical manner. That which is learned from the study of bacteria applies to man. But the problem of differentiation in embryonic development transcends the life processes of those organisms which undergo little or no differentiation.

It does not follow from the exciting advances of molecular biology that the older and more strictly descriptive aspects of biology which dealt with organisms and populations have lost their importance. Analytical biology must learn from descriptive biology what questions need to be examined; and descriptive biology can be alerted by analytical biology to look for details which had been overlooked previously. Furthermore, we must keep the wide aspects of descriptive biology in mind if we are not to lose sight of organisms as organisms, the very thing which as biologists we are concerned to understand. Let me make the point clear. When we analyze a whole into its parts, unless we are careful we lose sight of the whole and of the relation of the parts to the whole. For example, if we examine the stones of a mosaic, we lose the art of the mosaic. It does not exist at the level of the individual stones. When a phonograph record is analyzed, we discover a wavy groove. We may measure the amplitude, frequency, and wave pattern of the groove, but the music is no longer there. A book reduced to its letters is not literature. In the same manner, a many-celled organism which has been dissected to its cells, or a cell which has been analyzed to its molecules, is no longer an organism or a cell. The object which we desired to understand has vanished.

The chapters which follow are built around descriptive embryology as a core. We shall be concerned with the changes in structure which take place when an organism develops. But we shall also weave in at appropriate places some accounts of the experiments—physical, chemical, and surgical—which have been performed in trying to explain these changes. Explanations, however, are never complete. There will always be mystery. The more that science understands, the more surprising it becomes that so many complexities should exist in the universe. Especially, it is beyond our present comprehension how so many processes should work together in the production of a living creature.

Let us turn to a brief review of the history of embryology as an explanatory science.

AN HISTORICAL ORIENTATION

Epigenesis and Preformation

How did the embryologists of the past account for the origin of a new living being? Aristotle (384–322 B.C.), the "Father of Zoology," described the development of the chick in the egg as he and others had seen it with their naked eyes. The parts of the embryo, he said, arise in succession: first the heart; then the blood; then the blood vessels which come from the heart; and, lastly, the various organs of the body which form around the blood vessels by processes of condensation and coagulation. Aristotle called this process of the superposition of structure upon structure during development *epigenesis.*

But how can organized complex structure arise out of simplicity? Aristotle believed that there is a sort of soul, a nonmaterial purposive principle, which guides development. He called it "entelechy." He identified it with the "form" of the embryo as distinguished from the matter which composed it. He supposed that the soul or form comes from the father at the time of procreation and that the mother supplies the matter or soil in which the embryo grows.

Aristotle's doctrines were accepted as authority through most of the medieval ages. They appear in the writings of Fabricius (1537–1619) and Harvey (1578–1657), who described development as "an epigenesis through the superaddition of parts." The determining cause, or soul, so he thought, is present in the blood.

With the advent of the microscope, an alternative doctrine concerning the nature of development gained ascendency. In 1673 Malpighi presented a paper before the Royal Society of London on "The Formation of the Chick in the Egg," in which he asserted that with his own eyes he had seen the chick in the unincubated egg and, hence, that development is not a process of generation, but rather one of growth. This theory, originally termed unfolding (evolution), is now referred to as the *preformation* doctrine.

The theory was soon accepted by all biologists of note and dominated biological thought for the next hundred years. Some held it on the basis of supposed observations. Thus Swammerdam, the early student of insect structure, reported that the parts of the butterfly are discernible in the chrysalis (eggs and pupae were not clearly distinguished at this time), and Buffon asserted, "I have opened a large number of eggs at different times, both before and after incubation, and I am convinced by my eyes that the chick exists in the center of the cicatricule at the moment the egg leaves the body of the hen." Others held the theory of preformation as a matter of philosophy or theoretic necessity.

It was in the spirit of the day to be logical at all costs. Thus arose the "emboitement" or encasement theory of Bonnet and Swammerdam. According to this, successive generations of individual organisms preexist one inside the other in the germ cells of the mother. It was estimated that as many as 200 million years of human beings were present, already delineated in the ovaries of Eve. Those who believed that the miniature organisms were present in the egg were known as ovists.

But two views of preformation are possible. In 1677 Leeuwenhoek (or an associate) saw sperm cells for the first time in the seminal fluid, and Hartsoeker drew a figure of what he interpreted to be a miniature manikin in the head of a sperm (Fig. 1–1). The sperm was thought of as the seed (whence the name "sperm"), and the egg was the soil in which the seed was planted. Those who held this view were known as animalculists or spermists. The ensuing controversy lasted for a hundred years.

Recovery from the excesses of crude preformation began with the publication by Kaspar Wolff in 1759 of his "Theoria Generationis." First philosophically (he was moved by Leibniz's current discoveries of a mathematics of change), and then by careful observation, he demonstrated that the complete chick does not exist preformed in the unincubated egg, but that on the contrary the organs form successively in an epigenetic manner. His observations were abundantly confirmed by the great embryologists of the early nineteenth century, notably Karl Ernst von Baer

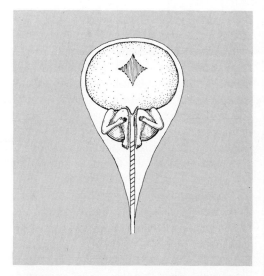

FIGURE 1–1   The human sperm cell as drawn by Hartsoeker, 1694, soon after the discovery of spermatozoa.

(1792–1876), to whom embryology owes a great deal. Today it is clear that development is epigenetic. There is no preformed organism in either the egg or sperm.

Mosaic versus Regulative Theories of Development

During the seventies and eighties of the nineteenth century, great progress was made in the techniques of hardening, sectioning, and staining tissues and in the knowledge of the structure and functioning of cells. The complex process of cell division by mitosis was worked out by Flemming, Strasburger, and others. It soon became evident that the chromosomes of the nucleus have an important part to play in the processes of heredity and development.

In 1883 August Weismann, impressed by the remarkable precision of mitosis, proposed the theory that early development involves the orderly unpacking of an embryo already pre-organized, if not actually preformed, in the chromosomes of the nucleus. He spoke of his theory as "the architecture of the germplasm." He postulated that there are units of heredity and development which he called "ids." The first cell divisions, he said, are differential; that is, they first separate the ids for the right side from those for the left, then the ids for the anterior end from the posterior, and so on for several divisions of the egg. After that, so he thought, interactions between parts and further differentiations make epigenetic development possible. Mendel's "hereditary factors"—we now call them genes—differ from Weismann's ids in that they do not segregate. Every cell during early development receives the complete set. Weismann at this time was not aware of Mendel's work.

In order to test Weismann's theory of the germplasm, Wilhelm Roux in 1888 performed a classic experiment which may be viewed as marking the beginning of the science of experimental embryology. He took a frog's egg

at the two-cell stage of cleavage and touched one of the two cells with a hot needle thus destroying the nucleus. He observed that the uninjured cell continued dividing and developed into what he interpreted to be a one-half blastula, a one-half gastrula, and ultimately a one-half embryo (Fig. 1–2). His work seemed to confirm the hypothesis that the early cell divisions of the egg are differential, a process of the orderly dispersal of the materials of the nucleus. Finally there results a ball of cells (blastula) in which each cell, so he supposed, possesses just its own potentialities for development. Roux sometimes observed a sort of reorganization of the embryo, but he dismissed this rather lightly as a secondary process of regeneration, or "post-generation," as he termed it. This general viewpoint of Weismann and Roux, namely, that parts self-differentiate, is known as the *mosaic theory* of development.

Three years after Roux's famous experiment on the frog's egg, another German scientist, Hans Driesch (1891), performed a somewhat similar experiment on sea urchin eggs. He put eggs at the two-cell stage in a vial of seawater and shook them sufficiently so that some of the cells broke apart. He then isolated the blastomeres in glass bowls and went to bed, expecting in the morning to see one-half blastulas and ultimately one-half larvae. But what was his surprise to discover that the one-half blastulas, which formed first, closed themselves into whole blastulas, and then became whole gastrulas and finally whole larvae of one-half size (Fig. 1–3)! He repeated the experiment with eggs at the 4-cell stage and even at the 8- and 16-cell stages. In some cases he obtained whole larvae of reduced size.

Driesch reasoned that, contrary to the

FIGURE 1–2 "Half embryos" of the frog which Roux obtained in 1888 by destroying the nucleus of one cell at the two-cell stage. (*From W. Roux, 1888, Virchow's Arch., 114:419–521.*)

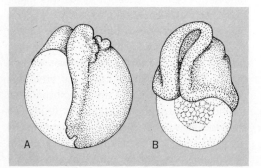

theories of Weismann and Roux, the early cleavages of the egg are equal (equational), "a quantitative division of homogeneous material." Hence at an early stage the blastomeres have equal potentialities. Their fate, he said, is determined by their position in the whole. Together they constitute an *"harmonious equipotential system."* Development such as Driesch observed in sea urchin eggs came to be called regulative development, and the eggs which were capable of performing thus were called regulative eggs. To account for the guidance of the course of development, Driesch resorted, as had Aristotle long before him, to the idea of a nonmaterial purposive principle, or entelechy. A vitalistic viewpoint such as Driesch's is, of course, a matter of faith; but so is the opposite viewpoint of mechanism.

Thus began a protracted controversy in biology between those who thought that mosaic development is primary and those who considered that regulative development is the basic method by which a new individual comes into existence. The controversy still has its repercussions.

Cell-lineage Studies

The controversy drove embryologists to their laboratories and to experiments on living eggs. It was soon discovered (first by Whitman on the leech, 1878) that many eggs cleave, that is, divide into cellular units, in so stereotyped a fashion that a "cell lineage" can be traced from cell generation to cell generation. Finally, it was discovered that a particular cell gives rise to a particular tissue of the embryo or larva.

In 1890 E. B. Wilson of Columbia University and E. G. Conklin, later of Princeton, were both working at Wood's Hole on Cape Cod. The former was working on the eggs of the marine

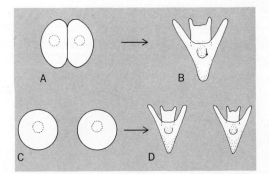

FIGURE 1–3 Driesch's experiment, 1891, in which he separated the blastomeres of sea urchin eggs at the two-cell stage by shaking them in vials of seawater. *A* and *B.* Controls—whole larva from unseparated blastomeres. *C* and *D.* Experimentals—two whole larvae of half size from separated blastomeres. (*Based on H. Driesch, 1891, Zeitschr. Wiss. Zool., 53:160–182.*)

annelid *Nereis,* the latter on those of the mollusk *Crepidula.* When they compared notes, they were excited to discover that these members of two entirely different phyla go through the same sequence of cell division and develop their several tissues from the same cell sources. The same symbols could be applied to designate the blastomeres (Fig. 1–4). Embryology had bridged a gap between two animal phyla which the study of adult structure had not spanned.

Cell-lineage studies became the vogue, and many who later became leaders in zoology earned their Ph.D. degrees by such investigations. For a time it appeared that the regularity of cleavage must have some important and necessary relation to normal development. However, there were notable exceptions, and it seemed necessary to classify eggs into those with determinate cleavage (mosaic eggs) and

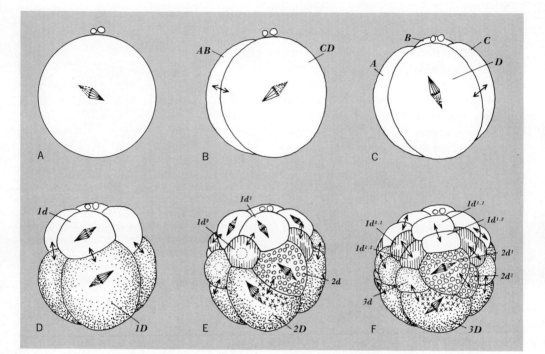

FIGURE 1–4 Diagram of cell lineage in the spirally cleaving eggs of primitive annelids and mollusks. Note that the mitotic cell divisions are alternately clockwise (dextral) and counterclockwise (sinistral). The conventional labeling of individual blastomeres is shown. In *E* and *F* the fates of blastomeres are indicated by symbols: anterior ectoderm, unmarked; posterior ectoderm, finely stippled; cells of the ciliated girdle, vertical lines; cells of the ventral plate which give rise to the nervous system, marked with circles; future mesoderm, marked with crosses; endoderm of the gut, coarsely stippled.

those with indeterminate cleavage (regulative eggs). (For a current viewpoint, see pp. 74, 75.)

Nuclear versus Cytoplasmic Theories of Development

Weismann elaborated his theories of development in terms of the nucleus, the central body which all cells possess. The early divisions of development, he said, are differential; that is, they are such that each body cell receives only that part of the nuclear "germplasm" which determines its fate. The cell divisions by which the future germ cells increase in number, so he said, are equational, for each germ cell retains the entire germplasm of the race. Weismann included this in a principle which he called "the continuity of the germplasm." The body "which bears and nourishes the germ cells is, in a certain sense, only the outgrowth of one of them." Weismann had trouble accounting for the facts of asexual reproduction and regeneration.

It was soon shown, however, that, contrary to the nuclear views of Weismann and Roux, it is the cytoplasm, namely, the protoplasm which is outside the nucleus, and not the nucleus which undergoes differentiation during early development. Driesch in 1892 compressed sea urchin eggs between glass plates and found that the direction of the cleavage divisions could be altered so that the 8- and 16-cell stages, instead of becoming balls, became flat plates. The distribution of the nuclei to the cells of the embryo had been changed, but the distribution of the cytoplasm was unchanged (Fig. 1–5). On release of the pressure, the plates of cells rounded up and developed into normal larvae. In 1896 Wilson performed similar experiments on the eggs of the annelid worm, *Nereis,* eggs with determinate cleavage, and often obtained larvae with only slight abnormalities.

In 1895 Morgan repeated Roux's experiment. He pricked one of the first two blastomeres of a frog's egg with a hot needle; but instead of leaving the egg undisturbed with the injured blastomere attached, he inverted the egg so that the cytoplasm of the uninjured blastomere flowed and rearranged itself. The uninjured half developed into an essentially normal whole embryo. Schultze found that if a frog's egg is inverted in the two-cell stage and held in the inverted position, a flow of cytoplasm takes place in each blastomere and a double-headed embryo may result. It must be concluded therefore, that it is the cytoplasm, and not the nucleus, which develops the pat-

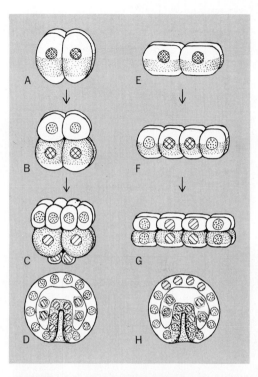

FIGURE 1–5 Driesch's experiment, 1892, in which he compressed sea urchin eggs between glass plates, thus altering the distribution of the nuclei to the blastomeres. The pattern of the cytoplasm remained unchanged. *A* to *D.* Controls. *E* to *H.* Experimentals. On release of pressure, the blastomeres rounded up and developed normally. *(Adapted from B. Dürken, 1928, "Lehrbuch der Experimentalzoologie," p. 332.)*

tern of the embryo. For a time it seemed that the nucleus played no part in development!

Organ-forming Stuffs

Strong support for the view that the cytoplasm, rather than the nucleus, is the basis of

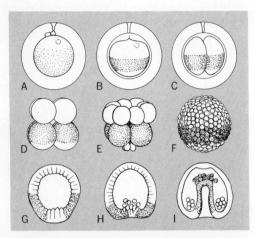

FIGURE 1–6 The early development of the sea urchin *Paracentrotus* as described by Boveri, 1901. A. Unfertilized egg. B. Egg after fertilization. C to F. Cleavage stages. G to I. Sections of blastula and gastrula. (*Adapted from W. Schleip, 1929, "Die Determination der Primitiventwicklung," pp. 384 and 390.*)

embryonic differentiation came from observations on the movements and final distribution of visible "stuffs" within the cytoplasm of various eggs. Boveri (1901) found that in the egg of the sea urchin, *Paracentrotus,* fertilization is followed by a flow within the cytoplasm. One result of this is that a broad superficial band of reddish-purple pigment collects just below the equator. (Most eggs have two opposite, unlike poles known as the animal and vegetal poles; the equator is midway between them.) Thus this egg is divisible into three regions: the animal hemisphere, which is pigment-free, and which becomes the outside cell layer (ectoderm) of the embryo; the subequatorial girdle with its superficial pigment, which becomes the gut and its derivatives (endoderm and mesoderm); and the small pigment-free area surrounding the vegetal pole, which gives rise

to loose mesenchyme cells. These migrate within the embryo and weave the larval skeleton (Fig. 1–6).

Conklin described five visibly different "organ-forming stuffs" in the eggs of the tunicate *Styela* (*Cynthia*) (Fig. 1–7). In the unfertilized egg these are not fully segregated, but after the entrance of the sperm, an active streaming takes place. The materials become arranged in a bilateral pattern which foreshadows the bilateral symmetry of the embryo.

This process of determining the fates of regions of an egg by events which take place within the cytoplasm has been termed *localization* (the restriction of developmental capacities to particular regions) or *segregation* (the separations of regions of unlike capacity as by cell boundaries). Experiments seemed to confirm the theory, for fragments of sea urchin and other eggs which possessed each of the visibly different regions were able to develop normally, while fragments lacking one or more of the stuffs commonly produced defective larvae.

At first it was thought that the pigment itself is an organ-forming stuff (morphogenetic substance). Then in 1908 Morgan subjected the eggs of another sea urchin, *Arbacia,* to a strong centrifugal force and discovered that the pigment could be displaced from its natural position to a new location in the egg (Fig. 1–8). Nevertheless, the manipulated eggs cleaved in accord with their original axiate pattern and developed quite normally, except that the pigment was now located in another part of the larva. These and similar observations on other eggs made it necessary to revise the original theory: It is not the visible stuffs (which may be wastes or by-products), but the invisible cytoplasmic substrate, or "ground substance," which is the basis of embryonic dif-

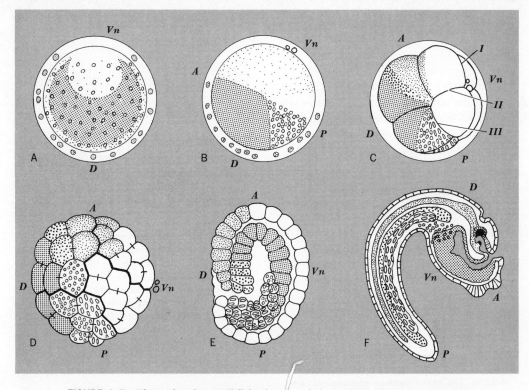

FIGURE 1–7 "Organ-forming stuffs" in the egg of the tunicate, *Styela*, as described by Conklin, 1905. From the right side. Vn, ventral; A, anterior; D, dorsal; P, posterior. *A. Unfertilized egg.* The large oocyte nucleus ("germinal vesicle") is filled with clear nuclear sap. It is close to the animal pole. Clear cytoplasm containing yellow granules (small circles) forms an outer layer surrounding a central mass of gray yolk. *B. Fertilized egg* just before the first cleavage begins. Polar bodies have been given off. A rearrangement of "stuffs" has taken place, including a gathering of the yellow yolk granules to form a yellow crescent on the future posterior side of the egg. *C. Eight-cell stage.* The first cleavage plane (I) divides the egg into symmetric right and left blastomeres. The second cleavage plane (II) divides anterior blastomeres from posterior blastomeres. The third cleavage plane (III) separates a ventral quartet of blastomeres around the animal pole from a dorsal quartet surrounding the vegetal pole. At this stage six regions can be distinguished. These are represented schematically thus: Unshaded—clear plasma, future epidermis; lightly stippled—bright gray plasma, prospective nerve cord; coarsely stippled—less brilliant gray, notochord; darkly stippled—slate gray yolk, endoderm; small circles—bright yellow, prospective mesenchyme; ovals—dark yellow, future muscle cells. *D. Sixty-four-cell stage.* The heavy lines mark the second and third cleavage planes. The short transverse lines connect daughter cells of the sixth cleavage. A slight flattening on the dorsal side indicates that gastrulation is about to begin. *E. Optical section through a gastrula.* Future neural and epidermal cells remain on the outside (ectoderm). Endoderm, notochord, mesenchyme, and muscle are inside. *F. Late embryo* showing the fates of the several formative "stuffs." *(A to E modified from E. G. Conklin, 1905, J. Acad. Sci. Philadelphia, 13:1–119, Figs. 1, 11, 30, 123, and 157.)*

ferentiation. Jacques Loeb expressed this viewpoint well when he referred to the cytoplasm of the uncleaved egg as "an embryo-in-the-rough." Recent work has emphasized the role of the outer cytoplasmic layer or *cortex* in embryonic differentiations (see pp. 50–52).

Embryonic Inductions

From the time of the first experiments concerning development, it has been evident that there are interactions between the parts of the developing embryo. Interactions assure that the parts will develop in their proper positions. Interactions also regulate their relative sizes. This is true even when the parts have been experimentally disarranged.

FIGURE 1–8 Morgan and Spooner, 1909, centrifuged uncleaved eggs of the sea urchin *Arbacia*. The heavy pigment granules (shown by stippling) were driven toward the centrifugal pole. Except for this dislocation of pigment, development continued normally. *(After T. H. Morgan and G. B. Spooner, 1909, Arch. Entwicklungsmech, 28:104f.)*

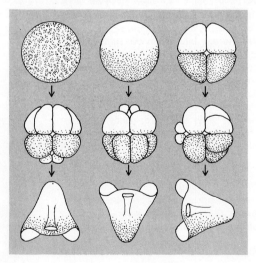

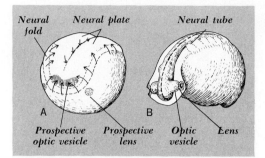

FIGURE 1–9 The development of the eye in amphibians. **A.** Open neural plate stage. The location of the prospective optic vesicles (future retinas) is shown by vertical lines, that of the prospective lenses by stippling. **B.** As a result of the folding of the neural plate to form the neural tube, the bulging optic vesicles come into contact with the epidermis and induce the production of lenses.

A classic example of interaction, or at least of action and reaction, was the discovery by Spemann in 1901 of the development of the lens of the vertebrate eye. The retina and lens come from separate embryonic sources: the retina from the sides of the forebrain, which in turn comes from the neural plate; the lens from head epidermis. When the neural plate folds to form the brain, the epidermis comes into contact with the bulges of brain tissue which are destined to form the optic cups (future retinas) (Fig. 1–9). What would happen if these two tissues should fail to come into contact with each other? Spemann and others studied this problem in amphibians by surgical procedures. They found that the epidermis of the future lens region possesses a competence to form a lens; but if no contact is made, a normal lens does not usually form. The optic cup is said to induce or evoke the epidermis to form a lens.

The area which possesses the competence

to form a lens is not sharply defined at first. Indeed, in some species epidermis from as far away as the sides of the trunk has some capacity to respond to the stimulus of the optic cup and form a lens. This has been shown by grafting an optic cup beneath flank epidermis and also by grafting flank epidermis to the head in place of head epidermis. Thus the competence of the epidermis and the inductive stimulus arising in the optic cup constitute a reaction system which assures that the lens will form just where it should, and that a unified eye will result. In the chapters which follow, we shall note many other examples of embryonic induction.

Organismic Viewpoints

The goal of experimental embryology has always been to give a causomechanical account of the processes of development. The earlier theories were predeterministic. Although they did not always postulate a "little man" in each germ cell, as did the theories of the early microscopists, yet they proposed an "architecture of the germplasm," or a precise sequence and pattern of cell divisions, or a cytoplasm that is an "embryo-in-the-rough." Even Driesch's nonmaterial, vitalistic entelechy was a purposive, predetermining principle. But all these older theories failed to come to grips with the first and basic problem of development; namely, whence comes organization in the first place? How does a new individual get its start? The trouble with the older theories was that they did not begin at the beginning. They begged the basic question of development by assuming that the embryo is represented in some manner in the germ. Some students of heredity still seem to talk this way, but as we soon shall see, heredity

is a very different thing from predetermination.

A completely new insight into this most fundamental of all embryological problems came indirectly through studies on the regeneration of lower animals. Regeneration has always proved a stumbling block to biological theorizing, especially to theories of the mechanistic type. If a part can form a whole—almost any part in the case of some of the lower organisms—then development must be completely epigenetic; the new individual cannot be predetermined in the part. And here vitalistic assumptions do not help us at all, for they are also doctrines of predetermination.

Roux (1885) tried strenuously to give a causomechanical account of development on an epigenetic basis, but he gave it up in favor of his mosaic theory. Morgan struggled with the problems of regeneration for years before he turned his attention to genetics in the hope that genes would supply the clue. Loeb clearly recognized that regeneration was the crucial problem in his attempt to give a mechanistic account of life.

A major breakthrough came as a result of the work of H. V. Wilson and C. M. Child. Wilson (1907) found that when sponges are pressed through bolting cloth into a dish of seawater, the individual cells become dissociated from one another and settle to the bottom. There they adhere together in clumps and shortly reorganize new and unified sponges. Similarly Child found that when he fragmented a large hydroid polyp (*Corymorpha*) by grinding it lightly in seawater with sharp ground glass, the isolated cells which still remained alive settled to the bottom, and there aggregated and reconstituted new individuals (Fig. 1–10). Here, beyond equivocation, were heaps of cells establishing a new pattern of wholeness. This

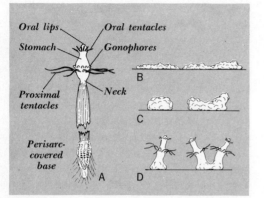

FIGURE 1–10 Reconstitution of a hydroid polyp as described by Child, 1928. A. Polyp of *Corymorpha* showing zones of the hydranth. B. Dissociated cells adhere. C. They draw together into clusters. D. They differentiate into new polyps.

is, indeed, epigenesis from a state near chaos. But of course it is not chaos, for the cells which are capable of such regeneration are organized living things which contain the capacities of their species. Moreover, there are different kinds of cells; and each cell, when it comes into contact with another cell, has the tendency to adhere to it and then to sort out with other cells of its own kind (page 54). Facts of this sort help explain how, out of a disordered mass, a new individual takes form.

On the basis of numerous experiments of many kinds, Child developed his theory of "axial gradients." He proposed that the first step in regeneration, as in all development, is the establishment within the developing cell, or group of cells, of some sort of quantitative differential from one region to the next. Between adjacent levels of the gradient, so he supposed, there is a relationship of "physiological dominance" and "physiological subordination." During regeneration (or development)

the dominant region, having become free from the control of adjacent regions, does what is its nature to do. In the case of the hydroid polyp it becomes oral lips. At the same time, under the influence of the dominant region, adjacent regions become in sequence other things. In the case of the marine polyp, *Corymorpha*, the subordinate regions produce in succession oral tentacles, the stomach, basal tentacles, the neck, etc.

Reproduction, then, in Child's view, starts with isolation—actual isolation in the case of development from fragments or eggs, but "physiological isolation" in the case of a budding polyp. That is to say, a region buds and develops a new whole when it escapes from the physiological dominance of its parent's body. The principal criticism of Child's theories has been that, in spite of all the evidence he accumulated concerning the many kinds of gradients in many organisms, he failed to discover a basic gradient of which all other gradients are but effects and expressions. But even if a basic physical or chemical gradient were to be identified—and today we may not be far from doing so—this in itself would not explain dominance and subordination. Such words do not explain anything. They serve to pinpoint what needs to be explained.

A concept which is similar in some ways to axial gradients is that of an embryonic field. This is, after all, just a word to remind us that patterns of a quantitative and geometric nature exist within the developing embryo, and that, in relation to these patterns, differentiations take place in an orderly manner. A field is not a mosaic work in which each part possesses its own independent capacity for self-differentiation. Rather, it is a whole in which each part differentiates according to its spatial (and temporal) relations to other parts. A field is something more than embryonic induction, or an interaction between parts, for it precedes the parts, and it is in relation to the field that the parts become different. Some prefer to speak of a field as a "pre-pattern." This adds nothing to our understanding.

Organismic views such as these are truly epigenetic. They do not assume a pattern or organization to begin with. Moreover, they are not deterministic, as are mechanistic and vitalistic hypotheses. They allow the regenerating part, or the developing embryo, to find its own way as it goes. It may develop as one whole; but if there is some interference with the gradient, or some distortion of the field, it may equally well give rise to an abnormal creature, or to a duplication of parts, or to twins.

The Nature of Differentiation

What is embryonic differentiation? It is, of course, the process by which a region or cell of a developing egg or embryo becomes demonstrably different from other regions or cells. It becomes a *rudiment* or *primordium* (German: *Anlage*). These terms are synonyms. But before visible differentiation has begun, some invisible process or event must have taken place.

In the older literature of experimental embryology this invisible process was described in negative and conceptual terms as a "segregation" or "localization" of the potentialities of development. Each region was thought of as being at first "totipotent," that is, as having the capacity of becoming any part of the embryo. Then, it was supposed that the potentialities became more and more restricted until finally no potentiality remained, except the potentiality to become the one thing which it was the region's fate to become.

This line of reasoning, however logical it may be, has proved sterile. Today, differentiation is described in positive and concrete terms as a straightforward coming-into-being of molecules which did not exist at first. Generally speaking, these molecules are macromolecules (nucleic acids, proteins, etc.). Differentiation, then, is the controlled process by which regions or cells of like hereditary capacities become different in their biochemical composition. The macromolecules include enzymes at the surfaces of which the chemical reactions of the cell take place.

The result of differentiation is *determination,* the decision as to what a given region will become. Determination is a progressive process. At first, it is "labile" and reversible; but with the passing of time it becomes increasingly stable and irreversible. Finally, the region or cell, having become fully determined, gives rise to the structure which it has acquired the power to become.

The Role of Heredity

Experiments on embryos have thus established that (1) development is epigenetic; (2) although it is regulated by the nucleus, it takes place primarily in the cytoplasm; (3) it involves interaction between parts; (4) the parts arise within gradient patterns or "fields"; (5) differentiation is in essence the development of the macromolecular pattern (including the distribution of enzymes) within the cell. (Much of the differentiation takes place before cleavage in mosaic eggs but after cleavage in so-called regulative eggs.) However, a basic question still eludes us: What supplies the *possibilities* of development? Why does a human egg, given a normal environment and opportunity, become a man and not a mouse? The answer, of course, is *heredity,* that precious something which is passed on to the offspring by its parents.

What is heredity? One of the triumphs of modern science is the discovery that what is passed on by heredity consists of an assortment of huge molecules—macromolecules—which are contained notably, but not exclusively, in the nucleus of every cell, and which are transmitted from one cell generation to the next. These essential macromolecules include proteins which, in their role as enzymes, control the chemical reactions of the cell. They also include nucleic acids. Some nucleic acids (DNA) are the genes which serve as patterns or templates for their own replication and for the production of other nucleic acids (RNA). The latter engage in the manufacture of the chains of amino acids (polypeptide chains) from which the proteins are formed. The nucleic acids pass on "information" in the form of an alphabetic "code" of four nucleotides. The nucleotides are arranged in a linear fashion like beads on a string. The sequence in which these nucleotides follow one another is significant, for the sequence of the nucleotides in the nucleic acids determines the sequence of the amino acids which make up the polypeptide chain (pp. 44, 45).

Note that the modern understanding of heredity is not a doctrine of preformation. It is not even a theory of predetermination, for an egg is subject to accidents and to the influences of its environment as it develops. What the egg is to become is not settled in advance. Heredity supplies only the *possibilities* of what it may become. Hereditary materials are blueprints, as it were, not of the individual that is to be, but of the developmental process which, under the influence of the circumstances which surround it, will bring the new individual into being.

In our enthusiasm for our new insights into the chemical basis of heredity, we must not lose sight of the fact that the organism as a whole nurtures and protects hereditary material. It is the entire organism which survives in the "struggle for existence." It is not the organism in any static sense which carries on. Rather, it is the entire *life cycle* as a system in both time and space—egg, embryo, larva, juvenile, adult—that endures. Every stage must be adapted to survive, or the race will come to an end. The proteins and nucleic acids are the thread of continuity which runs through the life cycle. These are the materials which are transmitted from one generation to the next.

Every life cycle is the product of a long evolutionary history and can be understood only when its past is taken into account. Just as the art of a mosaic, the music of a phonograph record, or the literature of a book is an achievement, so every life cycle is an achievement. It is the product of the creativity of nature, as truly as a painting or a symphony is the product of the creativity of a human mind. This is true whether we account for creativity in terms of time and chance (that is, by mutation and selection) or by the hypothesis of a Creator. The production of even the simplest life cycle has required untold millions of years. The macromolecules which are peculiarly of biological importance are those which have contributed to the continuation of living things on earth. They are "vital," not in the sense that they possess some nonmaterial principle or entelechy, but because they are the blueprints of successful living. As such they possess a significance which purely chemical systems do not possess.

Herein lies the distinction which sets biology apart from the physical sciences, and which we are apt to forget when we analyze an organism in terms of the elements which compose it and the physical processes by which it lives. The characteristics of chemical molecules such as salt, sugar, or a plastic can be predicted on the basis of the properties and arrangement of the atoms which compose them. But nucleic acids and proteins have properties (based on the sequence of nucleotides or amino acids) which can be accounted for only in terms of the long history of life on earth. Indeed, such words as "history," "heredity," "information," "function," and "survival" have no real meaning in physics and chemistry. They apply to self-reproducing systems which have had a past, and which, if all goes well, may be expected to have a future. Apart from a particular living system, a specific protein or nucleic acid would not even exist; much less would it have a "reason for being." These are matters which tend to be overlooked in discussions of the physics and chemistry of life.

How far embryology will ultimately go toward providing an account of the processes of development, only time will show. There is a sense in which it cannot be expected to go all the way, for the final outcome of development is a conscious, striving being. We know that this is true, each of us with respect to himself. We infer that it is true also of other men and, in varying degrees, of other organisms. But these are propositions which transcend science.

SELECTED READINGS

Oppenheimer, Jane M., 1955. Problems, concepts, and their history. In Willier, Weiss, and Hamburger (eds.), Analysis of Development. W. B. Saunders Company, Philadelphia. Pp. 1–24. —A scholarly review.

Gabriel, M. L., and S. Fogel (eds.), 1955. Great Experiments in Biology. Prentice-Hall, Inc., Englewood Cliffs, N.J. —Reprints historically important papers by Redi (1688), Spallanzani (1785), Newport (1854), Weismann (1889), Loeb (1899), Castle and Phillips (1909), Driesch (1891), Spemann (1928), and Mangold (1931).

Willier, B. H., and J. Oppenheimer (eds.), 1964. Foundations of Experimental Embryology. Prentice-Hall, Inc., Englewood Cliffs, N.J. —Includes classic papers by Roux (1888), Driesch (1892), Wilson (1898), Boveri (1902), Child (1914), and others to which reference will be made later.

THE
NUCLEUS
IN
DEVELOPMENT

Aliving cell is a partnership between the nucleus and the cytoplasm (Fig. 3–1). Now, although this concept has its limitations, it will serve as a basis on which to divide our subject matter. This chapter concerns the nucleus. The remainder of the book deals mainly with the cytoplasm.

The nucleus, with its store of hereditary factors, the genes, is the conservative member of the partnership. It controls the syntheses which go on in the cell. Hence it directs that most amazing of all syntheses, the course of development. Hence also it determines the species and individual characteristics which are the outcome of development. On the other hand, it is primarily the cytoplasm which develops and is the progressive member of the partnership. It begins in relative simplicity and then, by differentiation and molding, becomes the mature organism in all of its structural and functional complexity.

Both the nucleus and the cytoplasm are needed for the continued life and functioning of the cell. Many experiments have been performed in which the nucleus of a cell has been removed. For example, when an amoeba is cut into two parts, the part containing the nucleus heals and continues to live. The part which lacks a nucleus may carry on for a time, but ultimately it wears itself out and perishes, for it cannot continue to rebuild its substance. It is reported that a sea urchin egg from which the nucleus has been removed may cleave several times, but it will not continue to develop.

It is even more obvious that a nucleus cannot live and carry on its functions without the cytoplasm. In particular, the nucleus depends upon the cytoplasm for its raw material and energy. It is probable that only in the cytoplasm do the oxidations take place by which energy-rich molecules (ATP) are synthesized.

THE LIFE CYCLE OF THE NUCLEUS

The story of the nucleus as it is revealed by the microscope is a repetitive story, yet one that is full of interest. It begins at fertilization when the chromosomes of the two germ cells (gametes), an egg and a sperm, come together in the nucleus of the fertilized egg or zygote. The story ends and is ready to start again, at least as far as the contribution of the individual to the race is concerned, when the new individual has matured and has produced gametes of its own. Between fertilization and the maturing of the germ cells, the story of the nucleus is one of alternating growth and division by mitosis.

Fertilization

From the standpoint of the nucleus, fertilization is the coming together of the chromosomes of the egg and the chromosomes of the sperm in the same nucleus, the zygotic nucleus (Fig. 4–5, A to E). One set of chromosomes, the haploid number, is present in the egg. Another set is present in the sperm. This makes two sets, the diploid number, present in the zygote. In man the haploid number is 23. Hence the human diploid number is 46 (Fig. 2–1).

Each chromosome of a haploid set has its own individuality; that is, it is of its own kind. It becomes visible at the beginning of cell division and disappears at the close of cell division in its own characteristic manner. Very significantly, its genes are arranged within it in a definite sequence. Moreover, it affects development in its own specific way. Each chromosome of a diploid set usually has a mate. Yet, although the two mates are within the same nucleus, with rare exceptions each remains independent of the other throughout the cell divisions (mitoses) of development.

Mitosis

The coming together of the male and female nuclei in fertilization is followed by *mitosis,* the process by which one cell becomes two cells in so complicated but precise a manner that each daughter cell possesses just the same two sets of chromosomes that the mother cell possessed (Figs. 2–2 and 4–5, *E* to *I.* Several phases of the process are now described.

Interphase Let us begin our account with the interphase, that is, with a cell that is not dividing. Such a cell has commonly been called a "resting cell." Actually it is a busy cell doing everything which a cell does except divide. It is during the interphase that the genes self-duplicate and carry on their function of supervising syntheses.

Following self-duplication, the chromosomes continue to be stretched out (Figs. 2–2, *A,* and 2–3, *A*). Usually they are so attenuated that their boundaries cannot be seen with the light microscope. The extended chromosomes are able to carry on their chemical functions in an efficient manner.

"Nuclear sap" or karyoplasm fills the interstices between the chromosomes. One or more rounded bodies, the *nucleoli,* are usually present in the interphase nucleus. The visible boundary between the nucleus and the surrounding cytoplasm is the *nuclear membrane.* In the cytoplasm adjacent to the nucleus there is a body, the central body, which consists of two granules or, after each granule has replicated, of two pairs of granules, the *centrioles* (Fig. 2–2).

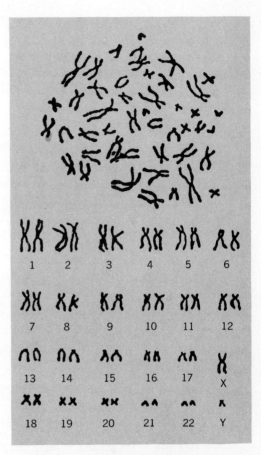

FIGURE 2–1 The 46 chromosomes of man, seen in polar view at the metaphase of a somatic mitosis. (Adapted from E. D. P. De Robertis, W. W. Nowinski, and F. A. Saez, 1960, General Cytology, W. B. Saunders Company, Philadelphia, 3d ed., from Ford et al.)

Prophase The events in the nucleus and in the cytoplasm which lead to the division of the cell constitute the prophase (Fig. 2–2, *B* and *C*). The following is a generalized account:

1 In the cytoplasm the pairs of centrioles move apart, and a sphere of gel rays, the *aster,* grows around each pair. Together the two

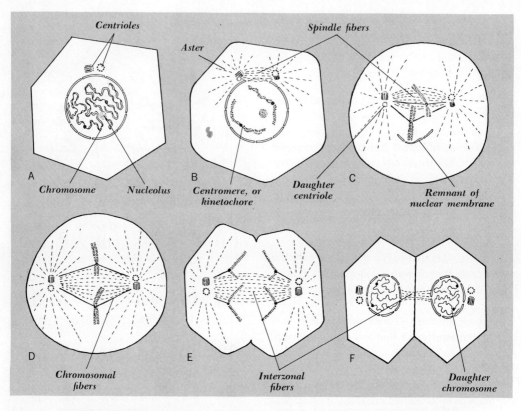

FIGURE 2–2 Diagram of mitosis showing one pair of chromosomes. *A.* Interphase after replication. *B.* Early prophase. *C.* Late prophase. *D.* Metaphase. *E.* Anaphase. (The telophase is not shown.) *F.* Early interphase before the chromatids, or "daughter chromosomes" have replicated. The centrioles have been drawn larger in order to show their structure. Replication of the centrioles is shown here beginning in the early prophase, but in some cases it begins as early as the preceding anaphase.

asters constitute the *amphiaster*. It is of interest that the centrioles are short, cylindrical bodies, each consisting of nine rods (or double or triple rods). In this they are structurally comparable to a cilium or flagellum. (They lack the two central tubules of a cilium.) In a typical case, the centrioles of a pair lie at right angles to each other and to the axis through the pairs.

2 The chromosomes condense into tight spiral coils (Fig. 2–3, *B*). Under the light microscope these appear to be rods. But when the material is suitably fixed and stained, each chromosome can be seen to be a double structure, each half of which is a chromatid.

3 A fusiform body, the *mitotic spindle,* now forms between the pairs of centrioles, sometimes in the cytoplasm, sometimes in the substance of the nucleus. The spindle's poles are anchored in the region of the centrioles of the cytoplasm. The electron microscope shows that it consists of hollow fibrils (microtubules). It elongates as the centrioles move apart. Possibly this is a result of imbibition of water, or incorporation of other molecules, or change in the shape of its constituent proteins. The amphiaster and the spindle together form the "achromatic figure," so-called because it does not stain deeply with basic dyes.

FIGURE 2–3 **The behavior of chromosomes in mitosis and meiosis. *A* to *F*. Mitosis, illustrated by a single chromosome: *A*, interphase after replication has taken place; *B*, prophase; *C*, metaphase; *D*, anaphase; *E*, telophase; *F*, interphase before replication has occurred. *G* to *K*. First meiotic division, illustrated by a homologous pair of chromosomes: *G*, leptotene stage; *H*, synaptene (zygotene) stage; *I*, pachytene stage; *J*, diplotene (at which time "crossing over" between adjacent segments of homologous chromosomes may take place) followed by diakinesis; *K*, anaphase. *L* to *O*. Second meiotic division, illustrated by one chromosome: *L*, metaphase; *M*, anaphase; *N*, telophase; *O*, interphase.**

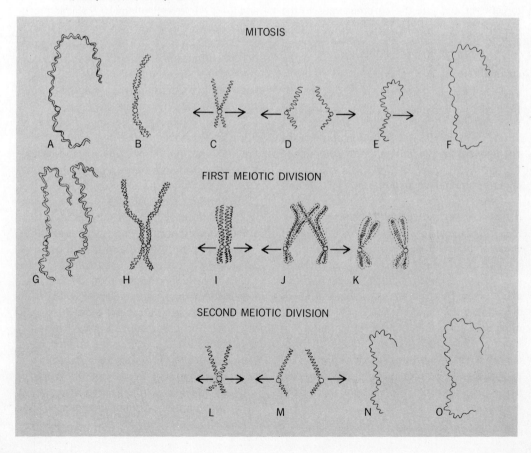

4 According to some accounts, chromosomal fibers grow out from particular locations in each chromosome until they reach the poles of the spindle. The locations from which they grow out are known as *centromeres* or kinetochores. The chromosomal fibers seem to consist of a contractile fibrous gel.

5 Toward the end of the prophase, the nuclear membrane disappears.

6 The nucleolus also disintegrates, and its substance, along with the nuclear sap, passes into the cytoplasm.

Metaphase The chromosomal fibers appear to contract and pull the chromosomes to the equator of the spindle (prometaphase) (Fig. 2–2, *C* and *D*). When thus spread out and seen in polar view, their number can be counted (Fig. 2–1).

Anaphase Continuing, or resuming, what appears to be contraction, the chromosomal fibers seem to drag each chromatid toward one pole of the spindle (Fig. 2–2, *E*). Actually two processes may be at work: a continuing enlargement and elongation of the spindle and a shortening of the chromosomal fibers. Since the fibers shorten without becoming thicker, the process probably involves the removal of water or other molecules from the fibers. As the chromatids move apart, they are often V- or J-shaped depending on the location of the centromeres to which the chromosomal fibers are attached. A band of "interzonal fibers" is often seen for a time after the separation has been accomplished, connecting the chromosomes which have pulled apart, and often including a remnant of the spindle. (The accompanying division of the cytoplasm will be described in Chapter 4.)

Telophase Having reached the poles of the spindle, the chromatids swell and disappear from view, as seen with the light microscope (Fig. 2–3, *E*). Although usually called "daughter chromosomes," they are actually single chromatids. The nuclear membrane reappears, and shortly one or more nucleoli re-form. Of the mitotic apparatus, little more than the centrioles of the cytoplasm remain.

Interphase The daughter cells have now returned to the interphase, or "resting stage," except for one feature: each chromosome still consists of only one chromatid (Figs. 2–2, *F* and 2–3, *F*). Investigations by a staining technique known as the Feulgen reaction (specific for DNA) have made it clear that the restoration of chromosomal material takes place during the interphase. At that time each chromatid becomes two chromatids and the double nature of each chromosome is restored.

Studies employing radioactive isotopes as tracers have added some important information. It is not the chromatids which replicate. It is half-chromatids, also termed chromonemata, which undergo self-duplication. Taylor (1957) placed bean seedlings in a solution containing radioactive (tritiated) thymidine (a substance which the cell uses to synthesize DNA). He left them in this solution until those cells which were in early interphase had each replicated once. He then cut the roots off and transferred them to a nonradioactive solution. In this they replicated a second time. Then, having killed the cells, he determined by autoradiography whether or not the chromosomes were radioactive. Now, if whole chromatids had replicated when they were in the first or radioactive thymidine solution, they would have made for themselves radioactive mates. After cell division half of the daughter chromosomes would have been radioactive, and half, namely, the original chromatids, would have been non-radioactive. But Taylor found that *all* the daughter chromosomes were radioactive! Why? The answer is that half-chromatids replicated, so that each chromatid now consisted of a new radioactive half-chromatid and an original non-radioactive half-chromatid.

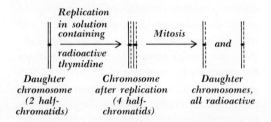

| Daughter chromosome (2 half-chromatids) | Chromosome after replication (4 half-chromatids) | Daughter chromosomes, all radioactive |

At the second replication, this time in a non-radioactive solution, each half-chromatid, whether radioactive or nonradioactive, made for itself a nonradioactive mate. Therefore, following the next cell division, the daughter chromosomes (chromatids) were half radioactive and half nonradioactive.

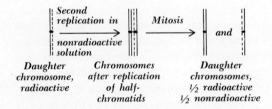

| Daughter chromosome, radioactive | Chromosomes after replication of half-chromatids | Daughter chromosomes, ½ radioactive ½ nonradioactive |

Similar results were obtained by Prescott and Bender, using cells of hamster tissues raised in tissue culture and also human leucocytes.

Experiments of this sort emphasize the remarkable stability of the half-chromatids. Although most of the macromolecules within the cell are in a state of flux, continually breaking down and being replaced, the half-chromatids are passed on intact from one cell generation to the next. And this has been going on for countless generations!

There is evidence that the centrioles may self-duplicate even earlier than the chromosomes, in some cases as early as the preceding anaphase (Fig. 2-2, *E*). At this time each centriole (there are two at each pole) makes for itself a partner, not usually by longitudinal splitting, however, as might be expected, but apparently by a process of lateral budding.

The introduction of mitosis in the history of life on the earth was one of the grand innovations of all time. Possibly it took place only once, for, with variations, it occurs in all plants and animals except bacteria and blue-green algae. Even in these there must be a process akin to mitosis, for without some such process the transmission of hereditary characters to each new generation in a balanced fashion would not be possible.

Meiosis

In 1884, Van Beneden described the processes of maturation, fertilization, and cleavage as they take place in the parasitic roundworm of the horse, *Ascaris megalocephala*. He demonstrated that the egg and sperm contribute an equal number of chromosomes to the offspring. This led, a few years later, to the discovery of *meiosis,* the process by which the number of chromosomes is reduced to one-half when the germ cells ripen (Fig. 2-4).

Each body cell and each unripe germ cell or *gonium* has two sets of chromosomes, the *diploid number*. Each of the chromosomes has a mate (except in the case of the unpaired sex chromosomes). When meiosis takes place, an oogonium or spermatogonium, as the case

FIGURE 2–4 Spermatogenesis and oogenesis. The black ovals represent paternal chromosomes (those derived originally from the sperm). The white ovals stand for maternal chromosomes (derived from the original egg).

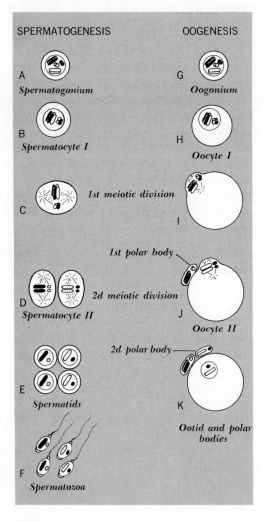

SPERMATOGENESIS   OOGENESIS

A
Spermatogonium

G
Oogonium

B
Spermatocyte I

H
Oocyte I

C
1st meiotic division

I

D
Spermatocyte II

1st polar body

2d meiotic division

J
Oocyte II

E
Spermatids

2d polar body

K
Ootid and polar bodies

F
Spermatozoa

may be, gives away one set of its two sets of chromosomes; that is, it gives away one chromosome of each pair of chromosomes. The result is that each ripe egg or sperm has left only one set of chromosomes, the *haploid number*.

Now, meiosis, unlike mitosis (1) takes place only in the gonads, (2) occurs only in those unripe germ cells which at the time are in the process of ripening, and (3) although it begins earlier, reaches completion only during the period of sexual maturity of the plant or animal. In most cases some unripe germ cells (residual gonia) remain in the gonad where they multiply by mitosis and so produce more gonia. In female birds and mammals, however, the period of multiplication of oogonia comes to an end about the time of hatching or birth. At this time several million gonia may be present; yet only a very few ever ripen into ova—about 400 in the case of women.

The gonia are thus lineal descendants, by mitosis, of the original fertilized egg or zygote. Hence they are cousins of the body cells, and, generally speaking, they possess the same two sets of chromosomes which the body cells possess.

Meiosis has been compared to two mitotic divisions, for two successive spindles are formed, and two separations of chromosomes take place. Yet there is only one complete prophase (Fig. 2–3, *G* and *H*), and this differs from a typical prophase in that it involves a pairing of the chromosomes side by side, each with its mate (Fig. 2–4, *B* and *H*). The process of pairing is known as *synapsis*. In a sense this pairing of homologous chromosomes completes the coming-together of the germ cells at fertilization; for all through the mitotic cell divisions of development the chromosomes of the egg and those which came from the sperm have remained separate. Now each chromosome joins its mate.

The stages of the prophase of meiosis have been given names: leptotene, when the chromosomes become threadlike; synaptene (or zygotene), when they come together in pairs; pachytene, when they contract and become tightly coiled; diplotene, when they begin to separate, except for locations where "crossing-over" between synapting chromatids has taken place; and diakinesis, when again they become compacted prior to separating. Some of these stages are illustrated in Fig. 2-3, *G* to *K,* and in part in Fig. 5-1 (see legends).

Note that at the beginning of meiosis each chromosome consists of two chromatids as in ordinary mitosis, so that, when pairing takes place, the result is bundles of four chromatids (Fig. 2-3, *I*). These are known as *bivalents* because each consists of two chromosomes. They are also known as tetrads because each is four chromatids.

During this period preparatory to the first meiotic division, the ripening germ cell is called a *primary gametocyte,* primary spermatocyte or oocyte, as the case may be. The diplotene stage of oocytes is of particular interest, not only because it is long drawn out (it may be as long as 40 years in the human female), but because it is a period of great growth (Fig. 5-1). The nucleus enlarges greatly due to the accumulation of nuclear sap and is often referred to as the *germinal vesicle.* The chromosomes become fuzzy objects which have been compared to lampbrushes (page 90). The growth of the cytoplasm is discussed in Chapters 4 and 5.

First meiotic division The stages of the prophase which have just been described end as the metaphase approaches. The bivalents move to the equator of the first meiotic spindle. At the metaphase they split; and at the anaphase each bivalent divides, and a *monovalent chromosome* or dyad (pair of chromatids) goes to each pole of the spindle (Fig. 2-3, *J* and *K*). The monovalents are actually the original chromosomes which united in synapsis, except that some crossing-over has usually taken place. The products of the first meiotic division are known as *secondary gametocytes.*

Second meiotic division Usually the telophase of the first meiotic division is brief and is quickly followed by the much reduced prophase of the second meiotic division. In some instances, the anaphase of the first meiosis leads directly into the metaphase of the second meiosis. In any case, the monovalent chromosomes (dyads, two chromatids each) take a position at the equator of the second metaphase spindle (Fig. 2-3, *L*). At the anaphase the chromatids or *daughter chromosomes* (monads) move apart to the poles of the spindle (Fig. 2-3, *M*). The resulting cells are known as *tids,* ootids or spermatids, as the case may be.

Note that a reduction in the number of chromosomes has taken place. The gonia had two sets (diploid number) of chromosomes. The primary gametocyte had one set (haploid number) of bivalent chromosomal pairs (tetrads). Secondary gametocytes had one set of monovalent chromosomes (dyads). Now the tids have one set of daughter chromosomes, that is, one set of chromatids (monads). But this is important: the set which each tid has is a complete set with one chromosome present to represent each pair of chromosomes of the original diploid set.

In the case of spermatogenesis, each primary spermatocyte divides twice equally and produces four equivalent spermatids (Fig. 2–4, *A* to *E*). In oogenesis, however, although the meiotic divisions of the nucleus are equal, those of the cytoplasm are grossly unequal (Fig. 2–4, *G* to *K*). At each division almost all the cytoplasm goes to only one of the daughter cells and thus is conserved to supply the substance of the embryo. The result is that the first meiotic division gives rise to one secondary oocyte and one small *first polar body*. The second meiotic division similarly produces one ootid and a small *second polar body*. The spermatids undergo metamorphosis and become sperm cells or *spermatozoa* (Fig. 2–4, *F*). The ootids, on the other hand, as a rule need only to burst from the ovary to cause them to ripen and become ready for fertilization.

It is a strange fact that the stage in meiosis which is attained before the fertilizing sperm enters the egg is not the same for all species (Fig. 2–5). In sea urchin eggs meiosis is complete, and both polar bodies have been formed before the egg is receptive to the sperm. This is rather rare. In vertebrates the first polar body has been given off and the second meiotic division has progressed to the metaphase before fertilization takes place. The extreme case is the parasitic roundworm, *Ascaris,* in which the sperm enters the cytoplasm of the egg before even the first meiotic spindle has formed. It remains inactive in the center of the cytoplasm while the egg completes meiosis. In most animal species, therefore, the meiosis which closes the nuclear cycle of the egg overlaps to a greater or lesser extent the entrance of the sperm which begins the nuclear cycle of the new individual.

Why does this complicated process of meiosis exist? The answer was pointed out in 1903 by Walter Sutton, who showed the parallelism between the story of hereditary factors (later termed genes), as worked out by Mendel and other students of plant and animal breeding, and the behavior of chromosomes in fertilization, mitosis, and meiosis. Meiosis, Sutton showed, is nature's way of reducing the number of chromosomes in each germ cell so that, at fertilization, the normal diploid number of chromosomes will be restored. But it is more than this: It is nature's way of seeing to it that the germ cells are all different. Let us assume that the chromosomes of different pairs are different and also that each chromosome has some mutant genes which make it unlike its mate. In nature this is probably always true. Since all the cell divisions during development are made by mitosis, every gonium will have the same two sets, or diploid number, of chromosomes and genes. Then, when meiosis takes place, each ripe germ cell retains only one chromosome of each pair of chromosomes, one gene of each pair of genes. Moreover, it is a matter of chance which chromosome of a given pair of chromosomes the cell retains.

Thus, from the standpoint of heredity, there can always be as many different genetic kinds, or *genotypes,* of ripe germ cells as there are possible combinations of chromosomes taking one chromosome from each mating pair of chromosomes. If there were just one pair of chromosomes in an unripe germ cell, there would be two kinds of ripe germ cells. If there were two pairs to begin with, there would be four possible kinds of germ cells. If there were three pairs, the number would be eight. In general, if there are N pairs of chromosomes before meiosis, there will be 2^N possible kinds of genotypes among the ripe germ cells. Now, man has 23 pairs of chromosomes. 2^{23} is 8,388,608. In other words, there is no more

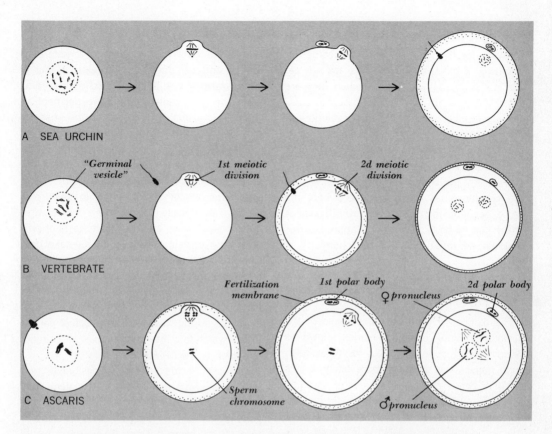

FIGURE 2-5 Fertilization (entrance of sperm) takes place at different stages of meiosis in different animals: *A*, after meiosis is complete in the sea urchin; *B*, midway in the second meiotic division in most vertebrates; and *C*, before meiosis in the parasitic roundworm *Ascaris*.

than one chance in over eight million that a man or woman will produce two germ cells with identical genotypes. (Crossing-over decreases the chance tremendously.) What is the maximum chance that two identical offspring will be born to the same parents by ordinary sexual reproduction? It is one chance in 8,388,608 times one chance in 8,388,608, or one chance in 70,368,744,177,664. Crossing-over at synapsis, mutations, and chromosomal aberrations greatly increase this variability.

THE ROLE OF THE NUCLEUS

The account of fertilization, mitosis, and meiosis which has just been given is similar to that usually found in textbooks of biology. It stops short, however, of explaining how the nucleus controls development. Indeed, it avoids the problem, for the compacted chromosomes of fertilization, mitosis, and meiosis are not active in development. It is only when they are expanded, as at the interphase, that they carry on their functions in growth and development. (The chromosomes of oocytes are expanded also during the long prophase.) Furthermore, the chromosomes presumably are alike in every cell of the body of an embryo. Now, how can chromosomes which are everywhere the same account for the origin of differences within the embryo? In order to approach this problem we must consider the chemistry of the nucleus.

The Chemistry of the Nucleus

The distinctive chemical substances of the nucleus are long chain molecules known as *deoxyribonucleic acid,* or DNA for short. Each such molecule is a gene; more likely it is a string of genes, the material units of heredity. A closely related type of nucleic acid is *ribonucleic acid,* or RNA. Both DNA and RNA are present in the nucleus. DNA is in the chromosomes. RNA is in the chromosomes, but especially in the nucleoli, and it is also present in the cytoplasm.

Now, nucleic acids have three unique and remarkable properties:

1 They are self-duplicating molecules. Possibly they are the only truly self-duplicating molecules. Centrioles and plastids are morphologically self-duplicating, but it is likely that, in this case, nucleic acids are involved.

2 Nucleic acids supervise the synthesis of proteins within the cell (pp. 44, 45). The proteins are the principal structural compounds

of protoplasm and are the chemical basis of its functioning. They include the enzymes which control the innumerable chemical reactions of the cell and also structural proteins. In fact, a gene (*cistron*) has been defined as the template for the production of one polypeptide chain. It is probably true that, in the absence of nucleic acids, no proteins are ever produced.

These statements are oversimplified. The self-replication of a nucleic acid requires the presence of specific enzymes (polymerases) and numerous other molecules in the protoplasm. The process is therefore a circular one: nucleic acids bring about the synthesis of proteins, and proteins are needed to make possible the replication of the nucleic acids. It is the entire protoplasmic system which endures.

Genes are not indivisible. Crossing-over of parts of genes may take place between synaptic mates during meiosis. The smallest part of a gene which may interchange thus with a mate has been termed a *recon.*

3 Nucleic acids (genes) occasionally mutate; that is, a short segment of a gene may change in composition. When this occurs, the gene replicates according to its new or mutant form. If this were not so, life on earth would never have evolved, and the ever-increasing adaptation of living things to their environments would not have been possible. The shortest segment of a gene which may mutate is probably a single nucleotide unit. It is known as a *muton.*

Great progress has been made in the last few years in the knowledge of the structure of nucleic acids and of the manner in which they duplicate. According to the Watson-Crick hypothesis, each DNA molecule is a long double chain of units known as nucleotides (Fig. 2–6, *A*). It may be compared to a rope ladder twisted into a helix. Each single nucleotide consists of a sugar, deoxyribose (S), a phosphate group (P), and one of four different "bases." The bases are adenine (A, a purine), guanine (G, also a purine), thymine (T, a pyrimidine), and cytosine (C, a pyrimidine). The single chain consists of alternating sugars and phosphates, with a base attached as a side group to each sugar. The precise sequence in which

the nucleotides are arranged in the long nucleic acid chain is the basis of the coded information which is passed on by heredity and which guides the course of development.

In the double chain of the DNA molecule, each base of one chain is linked crosswise, by weak hydrogen bonds, with its mate in the other chain. The relations are precise. Adenine is always bonded with thymine, and gua-

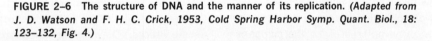

FIGURE 2–6   The structure of DNA and the manner of its replication. (Adapted from J. D. Watson and F. H. C. Crick, 1953, Cold Spring Harbor Symp. Quant. Biol., 18: 123–132, Fig. 4.)

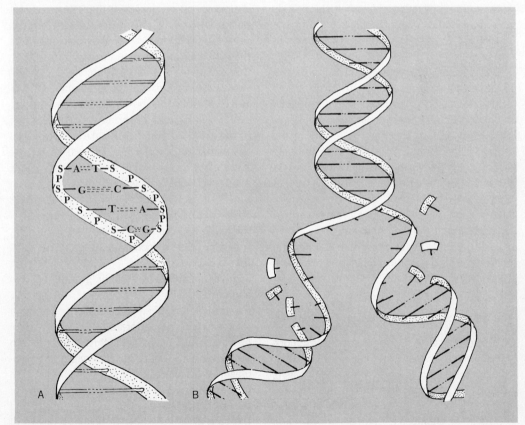

nine with cytosine. There are no other combinations. Hence the two chains fit each other exactly like a mold fits a model.

When a nucleic acid (DNA) molecule self-duplicates, the chains supposedly separate from each other (Fig. 2–6, *B*). Each single chain then makes for itself a replica of its former partner by attracting and assembling to itself appropriate nucleotides from the surrounding protoplasm. The result is that there are now two identical DNA molecules (double chains of nucleotides) where previously there was one. One notes at once the comparison between the replication of the half-chromatids (chromonemata), which takes place during the interphase of mitosis, and the replication of the single chains of DNA molecules.

The nucleus also contains proteins of two sorts: histones (proteins rich in basic amino acids), which are associated with the genes; and nonhistones, which compose the larger amount of protein and may be a part of the metabolic and mitotic machinery of the cell. The relation between histones and DNA is loose, yet it is so consistent that the combination is usually referred to as nucleoprotein. A small amount of nonhistone protein, termed residual protein, presumably holds the genes in order and organizes them into the definite bodies, the chromosomes. Lipids are also present in the nucleus.

The Influence of the Nucleus on Development

The genes (DNA) do not directly influence development. Instead, they "transcribe" their coded information to RNA, which then collects around the chromosomes and in the nucleoli. Apparently one of the two strands of a DNA molecule assembles ribonucleotides and makes for itself a complementary strand of RNA. (The other strand possibly self-duplicates.) The nucleotides (ribonucleotides) which compose the RNA, however, differ from those which compose DNA. The sugars are ribose instead of deoxyribose, and one of the four nucleotides is uracil instead of thymine (it lacks the methyl group of thymine). Although these differences from a chemical standpoint appear to be slight, yet they account for important differences in function. The RNA molecules are single chains and as such are not self-replicating. (RNA is self-replicating in some viruses.) They are dependent on DNA for their production. Moreover, the RNA molecules do not remain in the nucleus. From time to time they migrate into the cytoplasm where they function in the manufacture of proteins. Some of them (messenger RNA, see pp. 44, 45) serve as templates for the production of proteins, notably for the production of the enzymes which catalyze the chemical processes of the cell. Possibly this release of RNA is not a continuous process. An impressive release takes place at the prophase of the first meiotic division of oogenesis when the swollen oocyte nucleus (the so-called germinal vesicle) breaks down and its nuclear sap and the material of the large nucleolus mingle with the cytoplasm. A similar release occurs at the prophase each time a cell divides.

In very active cells, such as growing oocytes and certain cells of larval insects, the nuclear membrane has been shown by the electron microscope to be perforated. Nuclear material has been found to "bleb" through pores into the cytoplasm where it supervises the syntheses of proteins.

There may be a delay between the time of the formation of the RNA in the nucleus and the time when its influence becomes apparent in cytoplasmic processes. The clearest evidence that this is so comes from experiments on species hybrids, that is, on eggs which have been fertilized by sperm of another species. In this case the early events of development, for example, the pattern and rate of cleavage, are determined by the egg cytoplasm alone, presumably by the RNA that was synthesized while the oocyte grew in the mother's ovary and under the influence of the mother's genes. Generally speaking, the effect of the genes (DNA) brought in by the sperm is not apparent until about the time of gastrulation. But from then on, the genes of both the egg and the sperm control the course of development.

Even more crucial are certain experiments in which eggs were enucleated before they were fertilized by the sperm of another species, or in which the egg nuclei were destroyed immediately after the foreign sperm had entered. In such haploid eggs (termed androgenetic hybrids), the cytoplasm belongs to one species and the nucleus to another. Such combinations have not lived beyond the early embryonic stages, but the evidence indicates that, from the time of gastrulation onward, the influence of the chromosomes of the stranger nucleus is pronounced.

Do Nuclei Undergo Differentiation?

In the early days, Weismann and Roux believed that the nuclei of body cells become different during cleavage as a result of what were thought to be unequal divisions of the nuclei. The organization of the embryo, so they theorized, is the expression of the inherited organization within the chromosomes of the zygote nucleus.

This view has long since been given up. There is organization within the cytoplasm of the uncleaved egg. There is strong evidence

that the nuclei of every cell of the body contain initially the same set of chromosomes and genes, and that, during cleavage at least, mitosis is equal cell division in so far as the chromosomes are concerned. This fact was originally demonstrated in an experiment by Spemann in which he constricted an uncleaved amphibian egg by tightening a hair noose around it (Fig. 5–3). The egg was pinched into two halves, with only a narrow neck of protoplasm between them. One half contained the nucleus; the other half lacked a nucleus. After several cleavages had taken place in the nucleated half, one of the descendant nuclei, migrated across the isthmus of cytoplasm into the non-nucleated half (delayed nucleation). Cleavage followed, and both halves (provided the needed cytoplasm was present) became whole embryos of half size. This proves that a cleavage nucleus can do everything that the original zygotic nucleus can do. It is an exact copy of its original.

The fact that the daughter nuclei are equal during early development has been clinched by some remarkable experiments by Briggs and King. They removed the nucleus of an uncleaved frog's egg, and substituted a nucleus from a cell of a blastula or an early gastrula (Fig. 2–7). With refined techniques, many such operations were performed and were successful. Normal embryos developed. Here again, it is demonstrated beyond question that at this early stage of development when the future fates of the regions of the embryo are being determined, the nuclei, i.e., the genes of the cells of the different tissues, are equivalent. Since the nuclei are alike, it therefore must be the cytoplasms which are different.

Gurdon and others have extended this work and have found that some nuclei of stages as late as swimming larvae, when transplanted to

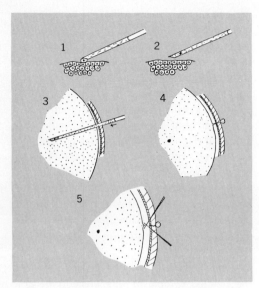

FIGURE 2-7 Method used by Briggs and King when they substituted nuclei from older embryos for the nuclei of uncleaved eggs of the frog. *(From R. Briggs and T. J. King, 1953, J. Exptl. Zool., 122:485–506.)*

enucleated eggs, are capable of supporting a complete development. Some of the tadpoles even live to metamorphose into frogs and become sexually mature.

This, however, is not the whole story. It is a common observation that the nuclei of cells of different tissues do differ in size and shape. Somehow during development nuclei do change. The cytoplasm, which supplies them with substance, energy, and necessary enzymes, also affects the visible characteristics which distinguish them. It does more than this: it affects their activity; and, at least in some cases, it affects their composition.

The nuclear changes which arise during development are of three sorts, and they arise in three distinct ways: (1) by the diminution of the chromatin, (2) by stable changes in the chromosomes, and (3) by the epigenetic control of gene activity by the cytoplasm.

1 Diminution of the chromatin This is a rare phenomenon. Boveri observed many years ago that a reduction of the substance of the chromosomes (diminution of the chromatin) takes place during the early cleavages of the parasitic roundworm, *Ascaris*. At the two-, four-, eight-, and 16-cell stages only one of all the blastomeres retains the typical four chromosomes of the zygote. In each case this cell is in the line of descent of the germ cells (germ line). The other cells cast off a part of their chromatin into the cytoplasm; that is, the tips of the chromosomes swell and disappear. At the same time the central part of each original chromosome breaks into several small segments which remain as separate chromosomes.

Other examples of the diminution of the chromatin in the development of body cells have been described, especially among the insects. Sometimes this process takes place early in development, sometimes late. But it never occurs in the germ line. We may conclude that, in the body cells of these embryos, the chromatin (DNA) has already performed its function by producing RNA. The RNA is stable and functions during development. The DNA (chromatin), no longer needed in the development of the body cells, is discarded.

Experiments have shown in *Ascaris* that a certain substance localized in the cytoplasm prevents the diminution of the chromatin. Normally, this material is passed on to only one blastomere during each of the first four cleavage divisions. But if, as a result of manipulation (centrifuging), this material becomes divided between two daughter cells, then diminution of chromatin does not take place in either of them. This is a clear case in which substances in the cytoplasm control the nucleus.

2 Changes in the chromosomes   Diminution of the chromatin is not known to occur in vertebrate development; yet there is evidence that changes in the genome or sum total of genes may take place. In their studies of transplanted nuclei, Briggs and King found that, when a nucleus from a late gastrula or neurula of an amphibian is substituted for the nucleus of an uncleaved egg, normal development does not usually take place, although it *may* take place. When examined under the microscope, the cells often do not have the normal chromosomal equipment. It is not clear whether changes of this sort take place in normal development. Possibly the nuclei of differentiating cells become increasingly sensitive to experimental manipulation. Possibly the transplanted nuclei are forced by the mitotic apparatus to divide before they are fully ready (page 99). Possibly, also, when cells differentiate, the mitotic cell divisions become slovenly. In any case, in these experiments the daughter cells did not always receive the normal complement of chromosomes.

The nuclear changes which take place in this manner are stable. Briggs and King have demonstrated that when the nucleus of a defective embryo is transplanted to an uncleaved egg, the result is an embryo with a defect similar to the defect of its nuclear parent. This transfer of nuclei may be repeated over and over with the same result (nuclear cloning).

For a time it was suggested that the changes in the genome which Briggs and King had discovered might be related to the normal differentiation of cells. For example, the nuclei derived from the endoderm seemed to favor the development of endodermal organs. But this interpretation has not been borne out. Defects of a similar sort have also been obtained when the nuclei are taken from the neural tube. On the other hand, some nuclei are capable of supporting a normal development even when taken from the gut of a swimming tadpole. However, one thing is certain: Any changes which may take place in the genome of differentiating cells always occur *after* the cytoplasm has already become specialized and committed to its fate.

3 The epigenetic control of the genes by the cytoplasm   Whether or not nuclear substance differentiates during development, it is certain that the same genes are not equally active in every cell and at every place, nor are they equally active at all times. Genes represent potentialities (in the sense of possibilities rather than powers) which may or may not have opportunities to express themselves. Which potentialities are realized depends upon the conditions within the cytoplasm which surrounds a given nucleus; these conditions change epigenetically as development progresses. It has been said that the cytoplasm turns the genes off and on like faucets. To change the metaphor, a nucleus is like a stockroom of a foundry full of patterns. From its vast store, patterns are requisitioned when, where, and as they are needed. No casting can be poured at the foundry for which a pattern is not provided, and, of course, the form of the casting depends upon the shape of the pattern. In the same way, no synthetic chemical process can take place in a cell unless a gene is present to supply the pattern (i.e., template) for the enzyme which mediates the process. Genes are templates in stock. But the decision as to which genes are active at a given time and place—this depends upon conditions in the cytoplasm. And these conditions change epigenetically as development goes on. It is a mistake to pass over lightly the epigenetic factors which turn genes on and off, for they, and not the genes, ultimately control the time and place of developmental processes.

Direct evidence that this is true is found in the behavior of the giant chromosomes of certain insect larvae as they approach metamorphosis. Such chromosomes consist of multiple strands of chromatids (so-called polytene structure). They are the result of self-duplication of the chromatids without accompanying separations (Fig. 2–8). For example, the chromatids of the salivary glands of fruit fly larvae multiply nine or ten times without separating, with the result that each giant chromosome consists of 512 to 1,024 chromatids. These compound chromosomes show bands (possibly representing genes) of DNA which stain darkly and are large enough to be seen under the light microscope. They also show swellings, or "puffs," at definite loci. These seem to be the result of RNA and protein accumulating and forcing the strands apart. The puffs may expand until they become what are known as Balbiani rings. Beermann has studied the puffs and rings in the larvae of the midge, *Chironomus*, and has shown clearly that they occur at different locations in the chromosomes of different tissues and at different times. The explanation seems to be that the puffs and rings are centers of intense synthetic activity in particular tissues, in preparation for the role which that tissue will play in the metamorphosis which is soon to follow. (See also page 98).

How are the Genes Controlled?

The analogy of a nucleus to a stockroom of patterns at a foundry has one obvious flaw. A stockroom has a stockkeeper who delivers the patterns as they are called for. The nucleus of a cell, however, is automated so that, cued by conditions in the surrounding cytoplasm, it delivers the templates when they are needed.

Lately developmental geneticists have been much interested in the problem of how the genes are activated. We shall make reference to the closely reasoned concepts of Jacob and Monod of the Pasteur Institute (Fig. 2–9), who studied enzyme induction in the bacterium, *Escherichia coli*. Normally this microbe produces no enzyme to metabolize the nutrient sugar lactose. Yet, when lactose is supplied in the nutrient medium, the enzyme β-galactosidase, which is able to metabolize it, appears in abundance. The presence of the substrate (substance acted upon) serves as an *"inducer"* and leads to the production of the enzyme.

The authors reasoned that no enzyme could have been produced if a "structural gene" for its production had not already been present in the germinal material. (Genes which de-

FIGURE 2–8 Polytene chromosome of the salivary glands of the midge, *Chironomis*, showing stages of "puffing." *(Adapted from figures by W. Beermann, 1963, Am. Zool., 3:23–32; and 1964, Sci. Am., 210 (April): 50–58.)*

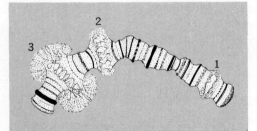

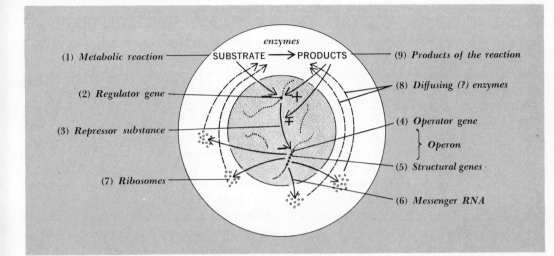

FIGURE 2–9 Schema illustrating Jacob and Monod's concept of nucleo-cytoplasmic interaction. 1. A substrate in the cytoplasm inhibits the activity of a regulator gene. 2. A regulator gene, the function of which is to produce a repressor substance. (When inhibited by the substrate it produces less repressor substance.) 3. Repressor substance which prevents the functioning of an operator gene in another chromosome. (When less repressor substance is present, the operator gene is "derepressed.") 4. Operator gene which activates a series of structural genes. 5. Structural genes which synthesize messenger RNA (and other sorts of RNA). 6. Messenger RNA which migrates from the nucleus into the cytoplasm where it associates with ribosomes in the production of specific enzymes (or other proteins). 7. Ribosomes scattered through the cytoplasm. 8. Enzymes which catalyze specific chemical reactions in which the substrate is altered. 9. Products of the reaction, which may augment the activity of the regulator gene or activate a repressor substance. In so doing the products act as inhibitors of the operator gene.

termine the structure of a protein are called structural genes.) If this is so, then why was the enzyme β-galactosidase not produced until lactose was added? Genetic studies by Jacob and Monod on mutant strains of *Escherichia coli* supplied them with the answer: The gene for the production of β-galactosidase is inactive because another gene, so-called "regulator gene," is present which produces a *repressor substance* that inhibits the activity of the structural gene. When the substrate lactose

is supplied, however, it inactivates the regulator gene (or the repressor substance which it produces). The result is that the structural gene, now freed from repression, gives rise to the enzyme β-galactosidase. The structural gene is thus activated by a process of *derepression*.

This statement is oversimplified. The repressor substance produced by the regulator gene acts upon an "operator gene" within a complex of structural genes. It is the operator gene which directly controls the structural genes, turning them on and off. An operational complex of genes of this sort was called an "operon" by Jacob and Monod. Such a complicated mechanism would be unbelievable if its several steps had not been experimentally verified. Generally speaking, a substrate acts as an inducer. Its presence leads to the production of the enzyme which metabolizes it. In the language of modern communication theory, this is a case of "positive feedback." In a similar manner, the products of a reaction are inhibitors which repress the production of the enzymes which produce them. This is a "negative feedback."

Embryologists are much interested in concepts such as this one of Jacob and Monod

because they suggest how epigenetic factors in the cytoplasm (represented in the above case by the substrate lactose) can control the activity of the genes. There is, however, an important difference between the actions of genes in bacteria and in the development of higher organisms. The genes of bacteria are immediately responsive to their environment, and the mRNA which they produce by transcription is immediately active in protein synthesis. It is also short-lived. The genes which control the development of higher organisms, on the other hand, commonly produce "masked" mRNA, which may be stored for a time until the time and place for its activity have arisen. These matters are discussed further in Chapter 3 (pp. 55, 56).

SELECTED READINGS

The Chromosomes of Man

Tjio, J. H., and T. T. Puck, 1958. The somatic chromosomes of man. Proc. Natl. Acad. Sci., **44**:1229–1237.

Ford, C. E., P. A. Jacobs, and L. G. Lajtha, 1958. Human somatic chromosomes. Nature, **181**:1565–1568.

Mitosis

Ris, H., 1955. Cell division. In Willier, Weiss, and Hamburger, (eds.), Analysis of Development. W. B. Saunders Company, Philadelphia. Pp. 91–125.

Mazia, D., 1961. How cells divide. Sci. Am., **205** (September): 100–120.

Taylor, J. H., 1958. The duplication of chromosomes. Sci. Am., **198** (June): 36–42.

———, 1962. Chromosome reproduction. Intern. Rev. Cytol., **13**:39–74.

Meiosis

Van Beneden, E., 1883. Researches on the maturation of the egg and fertilization. Translated and reprinted in Gabriel and Fogel (eds.), 1955. Great Experiments in Biology. Prentice-Hall, Inc., Englewood Cliffs, N.J. Pp. 245–248.

Sutton, W., 1903. The chromosomes in heredity. Biol. Bull., **4**:231–251. Reprinted in Gabriel and Fogel (eds.), 1955. Great Experiments in Biology, Prentice-Hall, Inc., Englewood Cliffs, N.J. Pp. 248–254.

Edwards, R. G., 1966. Mammalian eggs in the laboratory. Sci. Am., **215** (August): 72–81.

Levine, R. P., 1962. Genetics. Holt, Rinehart and Winston, Inc., New York. Chaps. 4 and 5.

The Chemistry of the Nucleus

Watson, J. D., and F. H. C. Crick, 1953. Molecular structure of nucleic acids. Nature, **171:** 737–738. Reprinted in Peters (ed.), Classic Papers in Genetics. Prentice-Hall, Inc., Englewood Cliffs, N.J. Pp. 241–243.

Crick, F. H. C., 1954. The structure of the hereditary material. Sci. Am., **191** (October): 54–61. Also **197** (September, 1957): 188–200; **207** (October, 1962): 66–74; and **215** (October, 1966): 55–62.

Levine, R. P., 1962. Genetics. Holt, Rinehart and Winston, Inc., New York. Chaps. 2 and 3.

The Role of the Nucleus

Spemann, H., 1928. The development of lateral and dorso-ventral embryo halves with delayed nuclear supply. Translated from the German and reprinted in Gabriel and Fogel (eds.), 1955. Great Experiments in Biology. Prentice-Hall, Inc., Englewood Cliffs, N.J. Pp. 215–219.

Gay, Helen, 1960. Nuclear control of the cell. Sci. Am., **202** (January): 126–136.

Callan, H. G., 1963. The nature of lamp-brush chromosomes. Intern. Rev. Cytol., **15:**1–34.

Beermann, W., 1963. Cytological aspects of information transfer in cellular differentiation. Am. Zool., **3:**23–32. Reprinted in Bell (ed.), 1965. Molecular and Cellular Aspects of Development. Harper & Row, Publishers, Incorporated, New York. Pp. 204–212.

Locke, M. (ed.), 1963. Cytodifferentiation and Macromolecular Synthesis. Academic Press Inc., New York.—This, the 21st Symposium of the Society for the Study of Development and Growth, includes important contributions by Grobstein, Yanofsky, Jacob and Monod, Markert, Gall, Lash, and others.

Nuclear Transplantation

Briggs, R., and T. J. King, 1952. Transplantation of living nuclei from blastula cells into enucleated frog's eggs. Proc. Natl. Acad. Sci., **38:**455–463.

King, T. J., and R. Briggs, 1956. Serial transplantation of embryonic nuclei. Cold Spring Harbor Symp. Quant. Biol., **21:**271–290. Reprinted in Bell (ed.), Molecular and Cellular Aspects of Development. Harper & Row Publishers, Incorporated, New York. Pp. 171–193.

Gurdon, J. B., 1964. The transplantation of living cell nuclei. Advan. Morphogenesis, **4:**1–43.

Gurdon, J. B., and D. D. Brown, 1965. Cytoplasmic regulation of RNA synthesis and nucleolus formation in developing embryos of *Xenopus laevis*. J. Mol. Biol., **12:**27–35.

The Control of the Nucleus

Jacob, F., and J. Monod, 1963. Genetic expression, allosteric inhibition, and cellular differentiation. In Locke (ed.), 1963. Cytodifferentiation and Macromolecular Synthesis. Academic Press Inc., New York. Pp. 30–64.

Bonner, J. 1965. The Molecular Biology of Development. Oxford University Press, Inc., New York. Chap. 10, Regulator, operator, effector.

chapter **3**

THE CYTOPLASM IN DEVELOPMENT

The genes of the nucleus control the course of development, but it is mainly the cytoplasm which develops. The genes are the "physical basis of heredity," but principally it is the cytoplasm which expresses this heredity insofar as it is expressed. There is interaction between the genes and the cytoplasm, but their roles are different. The genes largely control the species characteristics of the new organism and to a great extent its individual characteristics as well; but the characteristics which distinguish one tissue from another, or one organ from another *in the same individual,* are decided initially by processes which take place in the cytoplasm.

The history of the cytoplasm is indeed a true epigenesis; that is, it is a straightforward coming-into-being of molecules, structures, and functions which did not at first exist. It starts in the relative simplicity, or at least plasticity, of a single cell, the unripe egg. It moves forward irreversibly in ever-increasing complexity through the stages of the life cycle: embryo, larva (if any), juvenile, adult, and finally the senescent individual. The ultimate end of development is death unless, as in some of the lower organisms which are capable of regeneration, the story is able to begin again. Embryology is therefore, for the most part, a recital of processes which take place within the cytoplasm.

What is this cytoplasm which, synthesized under the direction of the nucleus, does and becomes so many different things? It is, to be sure, a colloidal aqueous system of proteins, lipids, sugars, and numerous other organic molecules plus various mineral salts. But cytoplasm is more than a chemical system, no matter how complex. It has a structure, an intimate, precise anatomy such as no system of a merely chemical nature ever had. Its proteins and lipids are organized into membranes, filaments, and granules, at the surfaces of which chemical reactions take place (Fig. 3–1). These are the organelles which carry on the many processes of living. The precise patterns of the protein surfaces are the principal basis of the specificity of the cell's chemical reactions, for the proteins include the enzymes which catalyze and control the rates of chemical processes. They also include nonenzyme proteins.

A very few years ago cytoplasm was described as a homogeneous, transparent gel or sol which was optically "empty." The various formed bodies which were seen under the microscope were spoken of as being within the "ground cytoplasm"; or they were interpreted as artifacts produced by fixation and staining. Now this picture has changed. The phase-contrast microscope, the electron microscope, and the ultracentrifuge have revealed an intricate morphologic structure in cytoplasm.

Although the cytoplasms of different kinds of cells have specialized functions (no single type of cell can be said to be "typical"), yet the cytoplasms of most cells posses certain organelles in common. These include: a plasma membrane (cell membrane or plasmalemma, the outer layer of the living cytoplasm); inner membranes, which form vacuoles, tubules, and lamellae; a nuclear membrane (actually a part of the cytoplasm); and formed bodies such as ribosomes and mitochondria. The central body with its centrioles was referred to in Chapter 2. Each of these organelles plays a role in the physiological life processes of the cell. Some of them presumably play a part in development, but it is not always clear what that part may be.

Specialized cells have filaments and granules

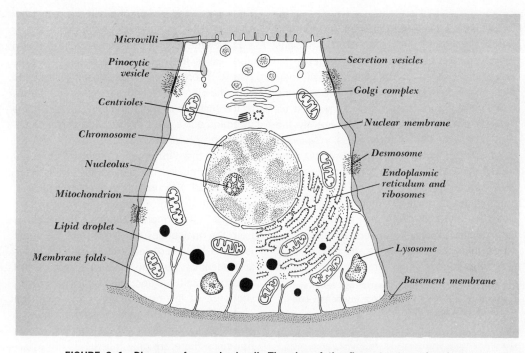

FIGURE 3–1 Diagram of an animal cell. The size of the finer structures has been exaggerated. *(Adapted in part from W. Bloom and D. W. Fawcett, 1962, "Textbook of Histology," W. B. Saunders Company, Philadelphia, 8th ed.)*

of various kinds which perform distinctive functions. Muscle cells have contractile fibrils; nerve cells have conductile fibrils; secretory cells have secretory granules; certain other cells produce pigment granules; while still others store reserve food materials such as yolk platelets, glycogen granules, and fat globules. The list could be multiplied. The embryologist sees these various structures appearing in the cytoplasm as development proceeds, each at the proper time and in the proper place. He has reason to believe that they result from the synthesizing activities of distinctive enzymes which are released in the cytoplasm, and that in turn each enzyme obtains the "information" (specific pattern) for its production

from a particular gene or genes in the nucleus. But why a particular gene is active producing a particular enzyme at a particular time and in a particular cell, he cannot as yet say. The answer to this most basic question of embryology is to be sought in the cytoplasm.

Ribosomes and the Synthesis of Proteins

A principal process in development is the synthesis of proteins. Our understanding of the manner in which this is accomplished has been one of the exciting areas of advance in modern biology. Beginning in 1941 and 1950 with the demonstration by Caspersson and Brachet that ribonucleic acid (RNA) is involved in the synthesis of protein, a great deal of information has been accumulated. As noted in the last chapter, the genes of the nucleus (DNA) "transcribe" their specific genetic "information," coded in the sequence of their nucleotides, to RNA molecules. These accumulate in and around the chromosomes and ultimately enter the cytoplasm. Apparently the DNA produces RNA of three sorts: *ribosomal RNA* (rRNA), *messenger RNA* (mRNA), and *transfer* or *soluble RNA* (tRNA). The ribosomal RNA associates basic proteins to itself and accumulates in nucleoli within the cell nuclei. Later this material (rRNA plus protein) enters the cytoplasm where it appears as ultramicroscopic bodies, the *ribosomes*. As much as 30 percent of the dry weight of cytoplasm may be

FIGURE 3–2 Schema illustrating how synthesis of a protein takes place under the control of messenger RNA.

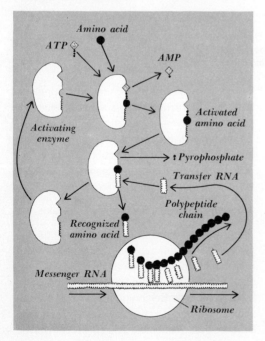

ribosomes, and about one-half of each ribosome is rRNA.

From time to time messenger RNA (mRNA) also leaves the nucleus and enters the cytoplasm. There it becomes closely associated with the ribosomes and serves as templates, at the surfaces of which the polypeptide chains of amino acids are built up. These chains become the proteins.

How is this "information" which is brought from the nuclei by the mRNA transferred ("translated") to protein molecules? It is carried in the form of an alphabetic code of four "letters," the ribonucleotides. These differ from the nucleotides of DNA in that the sugar is ribose, instead of deoxyribose, and the bases are adenine, guanine, cytosine, and uracil (which substitutes for thymine). The precise sequence of the nucleotides is the most important feature, for it determines the sequence of the amino acids in the polypeptide chain. But since there are only four kinds of nucleotides and there are twenty or more kinds of amino acids in a polypeptide chain, it must take a "word" or *codon* of at least three nucleotides to pick or choose one amino acid.

The third sort of RNA is transfer or soluble RNA (tRNA). It consists of smaller molecules, each of which bears, possibly at one end, a triplet of nucleotides which corresponds specifically to one amino acid. Since there are twenty kinds of amino acids, there must be at least twenty specific kinds of tRNA.

According to present evidence, the "translation" of information from mRNA to the polypeptide chain takes place mostly in the cytoplasm in the following manner: (1) First the amino acids in solution in the cytoplasm become activated with the aid of energy-rich molecules, such as ATP, and activation enzymes (Fig. 3–2). (2) Then the activated amino acids, still attached to the enzymes, are "recognized" by specific tRNA molecules. By recognition we mean that each amino acid becomes attached to its proper tRNA molecule. (3) Having accomplished this, the tRNA–amino-acid combinations separate from their enzymes, and each adheres one by one in sequence to a specific locus on a messenger RNA molecule as it moves across and past a ribosome. Thus, in turn, the amino acids, having been brought close together, link up, and a growing polypeptide chain is produced. A polymerizing enzyme is required for the linking-up process.

What is the role of the ribosomes in the synthesis of proteins? It seems not to be specific. Cell-free solutions in which mammalian mRNA have been combined with bacterial (!) ribosomes (plus a "pool" of amino acids, transfer RNA, and ATP) are capable of synthesizing mammalian proteins. One writer has referred to ribosomes as "workbenches" on which the polypeptide chains are forged. Ribosomes are said to "read" the message brought in by mRNA as it moves past. It may help us to visualize the process if we think of a ribosome as a sort of automatic machine which is controlled by a punched tape, the mRNA. The machine (ribosome) receives, as raw material, a mixture of activated and recognized amino acids (amino acids bound to tRNA molecules) and attaches particular combinations as the mRNA tape dictates. The amino acids link up, and an ever-lengthening polypeptide chain is produced. (The tRNA is released to be used again.) When the tape runs out, the polypeptide chain is complete and moves away. The entire process takes but little over one minute. The machine is also free to start over again with the same or another tape. Since the tape is long, it may be moving through or past several ribosomes at the same time, each of them in the process of producing a polypeptide chain. Such a series of ribosomes, connected by a thin thread of mRNA molecules, can be demonstrated with the electron microscope. They are known as *polyribosomes* (polysomes).

(There is some evidence that ribosomes are not all the same in their nucleotide composition, even that they differ in the cells of different tissues. Still, they may all be just workbenches.)

The "code" by which mRNA transfers its information to polypeptide chains has been broken. This was accomplished first by students of microorganisms. Bacteria were homogenized without destroying the ribosomes. The resulting cell-free system was supplied with amino acids, a source of energy (ATP), and artificially synthesized RNA chains of known composition. The polypeptide chains which were produced were then analyzed. The breakthrough came in 1961 when Nirenberg "fed" his system an artificial RNA consisting of the nucleotide urocil only. He obtained polypeptide chains composed of the amino acid phenylalanine. From that point on it was only a matter of time until the entire code was broken.

At first the ribosomes in the cytoplasm are free and scattered. Then, about the time that newly synthesized proteins begin to be secreted away from the cell, the ribosomes become attached to the endoplasmic reticulum (see page 49) within the cytoplasm.

This work on RNA and the ribosomes is of great interest to embryologists because it is possible that differentiation manifests itself first quantitatively in the distribution of the ribosomes. At any rate, gradients in the arrangement of ribosomes exist early in development, and any manipulation which disturbs their distribution, such as centrifuging, results in abnormalities.

Mitochondria and the Energy of the Cell

Whence comes the energy which makes these reactions possible? Proteins are highly ordered molecules, and their synthesis is an uphill (endergonic) process which requires a continuous source of free energy to make it go. If free energy were not provided, the living system would quickly revert to a more probable, that is, a more mixed-up state, namely, death.

Now, the development of an organism is, beyond anything else in the known universe, an improbable sequence of events. Only free energy, that is, ordered energy, can cause it to progress. Furthermore, an organism, and especially a developing organism, is a "dynamic system" like a flame or cyclone. Within it are gradients of substance and energy which would immediately flatten out and disappear if it were not that a continuous expenditure of free energy sustains them.

Where shall the needed order and energy be found? Heat cannot provide it, for it is only when heat is guided from a hotter source to a reservoir at a lower temperature, as in a heat engine, that some of it can be made to do work. Even then, a heat engine is inefficient. Since temperature differences do not exist in the cell, the cell cannot be a heat engine.

Green plants possess remarkable machinery, the chloroplasts, which capture the free energy of light and store it in the form of highly ordered organic molecules, primarily sugars. These molecules, built initially by green plants, are the ultimate source of the free energy which an animal uses in its development.

If the free energy of glucose (about 690,000 calories per gram-molecule) were to be released all in one step, as when sugar burns, most of the energy would become heat, and would be lost. But in the living cell, the energy is released stepwise in a complex series of controlled chemical reactions. In this way, a large part of the energy is transferred to the high-energy phosphate bonds of a ubiquitous substance known as adenosine triphosphate or ATP. Adenosine diphosphate (ADP) plus energy (approximately 12,000 calories per mole) becomes adenosine triphosphate (ATP). The energy-rich ATP can then migrate to where the energy is needed and, like a charged battery, give off its energy in the form of high-energy phosphate bonds, and so revert to ADP.

Perhaps the thing which, more than anything else, distinguishes living from nonliving matter, is this ability to transform and transfer energy in a controlled and useful manner. It is accomplished with the aid of enzymes, or rather, organized brigades of enzymes. One enzyme controls each step of the complex process.

Three successive chains of reactions are involved (Fig. 3–3):

Chain 1 The glycolytic sequence (cf. fermentation)   Glycolysis consists of a series of 14 chemical reactions by which glucose, glycogen, or other carbohydrate, is split and partially oxidized into a simpler chemical molecule, pyruvic acid. It takes a little energy (supplied by ATP) to start the process, much as it takes energy to start a stone rolling down a hillside; but the ATP which is used to start the process is more than restored by the ATP produced in the process. To control the 14 steps of glycolysis, at least 12 different enzymes are involved.

In the case of anaerobic organisms (those

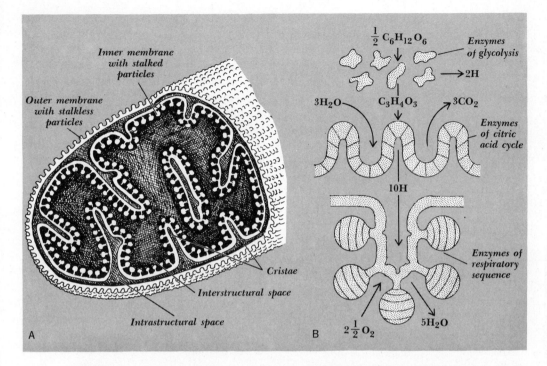

FIGURE 3–3 The mitochondrion as a converter of energy. *A.* The finer structure of a mitochondrion. *B.* Detail of the outer and inner membranes. The enzymes of glycolysis are probably in the cell solution. Those of the citric acid cycle may be in the outer membrane of the mitochondrion. The enzymes and coenzymes of the respiratory sequence probably occupy the inner membrane. *(Based in part on D. E. Green, 1964, Sci. Am., 210 (January): 63–74.)*

which do not use molecular oxygen), this is the end of the story, except that the resulting pyruvic acid is changed either to lactic acid, to acetic acid and CO_2, or to alcohol and CO_2, without much gain or loss of energy. But the process as it occurs in anaerobic organisms is incomplete, and over 90 percent of the energy of the glucose still remains in the pyruvic acid or related molecules.

Chain 2 The citric acid cycle Aerobic organisms, on the other hand, namely, those which ultilize oxygen (and this includes most de-

veloping eggs), continue the metabolic process by breaking down the pyruvic acid into CO_2 and H atoms. This is done in a closed series of reactions known as the citric acid cycle (of Krebs). It has appropriately been called "the metabolic mill." It is as though pyruvic acid were fed into a hopper and CO_2 and H atoms were ground out. Again specific enzymes and coenzymes are involved, and these act in sequence in an organized manner. Fats and certain protein fragments can also be fed into this same metabolic mill. The CO_2 which comes out is waste and must be discarded, but the H atoms are high in energy. Indeed, a single H atom (gram atom) represents some 52,000 calories of potential energy, four times the energy of an energy-rich phosphate bond.

Chain 3 The respiratory sequence The third chain of reactions is the one by which the H atoms that were produced in the first two chains are oxidized to water. The energy which is released is transferred to ATP. As before, the process does not take place in one step, for, if it did, a large part of the energy would be lost as heat. Instead, it takes place by a series of steps. At each step a special molecule accepts a pair of H atoms (or electrons—a H atom equals a H+ ion plus an electron) at a higher energy level. It then transfers them to another molecule at a lower energy level. At some of the steps, ATP molecules are synthesized. The final hydrogen acceptor is oxygen, and the end result is water.

The overall consequence of the three chains of chemical reactions is that, for each sugar molecule which is consumed, 6 molecules of CO_2 are given off, 6 molecules of H_2O are produced, and 38 molecules of ATP are restored. These represent 456,000 calories, or 66 percent of the potential energy of the glucose which was consumed. The remaining 34 percent of the energy is lost as heat.

The enzymes and coenzymes of the last two chains of metabolic reactions are organized in the form of certain ubiquitous organelles which are visible with the light microscope and which are known as *mitochondria* (Figs. 3–1 and 3–3). It is their business to oxidize pyruvic acid and restore ATP. This they apparently do after the manner of a bucket brigade at a fire; that is, they pass on pyruvic acid and its products from locus to locus until it is completely oxidized.

Mitochondria are present in abundance in all cells which are capable of utilizing oxygen in the release of energy. They are most concentrated in the more active cells, such as nerve, muscle, and gland cells, and notably in those regions of these cells where metabolic activity is greatest. A single cell may contain a thousand or more of these bodies. They may make up 20 percent of a cell's dry weight.

As pictured by the electron microscope, each mitochondrion consists of a double membrane surrounding a semifluid "matrix" (Fig. 3–3). The inside layer of the double membrane is thrown into internal folds or cristae which partially divide the cavity within. The enzymes of the citric acid cycle (chain 2) are thought to be in the outer layer, while those of the respiratory sequence (chain 3) are localized in the inner layer of the membrane of the mitochondrion (Fig. 3–3, *B*). Motion-picture microphotography shows the mitochondria to be in incessant motion, especially at the time of cell division.

Lysosomes

Lysosomes are cell organelles which contain powerful digestive enzymes (Fig. 3–1). Each is surrounded by a semipermeable membrane to prevent the enzymes from escaping and digesting the rest of the cell. Christian de Duve, a biochemist, was the first to recognize lysosomes as such. Later they were identified under the microscope. De Duve noted that when he homogenized rat liver with a Waring blender—a very severe treatment—enzymes were released which digested the organic molecules of the cell. When, however, he used a gentler method of homogenization, very little digestion took place. The enzymes were present, but they were held within the membranes of the bodies which enclosed them.

Lysosomes are especially important in white

blood corpuscles and other phagocytic cells (cells which devour). They constitute many of the granules of the cytoplasm. Cells of this sort take in foreign particles much as an amoeba engulfs a food particle. The process is called *phagocytosis* (Fig. 3–4). The vacuoles containing the food particles and the lysosomes move together, fuse, and their contents mingle. After a short time the vacuole becomes a digestive vacuole filled with the products of digestion. Absorption takes place, and the residual vacuole, now loaded with undigested debris, is ejected by the cell.

When, under certain circumstances, the membranes of the lysosomes become permeable, digestive enzymes are released into the cytoplasm, and the cell dissolves. This is known as *autolysis*. A controlled autolysis is not uncommon during development. For example, at the time of metamorphosis, the tail of a tadpole undergoes autolysis through the activity of enzymes released from lysosomes.

The Endoplasmic Reticulum

The electron microscope reveals the presence of a net of tubules and hollow lamellae within the cytoplasm of many cells; a light microscope is not powerful enough to reveal this. This net is the *endoplasmic reticulum*. The membranes of the net separate two liquid phases: an outer phase (the ground substance of the cytoplasm), which bathes the net; and an inner phase, which is contained within the net. In some cases, perhaps in all, the membrane of the net is continuous with the plasma membrane and with the nuclear membrane. Thus the fluid which is within the net has connection with the fluid outside the cell. Following cell division, the nuclear membrane seems to be restored from the endoplasmic reticulum.

Endoplasmic reticula are of two sorts:

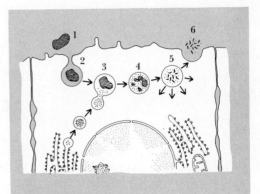

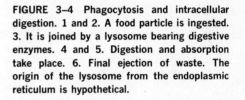

FIGURE 3–4 Phagocytosis and intracellular digestion. 1 and 2. A food particle is ingested. 3. It is joined by a lysosome bearing digestive enzymes. 4 and 5. Digestion and absorption take place. 6. Final ejection of waste. The origin of the lysosome from the endoplasmic reticulum is hypothetical.

"rough" (granular) and "smooth" (agranular). The roughness of the former is due to numerous ribosomes which adhere to and coat the surface of the net. Because the ribosomes stain with basic dyes, this material of the rough reticulum shows dark with the light microscope. It has been called *ergastoplasm* (Fig. 3–1).

Not all cells show a well-developed endoplasmic reticulum. In early development embryonic cells may possess an abundance of ribosomes, but these are mostly free, and the reticulum is limited in extent and largely smooth (Fig. 3–5, *A*). Later, especially in the case of those cells which secrete protein outwardly, the endoplasmic reticulum increases greatly and becomes granular (Fig. 3–5, *B*). The implication is that the reticulum is an organelle which receives the proteins that the adherent ribosomes have synthesized, concentrates them, and ultimately delivers them as secretion granules outside the cell. Ex-

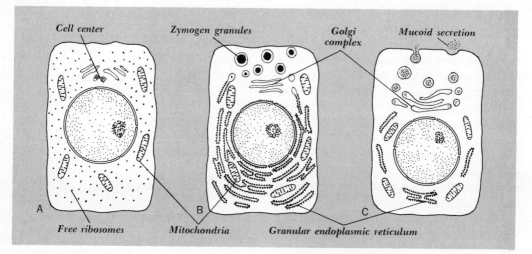

FIGURE 3–5 Three types of cells engaged in synthesis (idealized). **A.** A protein-retaining cell with free ribosomes. **B.** A protein-secreting cell with an extensive endoplasmic reticulum coated with ribosomes. **C.** A mucus-secreting cell with a well-developed Golgi complex. *(Adapted from E. H. Mercer, 1965, In DeHaan and Ursprung (eds.), "Organogenesis," Holt, Rinehart and Winston, Inc., New York, p. 42.)*

amples are the gland cells, which secrete zymogen granules, and the fibroblasts, which produce the structural proteins of connective tissue.

The Golgi Complex

Most cells possess in their cytoplasm a region of specialized membranes known as the *Golgi complex* or Golgi apparatus. In thin sections the membranes appear as irregular stacks of flattened vesicles (cisternae) often located not far from the centrioles (Fig. 3–1). The membranes are smooth; that is, they are devoid of a coating of darkly staining ribosomes.

Products of cytoplasmic synthesis accumulate in the Golgi apparatus, possibly having entered from the endoplasmic reticulum. Here

condensation and further synthesis takes place, with the result that secretory granules are produced which have carbohydrate (mucopolysaccharides) as well as protein in their composition. The granules collect in vacuoles which, in the case of gland cells, move to the surface of the cell and are discharged to the outside (Fig. 3–5, *C*). The process may be thought of as phagocytosis in reverse. Thus the membranes of the Golgi complex act as semipermeable barriers to the free diffusion of molecules, and they may be compared in this respect to the membranes of the contractile vacuoles of protozoa.

The Cell Cortex

The form and activity of a cell are intimately dependent on the properties of the outer zone (or zones) of the cytoplasm, namely, the *cell cortex*. The outer layer of the cortex is the *plasma membrane*. It directly relates the cell to its environment. Nutrients must enter through it, and unwanted substances must be kept out. Wastes are eliminated by it. Stimuli, except perhaps those of light, affect it first. The plasma membrane is of the order of 100 Å thick ($= 100 \times 10^{-7}$ mm). It has properties of semipermeability, but it must not be thought of as a passive sieve. On the contrary, it is an extremely active membrane which produces and retracts minute processes (microvilli) from its external surface, and forms microtubules and vesicles that push inwardly into the interior of the cytoplasm. By this means the cell feeds (phagocytosis) and drinks (pinocytosis). The particles and droplets which thus enter the cell do not directly penetrate the plasma membrane. Rather, the membrane moves before them and surrounds them as they advance into the cell. Later the membrane may dissolve, or the contents of the vacuole may be digested and absorbed into the cytoplasm (Fig. 3-4).

In addition to the plasma membrane, the cell cortex commonly includes a somewhat thicker layer of cytoplasm which is from 1 to 5 μ thick ($= 1$ to 5×10^{-3} mm). This layer, sometimes referred to as ectoplasm or plasmagel, is usually gel-like in its physical structure, although it may change its state from gel to sol and back again. When photographed with the electron microscope, the plasmagel often differs very little, if at all, from the more liquid internal cytoplasm (endoplasm, plasmasol). Indeed, it may not be present at all. But in other cases it is specialized by the presence of filaments or granules which are not found elsewhere.

The various responses by which a cell changes its shape are often cortical reactions. Development, also, is to a remarkable degree a matter of cortical response. This is obvious in the case of the extrusion of the polar bodies, the rising-up of the fertilization membrane, and the division of the cytoplasm during cleavage. It is not so obvious that the folding of the cellular layers by which the gut and later the nervous system are formed are also to a large extent cortical reactions.

Protozoologists long ago became aware that the form and much of the activity of one-celled animals are properties of the cell cortex. Consider, for example, the ciliate, *Stentor* (Fig. 3-6, *A*). The body of this animal is conical in shape. The sides are marked by a hundred or so longitudinal rows of cilia separated by intervening pigmented stripes. The base of the cone (anterior surface or "head") is a disc with spiral rows of cilia. The head is surrounded by an almost complete circle of membranelles. At the clockwise end of the arc of membranelles are the structures used in feeding, namely, the oral pouch, cytostome, and gullet. Nearby is the anal pore. The apex of the cone (posterior end) is specialized as an organ of temporary attachment, the holdfast. All these structures are specializations of the cell cortex.

Tartar has performed numerous surgical operations on *Stentor* and has repeatedly observed its extensive capacity for regeneration and reorganization. The controlling feature, he finds, is the presence and orientation of the cortex. A tiny bit of *Stentor*, only 1/123 the size of a normal animal, is capable of regenerating a whole *Stentor* provided that a little of the cortex and some nuclear material are present. But a far larger piece of *Stentor*, lacking a bit of cortex, does not regenerate even though adequate nuclear material is included. Undoubtedly the nucleus is necessary to con-

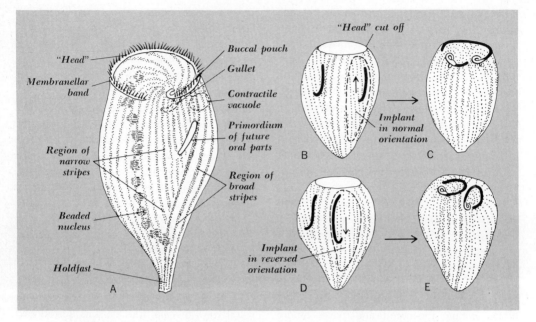

FIGURE 3–6   A. Schematic diagram of the ciliate protozoan *Stentor*. B and C. Tartar's experiment in which a region of the cortex with narrow stripes was implanted in a region of broad stripes in normal orientation. Result—a "doublet" *Stentor*. D and E. A similar experiment but with the implant revolved 180°. The implant developed a set of oral parts with reversed asymmetry. *(Based on V. Tartar, 1962, Advan. Morphogenesis, 2:1–26.)*

trol the processes by which substances of the cell are synthesized; but, Tartar concludes, the polarity and asymmetry of the body are transmitted from one generation to the next solely by the cortex. Two of Tartar's experiments are illustrated in Fig. 3–6.

In a somewhat similar fashion, the cortex of a metazoan egg transmits the pattern which largely controls the polarity, symmetry, and asymmetry of the future embryo. For a time these features are labile to varying degrees and subject to modification by the environment. Raven has studied the role of the cortex in the eggs of mollusks. When such eggs are centrifuged, the visible materials of the cytoplasm are displaced either toward or away from the axis of rotation. They are displaced toward the axis if they are less dense than the fluid cytoplasm which surrounds them, and away from it if they are more dense. In some cases, as we saw in Chapter 1, development proceeds normally except that the granules and droplets remain unnaturally distributed (see Fig. 1–8). In other cases, the displaced visible stuffs move back to their original locations. Now, since the internal cytoplasm is quite fluid and subject to flow (witness its movements during meiosis, fertilization, and cleavage), it must be the cortex which remains fixed and so transmits the pattern of the developing organism.

(Extreme centrifugation liquefies the cortex and results in abnormalities.)

The Adhesion of Cells

The cells of an embryo adhere to one another in varying degrees and in manners which are characteristic of the different tissues. At first the adhesions are weak. Cells easily dissociate when early embryos (blastulas and gastrulas) are placed in calcium-free solutions. As development proceeds, however, the adhesions between most cells become stronger. This is especially true in the case of epithelial cells, that is, in the case of cells which form layers and either bound the outside of the embryo or line cavities within it. Adhesion is especially close and strong immediately adjacent to where the contacts between cells border a cavity (Fig. 3–7, *A*). In effect, an epithelium presents a continuous outer surface—Holtfreter thought of it as a "coat"—which is semipermeable and which serves to separate the external medium which surrounds an embryo (or fills its cavities) from the internal medium which bathes its inner cells. Actually, it is not a continuous coat, but a network of close contacts between cells (often referred to as *terminal bars*). Except at these outer contacts pockets of liquid may separate the plasma membranes.

As development proceeds, adherent patches known as *desmosomes* develop between adjacent cells and strengthen the bonds between them (Figs. 3–1 and 3–7, *B*). The electron microscope sometimes shows filaments extending from each desmosome into the cytoplasms on both sides.

Other cells of the embryo, notably those known as mesenchyme, are less adherent to

FIGURE 3–7 Adhesion of epithelial cells. A. Early one-layered epithelium in which there is close contact next to the external surface. B. Later two-layered epithelium with desmosomes and a basement membrane. (Modified from E. H. Mercer, 1965, DeHaan and Ursprung (eds.), "Organogenesis," Holt, Rinehart and Winston, Inc., New York, pp. 32 and 33.)

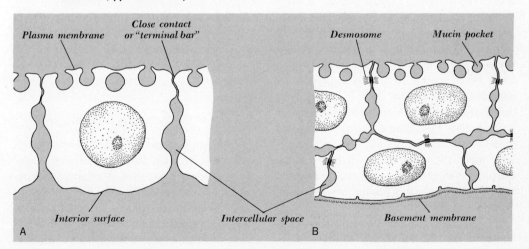

each other. Apparently they are free to wander independently; but they do not wander haphazardly. On the contrary, they move to appropriate positions in the embryo, and there they remain as if trapped by their surroundings.

What guides mesenchyme cells? What finally brings their wanderings to an end? In general, what holds embryonic cells to their places in the embryo?

Holtfreter has attempted to account for the direction of cell migration in terms of "contact guidance." Movement on a surface, he supposes, is controlled by the oriented macromolecular structure of the surface. He explains the final location of cells by the principle of "selective adhesion." That is, cells (or rather, cell surfaces) adhere most strongly to other cells for which they have an "affinity" (possibly a comparable surface pattern). When a cell makes contact with a congenial cell, its wanderings cease. Abercrombie terms the behavior "contact inhibition."

These responses of the cell surfaces are markedly specific. In some cases they appear to be species-specific, as when an egg cell reacts to sperm cells of its own species, or when lymphocytes react to foreign proteins (antigens) by producing antibodies. In other cases—and this is especially characteristic of embryos—the specificity of the cell surface seems to be tissue-specific. When an embryo is experimentally dissociated into its constituent cells and then the cells are permitted to reaggregate, they first adhere to each other and then tend to sort out and reform the tissues from which they were taken. Superficially, it seems as though each cell, when it contacts another cell, recognizes whether the other cell is of its own or another kind.

Steinberg offers a less mysterious explanation of contact guidance and selective adhesion. He explains segregation in terms of quantitative differences in the strength by which cell surfaces adhere. The principle is much the same as that which causes oil and water to separate; namely, the mean of the cohesive forces between like molecules (water to water and oil to oil) is greater than the adhesive forces between unlike molecules (water to oil). Similarly, intermingled cells sort out because like cell surfaces adhere to each other with more average strength than unlike surfaces adhere. Sorting out, therefore, is not so much a property of cells as individuals, as it is a property of cells in populations.

A corollary of this is that those cells of a mixed population which have the strongest mutually adhesive properties tend to pull together and become encapsuled within an outer mass of cells with less adhesive tendency.

The foldings and moldings of tissues during development (about which we shall have much to say in the following pages) have been interpreted in terms of programmed changes in the adhesive characteristics of cell surfaces. For example, a thickening of an epithelial layer of cells involves an increase in the areas of contact between cells. Presumably this results from an increase in the strength of their cohesiveness. A thinning of an epithelium is associated with a decrease in the areas of contact.

The Role of the Cytoplasm in Development

Many embryologists today are engaged in describing in detail the chemical and cellular changes which take place in an embryo as development proceeds. This is the problem of *cytodifferentiation*. Notable progress is being made with the aid of sophisticated physical and chemical techniques such as electron microscopy, radioactive tracers, and immunological methods. As a result, today we know something about the sequence of chemical events in the embryo and when and where new cellular structures appear and disappear. But it is one thing to *describe* the physical and chemical changes which take place during development, and quite another thing to give a *causal account* as to how these changes are brought about. The key question is: What controls differentiation?

There is no doubt today that the cytoplasm controls the course of cell differentiation. But there are two "levels" at which this control may be exerted: (1) It has been generally assumed that control takes place at the level of "transcription"; that is, that it takes place in the nucleus when DNA becomes activated and transfers its information to messenger RNA (pp. 36–38). Some genes are "turned on" at any given time and place, while other genes (probably most of them) remain repressed. (2) There is, however, increasing evidence that some control by the cytoplasm takes place at the level of "translation," namely, when mRNA supervises the synthesis of protein. In this case the mRNA produced in the nucleus remains "masked" until it reaches the proper time and place in the living system.

The classic example of control of differentiation at the level of translation is seen in the marine alga, *Acetabularia*. This is a giant cell, some two inches long, and shaped like a mushroom or parasol (Fig. 3–8). Its apical end spreads out radially like a hat or cap. Its basal end forms rhizoids (holdfasts), one of which contains the large nucleus. The entire life cycle of *Acetabularia* does not concern us now. Our present interest is in its capacity to regenerate a new cap when the old cap is cut off. Surprisingly, the regeneration of a cap in *Acetabularia*

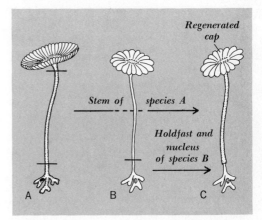
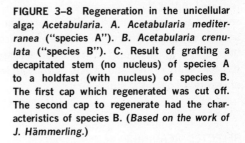

FIGURE 3–8 Regeneration in the unicellular alga; *Acetabularia*. A. *Acetabularia mediterranea* ("species A"). B. *Acetabularia crenulata* ("species B"). C. Result of grafting a decapitated stem (no nucleus) of species A to a holdfast (with nucleus) of species B. The first cap which regenerated was cut off. The second cap to regenerate had the characteristics of species B. (*Based on the work of J. Hämmerling.*)

takes place even when the nucleus has been cut away two or three weeks previously. Clearly the presence of nuclear DNA is not directly responsible for the restoration of lost parts.

Hämmerling, who studied these matters intensively, interpreted his findings in terms of gradients in the distribution of morphogenetic substances, i.e., a cap-producing substance with its highest concentration at the apical pole, and a holdfast-producing substance centered at the basal pole. His concept is that these gradients determine the location of the new parts.

The role which the nucleus plays in regeneration must be an indirect one. Presumably the genes of the nucleus produce the morphogenetic substances which then migrate within the cytoplasm and concentrate at the apex or base, as the case may be. There they stimulate regeneration and at the same time control the character of that which regenerates (Fig. 3–8). This interpretation is borne out by Hämmerling's observation that, when a piece of the stem of *Acetabularia mediterranea* (we shall call this "species A") is grafted by its basal end to a holdfast of *Acetabularia crenulata* ("species B"), the cap which regenerates resembles the caps of species A. He interprets this to signify that, at the start, regeneration is under the control of the morphogenetic stuffs which were emitted by nucleus A before it was removed. But if this first cap is cut away, the second cap which regenerates has the character of species B. This confirms the view that the nucleus, which is located at a distance from the apex both in time and space, is the source of the morphogenetic stuffs.

In terms of present theory we may identify the morphogenetic stuffs (as Brachet does) with messenger RNA. They are produced by transcription in the nucleus. Under the influence of the cytoplasm, the cap-forming stuff migrates and concentrates at the apical end of the cell. Here translation takes place; that is, the information brought by the mRNA is transferred *in situ* to newly synthesized enzymes and structural proteins. No theory of turning genes on and off is sufficient to account for this localized reaction.

What brings about the concentration of morphogenetic stuff at the apex? Since the inner cytoplasm of *Acetabularia* is engaged in active streaming, some factor in the cortex must be responsible. But it must be admitted that, as yet, we know very little of how this influence of the cortex originates or is exerted.

SELECTED READINGS

Ribosomes and the Synthesis of Protein

Brachet, J., 1961. The living cell. Sci. Am., **205** (September):50–61.

Ebert. J. D., 1965. Interacting Systems in Development. Holt, Rinehart and Winston, Inc., New York. Chaps. 5 to 8.

Hartman, P. E., and S. R. Suskind, 1965. Gene Action. Prentice-Hall, Inc., Englewood Cliffs, N.J. Especially Chaps. 7 to 9.

Scott, R. B., and E. Bell, 1964. Protein synthesis during development: control through messenger RNA. Science, **145**:711–713. Reprinted in Bell (ed.), 1965. Molecular and Cellular Aspects of Development. Harper & Row, Publishers, Incorporated, New York. Pp. 343–348.

Rich, A., 1963. Polyribosomes. Sci. Am., **209** (December):44–53.

Hurwitz, J., and J. J. Furth, 1962. Messenger RNA. Sci. Am., **206** (February):41–49.

Spiegelman, S., 1964. Hybrid nucleic acids. Sci. Am., **210** (May):48–56.

Spirin, A. S., 1966. On "masked" forms of messenger RNA in early embryogenesis and in other differentiating systems. In Moscona and Monroy (eds.), Current Topics in Developmental Biology. Academic Press Inc., New York. Pp. 1–38.

Collier, J. R., 1966. The transcription of genetic information in the spiralian embryo. In Moscona and Monroy (eds.), Current Topics in Developmental Biology. Academic Press Inc., New York. Pp. 39–59.

Mitochondria and the Energy of the Cell

Lehninger, A. L., 1960. Energy transformation in the cell. Sci. Am., **202** (May):102–114.

———, 1961. How cells transform energy. Sci. Am., **205** (September):62–73.

Giesy, A. C., 1962. Cell Physiology. W. B. Saunders Company, Philadelphia. 2d ed. Chaps. 5, 15, 16, and 17.

Green, D. E., 1964. The mitochondrion. Sci. Am. **210** (January):63–74.

Ball, E. G., and C. D. Joel, 1962. The composition of the mitochondrial membrane in relation to its structure and function. Intern. Rev. Cytol., **13**:99–134.

Lysosomes

De Duve, C., 1963. The lysosome. Sci. Am., **208** (May):64–72.

The Cell Cortex

Giesy, A. C., 1962. Cell Physiology. W. B. Saunders Company, Philadelphia. 2d ed. Chaps. 11 to 14.

Robertson, J. D., 1962. The membranes of the living cell. Sci. Am., **206** (April):64–72.

Curtis, A. S. G., 1963. The cell cortex. Endeavor, **22**:134–137.

Hokin, L. E., and M. R. Hokin, 1965. The chemistry of cell membranes. Sci. Am., **213** (October):78–86.

Tartar, V., 1962. Morphogenesis in Stentor. Advan. Morphogenesis, **2**:1–26.

Raven, C. P., 1964. Mechanism of determination in the development of gastropods. Advan. Morphogenesis, **3**:1–32.

The Adhesions of Cells

Moscona, A., 1957. The development in vitro of chimeric aggregates of dissociated embryonic chick and mouse cells. Proc. Natl. Acad. Sci., **43**:184–194.

———, 1961. How cells associate. Sci. Am., **205** (September):43–162.

Steinberg, M. S., 1963. Reconstruction of tissues by dissociated cells. Science, **141**:401–408. Reprinted in Bell (ed.), 1965. Molecular and Cellular Aspects of Development. Harper and Row, Publishers, Incorporated, New York. Pp. 64–76.

———, 1964. The problem of adhesive selectivity in cellular interactions. In Locke (ed.), Cellular Membranes in Development. Academic Press Inc., New York. Pp. 321–366.

Trinkaus, J.P., 1965. Mechanisms of morphogenetic movements. In DeHann and Ursprung (eds.), Organogenesis. Holt, Rinehart and Winston, Inc., New York. Pp. 55–104.

The Role of the Cytoplasm in Development

Brachet, J., 1964. The role of nucleic acids and sulphydryl groups in morphogenesis (amphibian egg development, regeneration in Acetabularia). Advan. Morphogenesis, **3**:247–300.

Ebert, J. D., 1965. Interacting Systems in Development. Holt, Rinehart and Winston, Inc., New York. See especially Chaps. 5, 7, and 8.

Gibor, A., 1966. Acetabularia: a useful giant cell. Sci. Am., **215** (November):118–124.

chapter **4**

GENERAL
FEATURES
OF
DEVELOPMENT

Development is the process by which a part of an organism becomes a whole organism. Developmental biology, therefore, includes more than the embryology of sexually reproduced individuals. It includes various forms of asexual reproduction, such as fission, budding, etc. It also includes regeneration and the healing of wounds. Wilhelm Roux and others compared organisms to candle flames. When one divides a flame in two, he immediately obtains two whole flames, not two half-flames. The process in organisms is less direct. The part, when it is isolated, begins processes of a complex nature which tend to restore wholeness. Some have called this "regulation." It is the essence of all development.

Development, however, is more than regulation, for an organism is not just a fixed pattern or ideal "whole." It exists in time as well as in space. An organism is a life cycle—egg, embryo, juvenile, adult, senescent individual. No stage in the life cycle is uniquely *the* organism.

How can one reconcile these two aspects of development, the regulative and the progressive or historical? A part of the answer is that the regulative aspect—the return of the individual to specific wholeness—depends on the integrity of the genetic factors which center in the nucleus. The progressive aspects of development, on the other hand, are best accounted for in terms of straightforward, apparently irreversible changes which take place in the epigenetic factors of the cytoplasm.

In this chapter we shall consider progressions of two sorts: *ontogeny,* the development of the individual, and *phylogeny,* the evolution of the race. In ontogeny the genetic factors presumably do not change. They are turned on and off. In phylogeny, on the other hand, as a result of mutations and selection, genetic factors have changed. A surprising proposition is

that, in evolution, epigenetic factors have changed relatively little. This is why embryologists can reason from the sequence of events in the development of one organism to the sequence of events in another.

ONTOGENY

The Ontogeny of a Sperm Cell

Actually there are two very different courses of development which a prospective germ cell may pursue under the control of its genes. Which of these it follows is determined by where it is located. If the immature germ cell, in this case a spermatogonium, is located in a tubule of a testis, it grows somewhat in size, its chromosomes pair in synapsis, and it becomes a primary spermatocyte. It then divides twice by meiosis and produces four spermatids. These metamorphose into four sperm cells (spermatozoa) of equal size.

This transformation of a spermatid into a spermatozoon is a process of differentiation, a true development (Fig. 4–1). It varies somewhat in detail in different species, but the basic features are the same. The genes themselves are not altered during the process, even though the nucleus becomes condensed in the head of the sperm and may assume different forms in different species.

The cytoplasm of the human spermatid shows at least three kinds of granules: (1) proacrosomic granules, (2) centrioles (two in number), and (3) mitochondria. During metamorphosis, the acrosomic granules (derived from the Golgi complex) gather toward one pole of the nucleus and condense into the form of a bead. The bead then spreads and forms a pointed cap or *acrosome* on one side of the

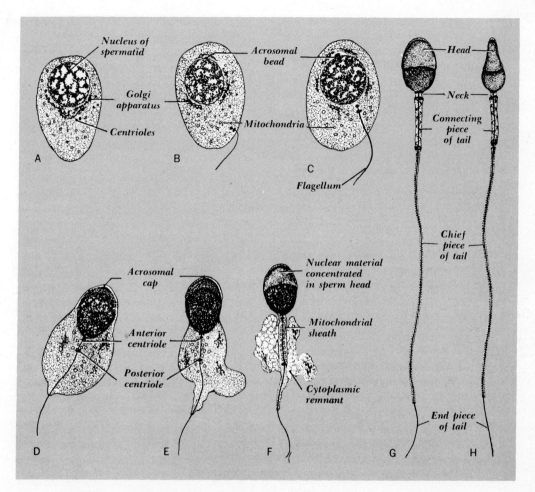

FIGURE 4-1 The development of the human sperm cell. (*From B. M. Patten, 1964, "Foundations of Embryology," McGraw-Hill Book Company, New York, 2d ed. After J. B. Gatenby and H. W. Beams, 1935, Quart. J. Microscop. Science, 78:1–29.*)

nucleus, the anterior side. At the same time, the two centrioles move toward the opposite side of the nucleus. Here the centriole nearest the nucleus spins a filament that grows outward and becomes the axis of the sperm's tail or flagellum. The centrioles then move apart. The proximal one remains close to the pos-

terior surface of the nucleus. The distal one forms a ring around the axial filament and recedes a short distance. Some of the mitochondria now gather around the axial filament in the space between the two centrioles and form the spiral of the middle piece.

Most of the cytoplasm, with its granules,

pinches away and is discarded; but a little remains as a delicate sheath over the head, middle piece, and tail, except at the tip. It is of interest that the immature sperm cells of vertebrates bury their heads in the cytoplasm of so-called Sertoli cells, which act as nurse cells. They remain attached in this manner until they are mature.

The Pre-embryonic Development of the Egg

If a prospective germ cell, in this instance an oogonium, is located within the cortex of an ovary, it grows greatly in size, undergoes synapsis, and becomes a primary oocyte. The growth is accomplished with the aid of surrounding follicle or nurse cells. Much of the increase is in the form of yolk platelets composed of lipoprotein and other nutrient materials. Immunological techniques have demonstrated that some of the proteins are synthesized elsewhere in the body and are transported to the oocyte by the body fluids. During oogenesis there is also a notable synthesis of ribosomal RNA. Furthermore, a sort of DNA appears in the cytoplasm of some growing oocytes. It is probably not genetically important but serves as a reservoir of nucleotides for use later in development.

Growth, as such, is an increase in size and amount of substance. It is not development, for development involves differentiation and the unfolding of a pattern. But development does in fact occur while the egg is in the ovary. The initial pattern (or pre-pattern) of the organization of the new individual is laid down. We may therefore speak of the period of the growth of the oocyte as *pre-embryonic development.*

The eggs of different animals vary greatly in the amount and distribution of the yolk which they contain (Fig. 4-2). Some, such as those of many annelids, mollusks, and echinoderms, are small and contain only a little yolk. If the yolk is evenly distributed they are called *homolecithal* or *isolecithal eggs.* Other eggs are larger and have a moderate supply of yolk. Of this sort are amphibian eggs and those of primitive fish. Still others, such as those of reptiles and birds, have a tremendous supply of yolk. Since, in these types, the yolk is concentrated toward one pole of the egg (known as the vegetal pole), these eggs are termed *telolecithal.* Arthropod eggs usually have the yolk massed toward the center, and hence are called *centrilecithal.*

The growth of the oocyte takes place under the influence and control of the genes (DNA) of the mother. Hence the characteristics which are acquired at this time are rightly designated as belonging to the *maternal inheritance.* They include, not only the basic stuffs from which the embryo is built, but also the formal distribution or arrangement of these stuffs, namely, polarity, sometimes bilaterality, and even asymmetry.

Polarity Ovarian eggs (oocytes), as is the case with all cells, are polarized. They have two unlike poles known as the *animal pole* and *vegetal pole,* respectively. Between the poles is the polar or egg axis. The nucleus is located nearer the animal pole; and it is here that the polar bodies are given off. (There are exceptions.) Usually the store of nutrient, mainly yolk, is more concentrated toward the vegetal pole. As a rule, this initial polarity becomes the outside-inside differentiation of the embryo; that is, mass movements of cells take place later by which the cells of the vegetal area sink in and become inner cells, while those of the animal area spread and cover the outer sur-

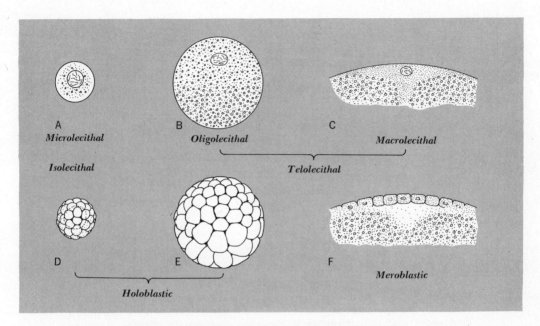

FIGURE 4–2 Terms used in describing eggs and cleavage.

face. Polarity may be present even in the eggs of some of the lower plants, although in some cases—for example, in the egg of the seaweed, *Fucus*—the polarity is for a time labile and subject to alteration by environmental influences.

Bilaterality A few eggs (insects, squids) develop a pattern of bilateral symmetry while still in the ovary. An egg of this sort has two differentiated axes: a primary polar (animal-vegetal) axis and a secondary dorsoventral axis. Only one plane will divide such an egg into symmetric halves. This is the median or sagittal plane which separates the right side from the left side. The halves are mirror images of each other. In most eggs bilaterality develops at the time of fertilization.

Asymmetry The cytoplasm of the ovarian eggs of certain freshwater snails (*Limnea*) possesses a third axis at right angles to the other two; that is, the right and left halves are not the same. Such eggs are therefore asymmetric. The evidence for this fact is not based on any visible differentiation within the cytoplasm, but rather on the inheritance of "handedness" in the spiraling of the embryos during later development (Fig. 4–3). Normally the embryos of this species coil to the right when looked at from either the apex or base, and hence are said to be dextral or right-handed (Fig. 4–3, *L*). But occasionally recessive mutants occur which coil to the left; that is, they are sinistral (Fig. 4–3, *A, F, G*). Now this is the significant feature: The direction of coiling is registered in the organization of the cytoplasm

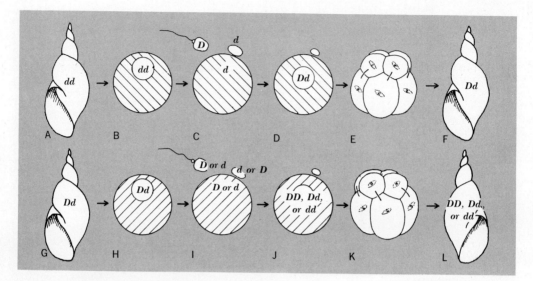

FIGURE 4–3   The inheritance of handedness in the freshwater snail *Limnaea. A to F.*
Parental cross. The ovum of a left-handed snail (dd) is fertilized by a sperm of a
right-handed snail (DD). According to Mendelian principles the offspring should be
right-handed because the gene for right-handedness (D) is dominant; but they are
left-handed because the cytoplasm was synthesized while the oocytes were under
the control of the mother's genotype. *G to L.* Filial cross. Irrespective of their genotype, the second filial generation of snails are right-handed because their cytoplasm
developed under the control of the dominant gene (D) which was present in all the
oocytes of the ovaries of the first filial generation. (*Schema illustrating the work
of A. E. Boycott and C. Diver as interpreted by A. H. Sturtevant, 1923. Science,
58:269–270.*)

that is formed while the oocyte is growing in
its mother's ovary and is under the control of
the mother's genes (Fig. 4–3, *B, H*). Asymmetry in these snails is therefore maternally
inherited, an expression of the mother's genotype. Thus a snail with a pair of recessive sinistral genes will produce eggs that give rise to
sinistral offspring, even though they are fertilized by sperm bearing a dominant dextral
gene. But the hybrids which result, although
they themselves are sinistral, produce dextral
offspring. It may well be that asymmetry of

this kind is present also in the cytoplasm of
vertebrate eggs.

What is the physical nature of the organization of the egg—its polarity, bilaterality, if any,
and its asymmetry? It has been suggested that
certain molecules of the ground substance of
the cytoplasm are oriented like the atoms in a
magnet, or the molecules in a crystal. This
might be true, but it is not the explanation of
asymmetry. If amphibian eggs are constricted
in the midplane at the two-cell stage, or even
at the blastula stage, they give rise to twins

which are identical in every respect except that sometimes the heart and viscera are mirror images of each other (page 154). It will hardly be argued that there are molecules in the twins which are mirror images of each other.

A more satisfactory theory is the gradient concept. According to this, eggs are polarized, bilaterally symmetrical, or asymmetric because there exists initially within them a pattern (or pre-pattern) of quantitative gradations of some sort, possibly in the distribution of ribosomal RNA or ribosomes. This is quite a different principle from that of an oriented intimate structure such as we find in a magnet or crystal. It must be emphasized, however, that patterns and gradients are descriptive terms which are still in need of analysis.

Gradients are labile. A continuous supply of energy is required to maintain them. Any influence which depresses the metabolism of an egg or embryo would be expected to level down the gradients and result in abnormal development. This, indeed, is what happens.

But whatever the nature of the initial gradient or pre-pattern, there soon arises within the egg a differential pattern with respect to the distribution of enzymes. Localized chemical reactions take place, and the regions of the egg become different chemically. Visibly different "stuffs" have played an important role in the history of embryology (Figs. 1–5 to 1–7). They have served as markers of the regions of cytoplasm which have different developmental futures. Some of the stuffs are yolk granules; others are wastes or by-products of metabolism. But the actual determining substances of the egg are most likely RNA and ribosomes. These reveal themselves in the nature and distribution of the proteins (enzymes and structural proteins) which they produce.

The Pre-fertilization Arrest

It is quite common for oocytes, as they approach maturity, to remain quiescent in the ovary for a long time with little or no further development. This is the case in echinoderms and amphibians. Finally, when the time for egg laying arrives, three events take place at approximately the same time: ovulation, further progress in meiosis, and developmental arrest.

Ovulation is the process by which the fully grown oocyte bursts from the ovary and, in most cases, enters the body cavity.

At about the same time meiosis is resumed. How far it gets before the egg is entered by a fertilizing sperm, varies in different species. When it is finally complete, however, a first and then a second polar body will have been given off by the oocyte with little loss of cytoplasm. A considerable rearrangement of the materials of the cytoplasm also takes place during meiosis. This is partly the result of the breakdown of the large oocyte nucleus (germinal vesicle) and the spread of its contents (nuclear sap) as a cap over the animal hemisphere (Fig. 1–7). The rearrangement is termed *ooplasmic segregation*.

Developmental arrest also occurs at approximately the same time that the egg leaves the ovary. After that, no new RNA is produced and no further development normally takes place until, and unless, the egg is released from its arrest by the entrance of a fertilizing sperm. In rare instances an arrest does not occur. Instead, the egg continues to develop and produce a new individual parthenogenetically. This is regularly the case in summer aphids and water fleas (*Daphnia*) and in the production of male ants and bees.

What brings this developmental arrest about? It is not likely that meiosis is a causal factor, for in some species (sea urchins, for example) the arrest does not take place until meiosis is complete (Fig. 2–5). In other species (for instance, the parasitic roundworm, *Ascaris*) the arrest precedes meiosis. In vertebrates the arrest comes during the process of meiosis. As a rule it takes place just after the first polar body has been given off and while the second meiotic division is at the metaphase. Possibly the arrest occurs because of some chemical change in the cortex of the egg. But at any rate the arrest has a chemical aspect, for it is accompanied by blockage of the synthesis of protein. Either the ribosomes are masked or the mRNA needed for protein synthesis is not available. Whatever its nature, the waiting egg is sensitive like a wound-up spring, ready to respond to a suitable stimulus, be it natural (the entrance of a sperm) or artificial (artificial parthenogenesis).

Activation

That aspect of fertilization by which an egg is released from its arrest and continues to develop is known as *activation*. The manner in which this is brought about has been studied most thoroughly in the eggs of marine invertebrates, notably in those of the sea urchin.

Frank Lillie was the first to show that a soluble substance is discharged by the newly laid sea urchin egg. It diffuses into the seawater and causes sperm cells in the neighborhood to become more active and to consume more oxygen. Tyler has found this substance in the jelly which surrounds the egg. Although there is no clear evidence that the sperm are attracted to the egg, the fact that they swim more rapidly when in the vicinity of an egg no doubt increases the likelihood of chance contact. The same secretion which activates the sperm causes the heads of the sperm to adhere together. Until recently, this has been considered a case of agglutination; but the electron microscope has revealed that, at least in some animals, when sperm cells are exposed to "egg water" (water in which ripe eggs have lain) or when they come into contact with egg jelly, the acrosomes of the sperm heads rupture and thrust out filaments (Fig. 4–4). It may be the tangling of these filaments that gives the appearance of agglutination. Lillie gave the name "fertilizin" to the substance produced by the egg, and he called the reacting substance in the head of the sperm "anti-fertilizin." The reaction between these substances is species-specific and in this respect has been compared to an antigen-antibody reaction. It helps to assure that an egg will be fertilized by a sperm of its own species.

The substance of the acrosomal granule is released when the acrosomal membrane ruptures. Apparently it is a lysin which dissolves a way through the jelly and membrane which surrounds the egg, thus permitting the acrosomal filament to come into contact with the plasma membrane of the egg. The eggs of some animals are surrounded by a thick envelope. In this case there is a special pore, the micropyle, which the sperm must find and penetrate.

Electron microphotographs of some marine animals, taken by Arthur and Laura Colwin, show that when the acrosomal filament from the head of the sperm comes into contact with the plasma membrane of the egg, it fuses with it, membrane with membrane (Fig. 4–4, *D*). Then the cytoplasm of the egg flows into the tubular filament, expanding it into a cone. As a result, the sperm is drawn toward and into the egg. A surface change quickly spreads over the egg which prevents other sperm from en-

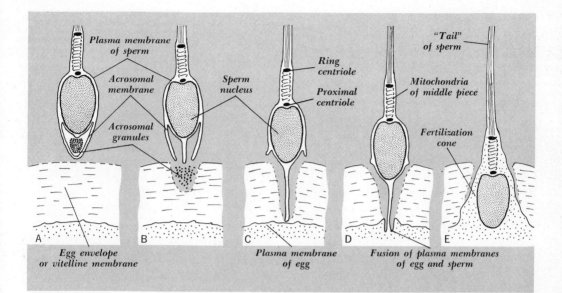

FIGURE 4–4 The entrance of a sperm into an egg (hypothetic). *A.* The sperm comes into contact with the egg envelope. *B.* The acrosomal membrane and plasma membrane of the sperm rupture and fuse; an acrosomal filament forms, and material of the acrosomal granule begins to dissolve a pathway through the egg envelope. *C.* The acrosomal filament contacts the plasma membrane of the egg. *D.* The plasma membranes of the sperm and egg fuse. *E.* A fertilization cone is formed through which the sperm then proceeds to enter the cytoplasm of the egg. *(Schematized and adapted from Arthur and Laura Colwin on the annelid worm Hydroides, 1964, in M. Locke (ed.), "Cellular Membranes in Development," Academic Press Inc., New York, pp. 233–279.)*

tering. At the same time, a membrane, often the vitelline membrane, now known as the *fertilization membrane* (Fig. 4–5), arises off the surface. Granules in the egg cortex discharge and form a thin jelly-like *hyaline layer* between the fertilization membrane and the living cytoplasm within (Fig. 4–5, *B*). Although no single account will apply to all the eggs which have been studied, it may be said that, in general, the new outer layer of the cytoplasm, namely, the new plasma membrane, is more permeable, more irritable, and exchanges oxygen and carbon dioxide with the environment at a greater rate than the older membrane did.

Almost at once, a radiating sphere of gel (cytaster) forms inside the cytoplasm of the egg, usually around the middle piece (centriole) of the sperm, as a center (Fig. 4–5, *C*). As the cytaster grows, it engulfs the sperm head (male pronucleus) and the nucleus of the egg (female pronucleus) and sweeps them together at its liquid center (Fig. 4–5, *D*). This coming together of the two gametes is known as *syngamy,* and the close association of the two pronuclei is called *karyogamy.* The

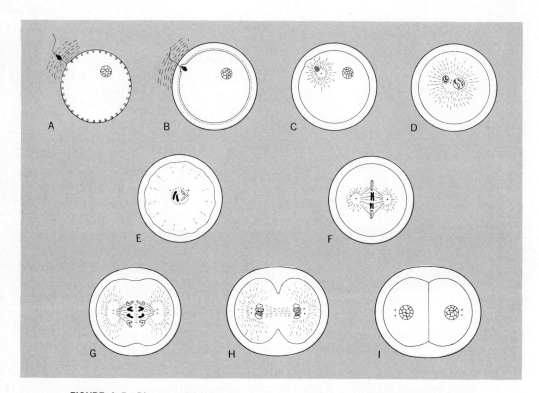

FIGURE 4–5 Diagram of fertilization and the first cleavage. Schematic figures based mainly on the sea urchin egg.

activation of the egg may now be said to be complete.

Activation is accompanied by a prompt and sharp increase in the synthesis of proteins. In some way, not as yet understood, the ribosomes and/or the mRNA already present in the oocyte are unmasked and begin to synthesize the polypeptide chains from which the new protein molecules needed in cell division are fabricated.

Syngamy and karyogamy are not necessary features in activation. Unfertilized eggs of various echinoderms, insects, and amphibians have been stimulated to development by artificial means without the presence of sperm cells. The phenomenon is termed *artificial parthenogenesis*. The unfertilized egg may be compared to a relaxed muscle cell or a resting nerve cell in that it is in a state of readiness to respond to any sufficient stimulus. The stimulus serves to release energies already present in the responding cell. The stimulus does not determine the nature of the response.

Very diverse stimuli have proved effective in activating eggs: heat and cold, electric shocks, ultraviolet radiation, pricking, friction, chemical changes in the surrounding medium, such as an increase or decrease in the salt content, acidity, alkalinity, change in ion content, alkaloids, and fat solvents. What do all these agents have in common? They irritate (stimulate) the plasma membrane of the egg so that it responds in its own characteristic way. It reorganizes the cortex and usually throws off a fertilization membrane.

Generally speaking, the result of artificial parthenogenesis is a weak development leading to the production of an abnormal embryo. Some embryos have been carried through to maturity. But why the weakness? Is it because the embryos are haploid and have an insufficient number of chromosomes (they are not always haploid), or is it because of the absence of the centriole that the sperm usually supplies; or is there some other explanation? As yet no clear answer has been found.

Cleavage

Activation is followed by cleavage, that is, by a rapid series of mitotic cell divisions in which the fertilized egg, or zygote, splits into cells of ever smaller size. The nuclear material (DNA and accompanying protein) is replicated after each cell division, but the cytoplasm does not increase. As a consequence, the ratio of the nuclear material to the cytoplasm augments until it is several times greater than it was at the start. Moreover, the ratio of the cell surface to the cell volume increases. (A cell which has half the diameter has a surface area one-fourth as great and a volume one-eighth as large. Hence the surface-to-volume ratio is doubled.) These facts, together with the fact that the nuclear substance becomes distributed throughout the embryo, may account, at least in part, for the increased activity of the cells which is observed as cleavage progresses.

Cleavage patterns Cleavage cells are known as *blastomeres*. They vary in size and content, principally by reason of differences in the amount and distribution of the yolk and other cytoplasmic inclusions which they contain. In those eggs in which yolk is meager in amount and fairly evenly distributed (isolecithal eggs), the cleavage cells are of approximately the same size. In the eggs of amphibians and primitive fishes, the yolk is more abundant and is concentrated toward the vegetal pole. In this case the cells nearer the animal pole tend to be smaller and are called *micromeres*. The usually larger cells nearer the vegetal pole are termed *macromeres*. But as long as the entire egg divides into cells, the cleavage is said to be complete or *holoblastic* (Fig. 4–2, *D* and *E*). In the case of sharks, reptiles, and birds, however, yolk is abundant and fills the egg, except for a thin disc of cytoplasm at the animal pole and an even thinner layer of cytoplasm around the periphery of the egg. In such eggs cleavage is incomplete and confined to the small area which surrounds the animal pole. The rest of the egg remains unsegmented. Such cleavage is termed incomplete or *meroblastic* (Fig. 4–2, *F*). In higher bony fish, although the eggs are small, the yolk is so highly concentrated that cleavage is meroblastic. In arthropod eggs, the yolk is concentrated toward the center (centrilecithal); and then, after the nucleus divides several times, the several offspring nuclei migrate to the periphery of the egg where meroblastic cleavage of the peripheral cytoplasm takes place.

There are two basic patterns of cleavage in the animal kingdom: radial cleavage and spiral

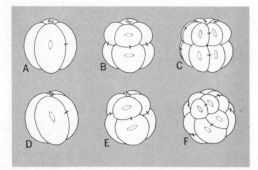

FIGURE 4–6 Types of cleavage. *A* to *C.* Radial cleavage; characteristic of echinoderms and chordates. *D* to *F.* Spiral cleavage as found in annelids and mollusks. (Compare Fig. 1–4.)

cleavage (Fig. 4–6). A third type, bilateral cleavage, is also described (ctenophores and tunicates).

In typical radial cleavage as found in echinoderms and amphibians, the first mitotic spindle elongates in a direction at right angles to the egg axis (Fig. 4–6, *A* to *C*). Hence the plane of first cleavage is meridional and includes the egg axis. The second spindle also elongates transversely to the egg axis and at right angles to the first spindle. The result is four meridionally arranged blastomeres. The third cleavage spindles elongate in a direction parallel to the egg axis, that is, vertically, if one thinks of the animal pole as upward. In consequence, the third cleavage planes are transverse (horizontal), and the eight-cell stage consists of four animal blastomeres and four vegetal blastomeres. But the cleavage pattern is not always simple. In the sea urchin, for example, the fourth cleavage planes are vertical in the animal hemisphere (Fig. 1–6, *E*). This results in eight similar animal blastomeres called "mesomeres" because they are intermediate in size. In the vegetal hemisphere the fourth cleavage planes are horizontal. They cut off four small vegetal "micromeres" from four larger subequatorial "macromeres".

In spiral cleavage, as found in annelids and mollusks, the mitotic spindles are oblique (Fig. 4–6, *D* to *F*). This is particularly obvious at the third cleavage, when the animal blastomeres, instead of being located directly above the vegetal blastomeres, are shifted to one side. The shift is clockwise (dextral) in most species, but counterclockwise (sinistral) in some others. Thus the quartet of animal blastomeres comes to lie in the notches between the vegetal blastomeres.

The spiral type of cleavage is the rule in annelids, mollusks, and in some flatworms and nemertean worms. So regular is this pattern of cleavage, in which the spindles slant alternately in one direction and then in the opposite direction, that embryologists have been able to designate each blastomere by a letter or number and to trace them to their certain fates in definite parts of the embryo or larva. Cell-lineage studies of this sort were referred to in Chapter 1 (Fig. 1–4).

In general, cleavage spindles elongate at right angles to the previous cleavage spindle. They also tend to elongate in the direction of the greatest protoplasmic mass. As a result, the cleavage planes cut across the smallest diameter of the cell and so divide the cell into more nearly spherical daughter cells than would otherwise be the result.

The mechanism of cleavage Anyone who watches cleavage as it takes place in a living egg will be moved to inquire as to what unseen forces are at work. How is the division of the cytoplasm accomplished? That energy is being expended is certain, for the division of a cell into two cells increases the surface area of the protoplasm and is therefore opposed by surface forces. The mechanism of cleavage in the sea urchin involves the elongation of the mitotic spindle and the growth of the asters at its poles (Fig. 4–5, *F* to *H*). These are spheres of gel with a radiating bristly structure like a burr. When the lengthening rays reach the cortex in the neighborhood of the poles of the spindle, the cortex expands, apparently by solation. At the same time the cortex at the equator of the spindle contracts as a gel ring and so starts the process by which the cell is pinched in two. This, however, is not the whole story, for the division of the cell continues to completion even after the cortex at the equator has relaxed and returned to its original position (Fig. 4–5, *I*). There is a sort of systole followed by diastole of the equatorial ring of cortex. The final separation of the daughter cells, at least in some instances, is accomplished by the formation of a cell plate at the equator of the spindle. In plants the formation of such a plate is the principal method of dividing the cytoplasm.

What holds the cleavage cells together? The truth is that they are not very firmly held together. In sea urchin blastulas the thin jelly (hyaline membrane) which was given off at the time of fertilization helps to keep the cells in contact with each other. If the membranes are removed—as can be done by placing the eggs in calcium-free seawater—the cells fall apart. However, by the time gastrulation has begun, the adhesive property of the cell surfaces has increased. If the cells are then drawn apart, microfibers which connect them can be seen.

The rate of cleavage At first the cleavage divisions take place rapidly and rhythmically. What determines their rate? Temperature has much to do with it, as indeed it has to do with the rate of all physical and chemical processes. But the rate is different for different species. Some species, even species of the same genus, cleave more rapidly than do others. It is of interest in this connection that when the eggs of one species are fertilized by the sperm of another species, cleavage takes place at the rate which is characteristic of the mother species. It is the cytoplasm which was synthesized in the ovary under the influence of the mother's genes, therefore, which determines the rate of cleavage.

The chemical events which accompany cleavage This brings up a fundamental proposition which has been verified biochemically. (Details must be left to biochemical accounts of development.) During the period of rhythmic cleavage, while the DNA of the nucleus is busy replicating itself, little or no ribosomal RNA is produced. Nucleoli, the temporary storage depots of ribosomal RNA, are absent from the nuclei of the cleavage cells, and ribosomes do not increase in the cytoplasm. Yet protein is being synthesized. What protein is it, and what RNA supervises its production?

Present evidence indicates that the proteins produced during cleavage are those which are used to build mitotic apparatus and to supply the framework of the multiplying chromosomes. Perhaps they are also needed for the increasing cell surfaces. The RNA involved in this synthesis, at least the ribosomal RNA, is the old RNA which was produced in the ovary during oogenesis under the influence of the mother's genes.

In some illuminating experiments, cleaving

eggs were exposed to actinomycin D, an anti-biotic which prevents transcription; that is, it prevents the production of new RNA at DNA templates. Yet, in spite of this "chemical enu-cleation" of the eggs, cleavage continued quite normally. It will be recalled that sea urchin eggs from which the nucleus has been re-moved continue to cleave. This seems to in-dicate that, not only is ribosomal RNA not produced during cleavage, but new messenger RNA is not required. When, however, the cleav-ing eggs are treated with puromycin, an anti-biotic which inhibits translation, i.e., the pro-duction of protein by RNA, cleavage comes immediately to a halt.

As the period of cleavage draws to a close, the cell divisions become less frequent and more irregular. Most of the cells at a given moment are in interphase, and their nuclei are engaged in the production of mRNA and tRNA. Nucleoli reappear in the nuclei. Mes-senger RNA, migrating into the cytoplasm, joins the ribosomes already present in the synthesis of new protein. Soon active differ-entiation gets under way. (The account of chemical changes following cleavage is con-tinued in Chapter 5.)

The fate of cleavage cells   What determines the fates of the blastomeres? Generally speak-ing, there are two possible answers: (1) the fate of a blastomere is decided by the "stuffs" which it contains; or (2) the fate of a blasto-mere is determined by its location with respect to other blastomeres.

Experiments on the eggs of annelids and mollusks seem to indicate that the first answer applies (Fig. 1–4). If a blastomere, or group of blastomeres, is isolated, it becomes just what it would be expected to become by reason of the material within it. Such eggs, made up

of self-differentiating parts, are called mosaic eggs. The egg of the toothshell, *Dentalium,* may be cited as an example. E. B. Wilson (1904) found that it possesses, near its vegetal pole, a stuff which ultimately becomes in-corporated into the mesoderm of the larva (Fig. 4–7). At the first cleavage this material bulges out as a first "polar lobe" and is passed on to only one of the first two blastomeres, the blastomere designated CD. The lobe is then withdrawn. At the second division a sec-ond polar lobe is formed and passed on to one of the four resulting blastomeres, the blastomere labeled D. Now, if at either stage the lobe is cut away, an imperfect larva is formed which lacks mesoderm. The entire posterior portion of the embryo is missing, and the anterior part is abnormal. Observations such as these by Wilson, Conklin, and others laid a firm foundation for the mosaic theory of development (pp. 5–7).

In the eggs of echinoderms and vertebrates, on the other hand, the fate of a blastomere is decided, not only by its stuffs, but also by the position which it occupies with respect to other blastomeres and their stuffs. Sven Hörstadius has found it possible, by careful hand-dissec-tion with finely drawn glass needles, to sep-arate the individual blastomeres or groups of blastomeres of sea urchin eggs and then to recombine them in various combinations. Now at the 16-cell stage the sea urchin egg con-sists of eight mesomeres surrounding the animal pole, four macromeres below the equa-tor, and four micromeres at the vegetal pole (Fig. 4–8, *A*). In normal development, the eight mesomeres become ectoderm and pro-duce a small apical tuft of cilia, a ciliary band, and a stomodaeum (the oral cavity lined by ectoderm). When, however, the mesomeres are cut away and reared as an isolated group,

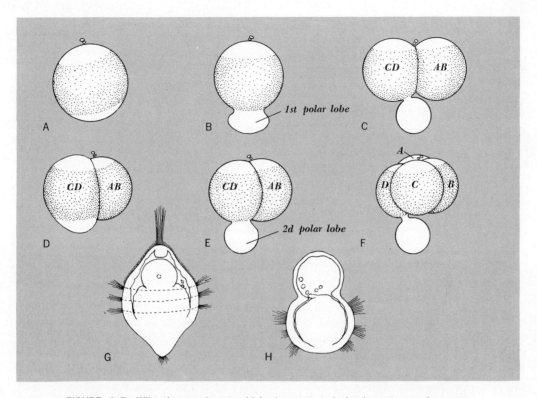

FIGURE 4–7 Wilson's experiment which demonstrated the importance of a cytoplasmic "stuff" in the development of the tooth shell *Dentalium. A to F.* The first two cleavages, showing the first and second polar lobes. *G.* A normal larva. *H.* An abnormal larva which developed when a first polar lobe was cut off. The result of removing the second polar lobe is similar. (*After E. B. Wilson, 1904, J. Exptl. Zool., 1:1–72.*)

they become a ciliated ectodermal vesicle with a greatly exaggerated apical tuft of cilia but with no ciliary band or stomodaeum (Fig. 4–8, *B*). But if micromeres are grafted to the group of mesomeres, the combination may become an essentially normal larva. In this case the mesomeres contribute to the gut endoderm, a thing they normally never do.

The cells of the vegetal hemisphere (macromeres and micromeres), cut off by themselves,

give rise to an unbalanced larva with very little ectoderm but with a huge gut (endoderm). The gut may be unable to cave in, as it normally does, by the process of gastrulation. Instead, it may bulge outward (exogastrulate) (Fig. 4–8, *C*).

Now a strange chemical effect has been discovered: If sea urchin eggs are treated with sodium thiocyanate, zinc or mercury salts, or trypsin or various other agents, they become

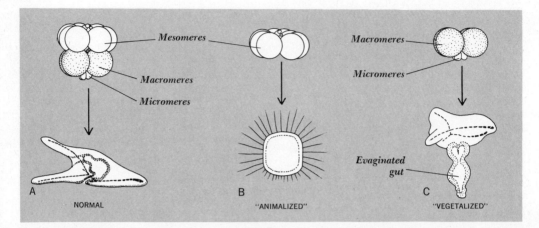

FIGURE 4–8 Examples of "animalized" and "vegetalized" embryos of the sea urchin. *(Adapted from illustrations by Sven Hörstadius.)*

"animalized"; that is, they develop into ecto-dermal vesicles comparable to those which are produced by isolated animal mesomeres. There is an exaggerated apical tuft of cilia, but no ciliary band, no stomodaeum, and no mesenchyme. The cells which normally would have become endoderm and mesoderm be-come ectoderm. If, on the other hand, sea urchin eggs are exposed to lithium salts, sodium azide, or dinitrophenol, they tend to be "vegetalized." They develop like isolated vegetal halves (macromeres and micromeres). Only a small region at the animal pole becomes ectoderm.

What would happen if an isolated animal half (mesomeres only) of an egg were to be treated with a vegetalizing chemical? This ex-periment was performed by von Ubisch, who found that in favorable cases abnormalities are avoided and a small but fairly normal larva may result!

These and other observations have led to speculation concerning a functional gradient pattern in the sea urchin egg: an animal gradi-ent with its center at the animal pole, and a vegetal gradient centering in the micromeres at the vegetal pole. According to another in-terpretation, there is only a single bipolar gradient, one which produces unlike effects at its opposite poles. In any case, normal devel-opment requires a proper balance between what seem to be two influences. Animalizing agents act by suppressing the vegetal influ-ence. Vegetalizing agents suppress the animal influence. What a cell becomes, therefore, is determined not only by the stuffs in its cyto-plasm, but also by its position in the functional gradient pattern of the whole.

How does one resolve the apparent conflict between experiments on mosaic eggs and ex-periments on regulative eggs? The difference in part is one of timing, and in part it is a ques-tion as to which morphogenetic factors are emphasized. In mosaic eggs (e.g., annelid and mollusk eggs) a considerable amount of re-gional differentiation (localization) takes place within the functional gradient pattern of the egg as a whole before cleavage begins. According to current theory, the now-differentiated stuffs become segregated by cleavage planes and then turn appropriate genes on or off in their respective regions of the embryo. In regulative eggs (such as those of echinoderms and vertebrates) a lesser amount of differentiation precedes cleavage. Relatively more differentiation follows cleavage and results from inductive interactions between the germ layers. Actually, the contrast between mosaic and regulative eggs has been grossly exaggerated. There is localization and interaction in both.

Gastrulation

The result of cleavage is a *blastula,* typically a ball of cells surrounding a central cavity, or *blastocoel* (Fig. 1–6, *G*). Cell divisions by mitosis continue after the blastula is formed, but, generally speaking, the divisions alternate with intervals of cell growth. Hence no further decrease in the size of cells takes place. Instead, mass movements of cells commence by which the single-layered ball of cells, the blastula, becomes transformed into a two- or three-layered ball of cells (two-layered in coelenterates), the *gastrula* (Fig. 1–6, *G* to *I*). The process is known as *gastrulation,* and the layers of cells are called *germ layers.*

We shall first describe gastrulation as it occurs in echinoderms and amphibians. One side of the blastula folds inward and partially obliterates the original cavity or blastocoel (Fig. 4–9, *A* to *C*). A new cavity open to the outside is formed. It is known as the *archenteron* or primitive gut. The opening from the archenteron to the outside is the *blastopore* (primitive pore). The cells which border the

blastopore constitute the *blastoporal lips.* The outer germ layer of the gastrula, when gastrulation is finally complete, is the *ectoderm.* The inner germ layer is the *endoderm.* The third germ layer, which in vertebrates includes the *notochord,* is the *mesoderm.* It occupies a position intermediate between the ectoderm and endoderm. Loose cells which are commonly present between the germ layers are termed *mesenchyme* (Fig. 4–10, *A*). They are usually thought of as mesoderm, although they may arise from any of the three germ layers.

In echinoderms and amphibians, gastrulation is essentially an inward folding, and is accomplished by three easily defined processes: invagination, involution, and expansion (Fig. 4–9, *B*).

1 *Invagination,* or "caving in," is the process by which cells of the vegetal area pocket inward to form the archenteron. Note that the invaginating layer of cells is concave when looked at from the outside.

2 *Involution,* or "rolling under," is the process by which the cells which border the vegetal area move toward the blastopore and progressively roll around and beneath the lips of the blastopore, and so come to lie within the egg. Note that during involution the layer of cells which is in process of passing around the lips of the blastopore is convex when seen from the outside.

3 *Expansion* is the process by which the cells of the animal hemisphere extend and converge toward the blastopore. As a result of the coordination between these three processes, an echinoderm or amphibian egg during gastrulation does not cease to be a sphere.

In echinoderm eggs, such as those of a starfish or sea urchin, the movements of gastrulation are fairly equal on all sides (Fig. 1–6). The newly formed archenteron lies near

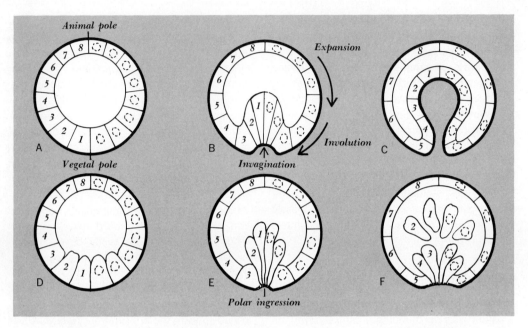

FIGURE 4–9 Schema illustrating types of cell movements which take place during gastrulation and mesoderm formation. The heavy line indicates the original cortex. A to C. In invagination and involution the cells retain connection with the cortex. D to F. In polar ingression, attachment to the cortex is broken.

the center of the vegetal area. In vertebrates, however, because of the greater mass of yolk in the vegetal hemisphere, the process of gastrulation takes place predominantly on the dorsal side; involution is mainly around the dorsal and dorsolateral lips of the blastopore, and expansion is more of the dorsal than of the ventral surface (Fig. 5–10). As a result, the archenteron is dorsal in position. It is almost as though the future endoderm of the vegetal area swings inward upon the ventral lip of the blastopore as though on a hinge, and so comes to lie within the ventral half of the gastrula.

There are, however, other methods of gastrulation in the animal kingdom than the

method just described. In coelenterates, the stage which corresponds to the gastrula is usually a free-swimming larva known as a *planula*. It consists of an outer layer of ectoderm and an inner solid core of endoderm. A cleft later appears in the endoderm (cf. the archenteron) which forms a mouth by breaking through the ectoderm to the outside (cf. blastopore).

In annelids, mollusks, and certain other animals (mainly the animals which undergo spiral cleavage), there is an inward migration of cells at the vegetal pole (Fig. 4–9, *D, E*). This process is termed *polar ingression*.

In the yolk-rich eggs of reptiles and birds, the endodermal cells appear to split away from the inner surface of the outer layer by a process known as *delamination*. The same process takes place in the eggs of mammals which behave in gastrulation as though yolk were still present. There is some question, however, as to just what the cells which delaminate represent (see Chap. 9).

Now it is quite possible that these various processes are modifications of a single type of gastrulation, although it is not easy to understand how this could be so.

The mechanisms of gastrulation Because of the variety of processes of gastrulation, the following account is based largely on the sea urchin. During gastrulation a ball of cells folds and differentiates into the first tissues of the embryo. How is this accomplished? It seems to be a movement of groups of cells. Actually it is, at the start, the result of individual cells changing their shape. This in turn may well be the result of changes in the adhesive properties of the cell surfaces. It is not an osmotic

FIGURE 4–10   The development of the pluteus larva of the sea urchin. This figure continues the story of development begun in Fig. 1–6.

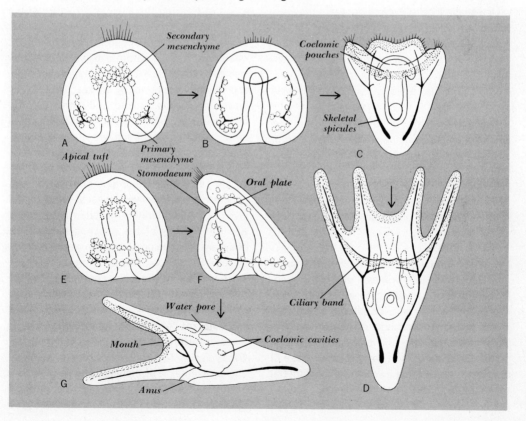

phenomenon, that is, a caving-in of the blasto-coel due to the withdrawal of fluid from the cavity. This is proved by the fact that it will take place even when the blastocoel is opened. In some cases, small groups of cells will perform their characteristic movements even when separated from the rest of the egg.

The outer layer of the blastula (cortex) plays an important part in gastrulation. The cells of the vegetal pole become wedge-shaped by the withdrawal of most of their cytoplasm inward, while at the same time retaining connection with the surface by a narrow neck of cortex (Fig. 4–9). The result is invagination and the formation of the archenteron. When next these wedge-shaped cells expand to their previous cuboidal shape, the gut is enlarged. If for any reason invagination fails to complete itself, then, when the wedge-shaped cells regain their cuboidal shape, an outpocketing (exogastrulation) instead of an inpocketing may occur.

Various explanations have been suggested for the changes in shape which cells undergo. Surface tension is not strong enough to account for the movements. Changes in the adhesiveness of cell surfaces are probably most important. Solation (i.e., decrease in viscosity) and the contractile properties of gels also may be involved.

Mesoderm formation   Mesoderm formation is one aspect of the process of gastrulation. Mesoderm-forming cells may enter within the egg either before, with, or after the invagination of the definitive endoderm. Because of the variation in the time and place of mesoderm origin, the inner cells at the two-layer stage of development are often given the noncommittal name of *hypoblast,* and the outer cells are called *epiblast.*

In sea urchin development certain loose cells (primary mesenchyme) wander inward from the vegetal pole in advance of the invagination of the future gut endoderm (Fig. 1–6, *H*). Unlike the wedge-shaped cells of invaginating endoderm, these do not remain attached to the surface coat and to each other, but instead they move independently. They throw out filamentous pseudopods which attach to the inside of the wall of the blastula and then contract. In starfish development similar cells separate from the tip of the gut after it has formed (Fig. 4–11, *D*). In both cases the inwandering cells take a position in the space between ectoderm and the gut. Here they play a part in weaving the skeletal structure of the larva (Fig. 4–10). It is an impressive fact that these cells, although apparently free to wander wherever they will within the larva, nevertheless go to the proper place, at the proper time, and do the proper thing. In sea urchins they form a ring at the equator and then break into two groups of cells, a group on the left and a group on the right. Each group becomes triangular, and beginning at its center calcareous spicules form.

Later in the development of echinoderms, pouches form from the sides of the gut and give rise to the definitive mesoderm (Fig. 4–11, *E*). The cavities of the pouches become the body cavities (coeloms) and the water-vascular system. Similar pouches give rise to the mesoderm in some of the lower chordates. In these the roof of the gut between the pouches becomes the notochord.

Mesoderm forms in a different manner in annelids, arthropods, and mollusks (Fig. 4–11, *B*). Typically, two cells, known as *mesoblasts,* one on each side of the vegetal pole, enter the blastocoel, proliferate, and form two longitudinal bands of mesoderm. As the bands grow forward they segment and produce a series of clefts which become the coelomic cavities.

Formation of Mouth and Anus

There are two ways in which mouths develop in the animal kingdom. In echinoderms and chordates, the mouth is a new formation (Fig. 4–11, *D* to *F*). It is produced when the anterior tip of the gut grows forward and comes into contact with a shallow pocket of ectoderm (stomodaeum) on the ventral surface. The

oral plate, which is thus formed by the fusion of ectoderm and endoderm, breaks through so that the mouth opens into the gut. Those animals which form their mouths in this fashion are known as *deuterostomes*. They are, generally speaking, the same animals which cleave in radial fashion and form their mesoderm from the sides of the gut.

In annelids and mollusks, on the other hand, the blastopore becomes the mouth (Fig. 4–11, *A* to *C*). The anus is either a new formation, or else both mouth and anus are formed from

FIGURE 4–11 The development of protostomes and deuterostomes (idealized). *A* to *C*. Gastrulation, mesoderm formation, and the trochophore larva of a protostome. *D* to *F*. Gastrulation, mesoderm formation, and an echinoderm-type larva of a deuterostome.

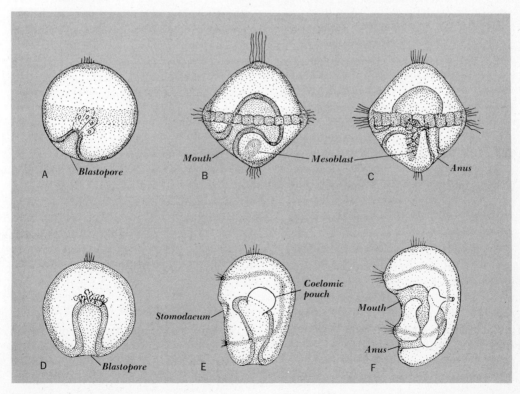

the blastopore by the coming together of the lateral lips of the blastopore. (In some cases the mouth closes and later reopens as an apparently new formation.) Animals in which the mouth is formed from the blastopore are termed *protostomes*. They are, generally speaking, the ones whose eggs cleave spirally and which form their mesoderm from posterior mesoblasts.

Embryos and Larvae

At every stage of development, an organism must have a source of energy. If the egg is very small and there is little energy stored away as yolk, then the organism must hatch while yet it is very immature, and must begin to feed for itself. Such an immature stage, specialized for feeding (or for finding a new location where it can feed), is known as a *larva*. Larvae are nature's way of bridging the gap between a tiny egg and a complex adult. The life of a larva is always precarious. The chances that it will become a morsel of food for another animal are tremendous. Hence, those animals which produce small eggs and larvae always produce them in great numbers. Often a profound metamorphosis takes place between the larval stage and the young adult stage, during which the organism undergoes a radical change in its structure and mode of life.

One direction of progress in the animal kingdom has been toward providing better methods for the nourishment and protection of the young. Stages in development, which in more primitive creatures are passed as larvae, are in these more advanced forms passed as embryos. In this case hatching does not take place until the young organism is far enough advanced to be termed a juvenile. Take the crustaceans as an example. The lower members of this class have a larval stage known as a nauplius in which there are three pairs of bristly appendages for swimming and chewing (Fig. 4–13, *C*). In the higher crustaceans (lobsters, crabs, etc.) the "nauplius stage" is passed over rapidly while the organism is yet a yolk-rich embryo. Similarly, the frog hatches as an immature tadpole, while the bird and mammal go through comparable "tadpole stages" rather early in their embryonic development.

The opposite evolutionary trend may have taken place in the case of the higher insects. Instead of larval stages becoming embryonic stages, embryonic stages may have become larval. It is possible to think of caterpillars, maggots, and grubs as embryos which have adapted to free life in order to make use of certain rich sources of nutrition. If this interpretation is correct, then the profound metamorphosis which these insects undergo is in truth a continuation of their embryonic development.

PHYLOGENY

Discussions such as these lead at once to the subject of phylogeny, the history of the race. Embryology has much to say concerning evolution and the interrelations of the great animal phyla. It supplements the record of the rocks where this record is weak.

Von Baer's Principle and Haeckel's Law

Von Baer, the great embryologist of the early nineteenth century, called attention to the fact that those characters which define a more inclusive group of animals, such as a phylum, develop earlier than the characters which distinguish a group of lesser rank, such as a class. Thus the notochord, dorsal neural tube, and gill clefts, which define the phylum Chordata, appear earlier in development than the jaws and limb buds, which distinguish most of the vertebrate classes. And those characters which distinguish species are ordinarily the last to appear. Indeed, in some instances, species differences do not become apparent until the animal has reached maturity.

Haeckel, in 1868, incorporated Von Baer's principle into a broad generalization which he called "the biogenetic law." Today it is called the "recapitulation theory." Those characters which are common to a large group of animals, he reasoned, must also have been the characters of the remote common ancestors of the group. Now, since they appear earlier in development, and the characters which define lesser assemblages appear later, it must be true that an organism, in its development, recapitulates the history of its race.

Haeckel's law awakened tremendous enthusiasm as well as controversy among biologists. Many were moved by the thought that here right before our eyes, and in the course of a few days or weeks, there is being reenacted the long history of the race. Embryos and larvae recall ancestors, or at least so it seemed. The uncleaved egg recalls the one-celled protozoan ancestor. Because most animals pass through a one-layered (blastula) and then a two-layered (gastrula) stage, Haeckel imagined hypothetical ancestors which he named "blastaea" and "gastraea." Coelenterates, although they may become complex in other ways, have not gone beyond the gastraea stage.

The principal weakness of the theory of recapitulation is that it overlooked the great degree to which larvae and embryos have acquired characters which adapt them to their special larval and embryonic environments. The fetal membranes of a reptile or bird, for example, have evolved as adaptations to development on land. They are of more recent origin (cenogenetic), and they are not ancestral (palingenetic). In addition the placentas of mammal embryos are purely embryonic organs, which permit the embryo to exchange substances with the uterus of its mother. It is obvious that evolution has progressed by modifying stages of the life cycle, even early stages, and not solely by adding new stages to the end of previous life cycles. Hence, it is simply not true that embryos recapitulate ancestral history in any necessary or slavish way. Yet in spite of the weaknesses of the theory of recapitulation, it is certain that embryology has much to tell us about the history of life on the earth and the interrelations of the phyla of the animal kingdom.

Protostomes and Deuterostomes

We have already noted that many-celled animals above the coelenterates develop in two diverse manners. One assemblage, which includes the annelids, arthropods, mollusks, and several phyla of lesser rank, is known as the protostomes (Fig. 4–11). Their eggs cleave in spiral fashion; they form their mesoderm as strands which bud off from mesoblast cells at the posterior end; and their blastopores become a mouth (or mouth and anus). When they produce skeletons, these are mostly cuticles or exoskeletons of ectodermal origin.

The other assemblage is the deuterostomes, to which belong the echinoderms, chordates, and a few minor phyla. Their eggs cleave in a radial or even an irregular manner; they form mesoderm (theoretically at least) from paired pouches at the side of the gut; their mouths are not derived from the blastopore but instead are new formations; and when they produce skeletons, these are for the most part internal skeletons of mesodermal origin.

Trochophores and nauplii The typical larvae of protostomes are the *trochophores* of annelids and mollusks and the *nauplii* of crustaceans. Trochophores are essentially spheres with a sensitive tuft of cilia at one pole (apical

FIGURE 4–12 Further development of a trochophore, a type of larva found in both annelids and mollusks. A and B. In annelids the trochophore strobilates (segments transversely) and becomes a segmented worm. A, C, and D. In mollusks a similar trochophore develops a "foot" and a mantle which usually secretes a shell.

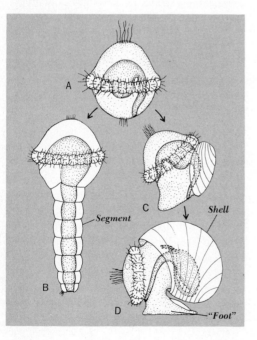

organ) and a single or double girdle of cilia (prototroch) around the equator (Figs. 4–11 and 4–12). A mouth is present on the ventral side, just below the ciliated girdle, and an anus occupies a position near the pole opposite the apical tuft. Generally the whole trochophore rotates like a wheel as it swims—hence the name. The trochophores of mollusks develop a muscular region (foot) on the ventral surface between the mouth and the anus and a shell on the dorsal surface between the girdle and the anus (Fig. 4–12, *A, C, D*). Annelids, on the other hand, although they may start with an almost identical structure, constrict off a series of body segments anterior to the anus, and thus they elongate into segmented worms (Fig. 4–12, *A* and *B*).

The crustaceans pass through a stage comparable to a trochophore while they are yet embryos. When the more primitive crustaceans hatch, they are nauplii larvae, which may be compared to annelid larvae that have formed three body segments, and in which three pairs of lateral appendages have already been developed (Fig. 4–13, *C*). The nauplius, however, has no cilia. Instead, it swims with a jerky motion produced by the strokes of its appendages. Facts such as these seem to link the annelids with the mollusks, and suggest—but do not prove—that the arthropods are descended from specialized annelid-like ancestors. The similarities could, of course, be a result of convergence in evolution.

The dipleurula The hypothetical larva of ancestral deuterostomes, namely, echinoderms and chordates, has been visualized as something like the larvae of present-day echinoderms (Fig. 4–11, *F*). A generalized larva of this sort has been called a *dipleurula*. It has been pictured as a sort of common denominator of the bipinnaria of a starfish, the plutei of sea urchins and brittle stars, and the auricularia of sea cucumbers. Like these, the blastopore becomes the anus, and the mouth would be a new formation on the ventral side. Hypothetically it possessed a band of cilia; but instead of girdling the larva, the band formed a border between the dorsal and ventral surfaces. The dipleurula presumably glided smoothly through the water, sweeping plankton into its mouth by means of currents set up by the ciliary band.

The echinoderms seem to have evolved from such an ancestral type by attaching themselves to some solid object by their anterior end, then undergoing a torsion of their bodies (possibly originally so as to bring their mouth and anus closer together), and finally developing outgrowths from their body for "arm feeding." Particles which came into contact with the arms were entrapped in mucus and then driven by cilia along grooves of the arm into the mouth. The most ancient echinoderms, which are known only by their fossils, were quite certainly of this sort. Indeed the most primitive group of echinoderms living today, namely, the crinoids or sea lilies, are, for the most part, sedentary arm feeders. Even present-day free-living echinoderms—starfish, sea urchins, and sea cucumbers—show the marks of their sedentary ancestry by their centrally located mouths and general radial symmetry. Radial symmetry is an adaptation which is evolved when an animal uses all its sides alike.

Some of the starfish in their development actually recapitulate the history of their race— or at least this is a possible interpretation. First they are free-swimming, bilaterally symmetrical larvae known as *bipinnaria* (Fig. 4–14, *A* to *C*). These then settle down, attach

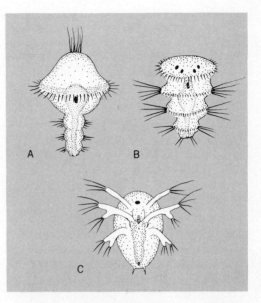

FIGURE 4–13 The nauplius larva of a crustacean compared to an annelid larva of three segments. *A.* Larva of a primitive annelid. *B.* Larva of the clamworm *Nereis*. *C.* Nauplius larva of one of the lower crustaceans.

to the substrate by their anterior ends, and become sedentary larvae known as *brachiolaria* (Fig. 4–14, *D*). A most remarkable metamorphosis now takes place. The anterior region of the larva regresses. The left side turns toward the substrate and surrounds the mouth. The right side turns away from the substrate. The midregion of the left coelom gives rise to the water-vascular system. Starfish arms grow out which have no relation whatever to the original bilateral symmetry of the larva. Finally, the stalk by which the brachiolaria was attached to the substrate regresses, and the starfish becomes free to approach food or retreat from danger by moving in any direction.

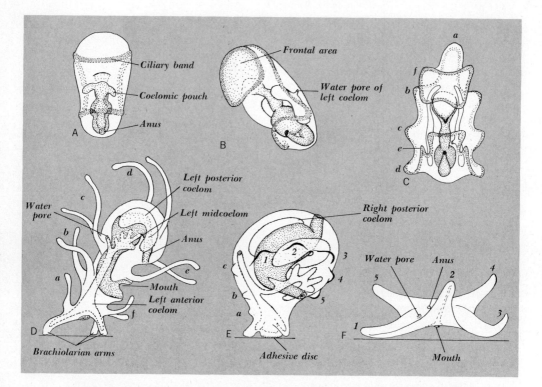

FIGURE 4–14 Later development of the starfish. (Earlier stages in the development of the starfish are illustrated in Figs. 4–6, *A* to *C*, and 4–11, *D* to *F*.) *A* to *C*. Bipinnaria larvae, which are bilaterally symmetrical except for the water pore on the left side. *D*. Brachiolaria larva from the left side. At this stage the larva is attached to a surface. *E*. Stage in metamorphosis. *F*. Mature starfish. The letters *a* to *f* label the ciliated arms of the larva. The numerals *1* to *5* indicate the arms of the adult.

The ancestral chordate   We may suppose that the first step toward the chordate type was the development of a dipleurula-like creature with gill clefts at the sides of the gut (pharynx) just posterior to the mouth (Fig. 4–15, *B*). Some have speculated that the gill clefts indicate a sedentary stage in the ancestry. It is true that one group of primitive chordates, the tunicates, are sedentary and have a remarkable develop-

ment of gill clefts (Fig. 4–16). But their larvae are typically chordate and form gill clefts before they settle down. In any event, it is from forms like the free-swimming larvae, not from sedentary adult ancestors (if such ever existed), that the vertebrates are descended. (Reproduction by larval stages is known as neoteny or paedogenesis.) From the first, the gill clefts were important for plankton feeding, that is, for filtering out particles of food from the water about them. Their role in respiration came later.

The primitive chordate larva has been pictured by Garstang as a modified dipleurula (Fig. 4–15). The ciliary band which surrounds the dorsal area is envisioned as having become the neural folds, the dorsal area as the neural plate. According to this hypothesis, the ciliated bands rose up and met dorsally so that the neural plate became transformed into the neural tube, just as it is transformed in the development of most vertebrates. The posterior ends of the neural folds enclosed the blasto-

pore. As they came together, they supposedly formed a canal (neurenteric canal) which connected the cavity of the neural tube with that of the gut. This also occurs in the development of vertebrates. Both mouth and anus in this case were new formations. According to Garstang's theory, the notochord, so characteristic of chordates, was a flexible, longitudinal rod formed from the top of the primitive gut, just as it is in vertebrate embryos today. It served to appose the muscle fibers of the right side of the body to those of the left and to stiffen the larva so that by undulatory movements it could dart tadpole-like through

FIGURE 4–15 Diagrams which illustrate Garstang's hypothesis of the origin of the chordate body plan.

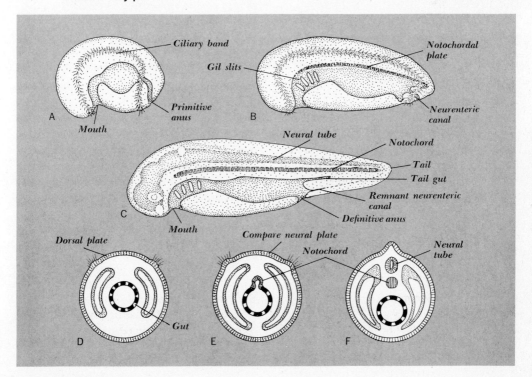

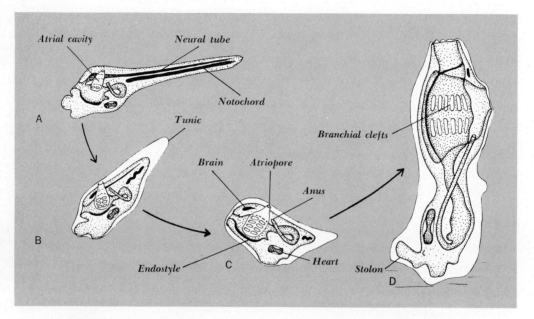

FIGURE 4–16 Metamorphosis of a tunicate. *A*. The "tadpole larva" of a tunicate showing its chordate characteristics. *B* to *D*. The larva attaches to a surface by its "chin," develops a tunic around itself and a large gill basket with numerous branchial clefts. It then proceeds to lose its tail, notochord, and neural tube, except for a tiny "brain."

the water and filter-feed on the planktonic organisms which float in the sea.

Such evolutionary speculations do not assume that a larval or embryonic stage represents an adult ancestor. There is really no way of knowing what an ancestor looked like. But it is good sense to suppose that if two present-day animals have similar larvae, then they may have had a common ancestor with a similar larva. And may we not go further and speculate that the common ancestor was not specialized in the direction of either of these descendants?

SELECTED READINGS

Ontogeny of the Sperm

Balinsky, B. I., 1965. An Introduction to embryology. W. B. Saunders Company, Philadelphia. 2d ed. Chap. 3, Spermatogenesis.

Burgos, M. H., and D. W. Fawcett, 1955. Studies on the fine structure of the mammalian testis. I. Differentiation of the spermatids of the cat. J. Biophys. Biochem. Cytol., **1**:287–300.

Oogenesis

Bloom, W., and D. W. Fawcett, 1962. Textbook of Histology. W. B. Saunders Company, Philadelphia, 8th ed. Pp. 584–593.

Balinsky, B. I., 1965. Introduction to Embryology. W. B. Saunders Company, Philadelphia. 2d ed. Pp. 56–76.

Anderson, E., and H. W. Beams, 1960. Cytological observations on the fine structure of the guinea pig ovary with special reference to the oogonium, primary oocyte, and associated follicle cells. J. Ultrastruct. Res., **3:**432–446.

Boycott, A. E., C. Diver, S. L. Garstang, and F. M. Turner, 1930. The inheritance of sinistrality in Limnea peregra. Phil. Trans. Roy. Soc. London, Ser. B, **219:**51–131.

Clavert, J., 1962. Symmetrization in the egg of vertebrates. Advan. Morphogenesis, **2:**27–60.

Fertilization and Artificial Parthenogenesis

Balinsky, B. I., 1965. Introduction to Embryology. W. B. Saunders Company, Philadelphia. 2d ed. Chap. 6, Fertilization.

Ebert, J. D., 1965. Interacting Systems in Development. Holt, Rinehart and Winston, Inc., New York. Pp. 27–34.

Allen, R. D., 1959. The moment of fertilization. Sci. Am., **201** (July):124–134.

Tyler, A., 1954. Gametogenesis, fertilization, and parthenogenesis. In Willier, Weiss, and Hamburger (eds.), Analysis of Development. W. B. Saunders Company, Philadelphia. Pp. 170–212.

Colwin, A. L., and L. H. Colwin, 1964. Role of the gamete membranes in fertilization. In Locke, M. (ed.), Cellular Membranes in Development. Academic Press Inc., New York. Pp. 233–279.

Metz, C. B., 1962. Fertilization. In Moment (ed.), Frontiers of Modern Biology. Houghton Mifflin Company, Boston. Pp. 132–142.

Cleavage

Balinsky, B. I., 1965. Introduction to Embryology. W. B. Saunders Company, Philadelphia. 2d ed. Chap. 7, Cleavage.

Ebert, J. D., 1965. Interacting Systems in Development. Holt, Rinehart and Winston, Inc., New York. Chap. 3, The shape of things to come.

Wilson, E. B., 1892. The cell lineage of Nereis. J. Morphol., **6:**361–480.

Conklin, E. G., 1897. The embryology of Crepidula. J. Morphol., **13:**1–226.

Costello, D. P., 1954. Cleavage, blastulation, and gastrulation. In Willier, Weiss, and Hamburger (eds.), Analysis of Development. W. B. Saunders Company, Philadelphia. Pp. 213–229.

Wolpert, L., 1960. The mechanics and mechanism of cleavage. Intern. Rev. Cytol., **10:**164–216.

Cell Differentiation

Wilson, E. B., 1925. The Cell in Development and Heredity. The Macmillan Company, New York. 3d ed. Especially Chap. 13.

Hörstadius, S., 1950. Transplantation experiments to elucidate interactions and regulations within the gradient system of the developing sea urchin egg. J. Exptl. Zool., **113:**245–276.

Ranzi, S., 1962. The proteins in embryonic and larval development. Advan. Morphogenesis, **2:**211–257.

Lallier, R., 1964. Biochemical aspects of animalization and vegetalization in the sea urchin embryo. Advan. Morphogenesis, **3:**147–195.

Gastrulation

Balinsky, B. I., 1965. Introduction to Embryology. W. B. Saunders Company, Philadelphia. 2d ed. Chaps. 8 and 9.

Costello, D. P., 1954. Cleavage, blastulation, and gastrulation. In Willier, Weiss, and Hamburger (eds.), Analysis of Development. W. B. Saunders Company, Philadelphia. Pp. 213–229.

Moore, A. R., 1941. On the mechanics of gastrulation in Dendraster eccentricus. J. Exptl. Zool., **87:**101–111.

Gustafson, T., and L. Wolpert, 1963. The cellular basis of morphogenesis and sea urchin development. Intern. Rev. Cytol., **15:**139–214.

chapter **5**

THE
EARLY
DEVELOPMENT
OF
AMPHIBIANS

An understanding of amphibian development is fundamental to the interpretation
of chick and mammal development. Amphibian
eggs are easily obtained and may be studied
in the laboratory from the moment of fertilization onward. This is not so easily true of the
higher vertebrates. The amphibian embryos,
also, are relatively straight, not coiled. This
makes description easier. Frog eggs are generally used, having been either collected in
freshwater ponds in the early spring or obtained by means of induced ovulation at other
seasons of the year. Some of the tailed amphibians, however, have eggs which are more
satisfactory for observation and experiment
than frog eggs, although they are less readily
provided. Both frog and salamander eggs are
considered in this chapter.

THE DEVELOPMENT OF THE OOCYTE IN
THE OVARY

The pre-embryonic development of the eggs
of a frog or salamander is a slow process. It
requires three seasons during which time it
passes through three phases: multiplication,
growth, and retention.

Multiplication of Oogonia

During the first or multiplication phase, the
oogonia divide several times by mitosis and
give rise to "nests" of cells in the wall of the
ovary. The nests lie in mesenchyme, sand-
wiched between the outside epithelium of the
ovary (peritoneum) and the inner epithelium
which lines the cavity of the ovary. (Birds and
mammals do not have an ovarian cavity.) The
oogonia are small (about 20 μ in diameter) and

scarcely to be distinguished from the spermatogonia of the male.

Meiosis and Growth

When the time comes for meiosis to begin,
some of the oogonia move apart and begin to
be surrounded by an envelope of follicle cells.
This takes place in the tadpole and annually
thereafter in the case of older frogs. As the
oogonium becomes an oocyte, its chromosomes elongate into threads (leptotene stage)
and come together in pairs (synaptene stage)
(Fig. 5-1, A). In some amphibians the clusters
of chromosomes appear as "bouquets" with
their ends all turned toward one pole of the
cell.

The second phase involves a process which
is unique to oogenesis. Instead of contracting
and remaining contracted (as they do in spermatogenesis), the chromosomes expand and
develop a fuzzy appearance (diplotene stage).
What actually happens is that the chromosomes lengthen greatly and develop pairs of
lateral loops, so that each chromosome has
the appearance of an old-fashioned lampbrush
(Fig. 5-1, B to D). The oocytes remain at this
stage (diplotene) for more than a year, during
which time they grow tremendously. What is
the meaning of the lateral loops? The presence
of RNA and protein around them indicates
clearly that they are an adaptation to facilitate
transcription and synthesis.

During this phase the nucleus of the oocyte
enlarges mostly by accumulating liquid (nuclear sap). No doubt the liquid is a "pool" rich
in nucleotides, amino acids, and other substances used in synthesis. In its inflated condition it is known as a *germinal vesicle*. Nucleoli increase in number within the nucleus

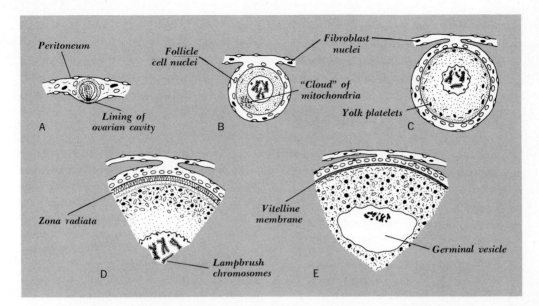

FIGURE 5–1   Diagram of oogenesis in amphibians. **A.** Leptotene and synaptene stages of oocyte, showing the "bouquet" of chromosomes. × 220. **B to D.** Diplotene stages, showing lampbrush chromosomes. × 130, × 65, and × 65, respectively. **E.** Diakinesis. × 44. The cytoplasm of the oocyte is stippled; that of other cells is not shown. *(Based mainly on S. Wischnitzer, 1966, Advan. Morphogenesis, 5, pp. 135 and 161.)*

and come to lie next the nuclear membrane. Some of them have been observed to burst through the nuclear membrane and discharge their contents into the surrounding cytoplasm. At the same time rRNA is synthesized and ribosomes increase in the cytoplasm, so that the oocyte cytoplasm comes to stain darkly with basic dyes. (It is of interest that these ribosomes of the oocyte are conserved and continue to serve the developing embryo at least until the tail-bud stage.)

Mitochondria also increase in the oocyte cytoplasm. At first they form a "cloud" in the region of the centrioles (Fig. 5–1, *B*). Then they scatter and move closer to the periphery.

Vesicles appear in the cytoplasm, possibly having derived from the Golgi complex.

Yolk granules (platelets) are not present in the cytoplasm until about the middle of the second summer. By this time the oocyte is about 350 μ in diameter (Fig. 5–1, *C*). At first they are located close to the plasma membrane; but as they increase in number they encroach nearer to the nucleus until finally, when the growth of the oocyte is complete, they are present throughout the cytoplasm.

Where do the several proteins and lipoproteins that constitute yolk come from? Radiotracer experiments and experiments using immunological techniques have made it clear that most of them are synthesized in the liver. They circulate in the blood, penetrate past the follicle cells—possibly with their help—and are deposited as organized platelets in the oocyte cytoplasm. The mitochondria and Golgi bodies have something to do with the production of the platelets, but the details are not entirely clear.

The yolk platelets are large and densely packed in the vegetal hemisphere of the oocyte. They are smaller and less consolidated in the animal hemisphere, especially in the region of the nucleus. A thin layer of dark pigment is laid down in somewhat similar fashion beneath the plasma membrane of the animal hemisphere.

The Envelopes of the Oocyte

Three layers of cells enclose the oocyte in the ovary. The inner layer, already referred to, is the epithelium of follicle cells. The other two layers develop in the following fashion: As the oocyte grows it protrudes inwardly into the cavity of the ovary (Fig. 5-1, *B* to *E*). In doing so it pushes a thin mesenchymal stroma (including fibroblasts, collagen fibers, and blood capillaries) and the lining of the ovarian cavity before it. Finally the oocyte becomes completely enclosed in the three membranes, except at one pole where it retains an attachment to the wall of the ovary. (At ovulation only the follicular epithelium and the peritoneum rupture. The mesenchyme and the lining of the ovarian cavity remain intact.)

A noncellular layer known as the *zona pellucida* gradually forms between the follicle cells and the plasma membrane of the oocyte (Fig. 5-1, *D*). Microvilli grow into it from both the follicle cells and the oocyte, giving it a cross-striated appearance when seen in section.

Hence it is also called the zona radiata. The interdigitating villi presumably facilitate the transfer of nutrients to the growing egg. As the oocyte attains its full size (1,500 to 2,000 μ) the villi are withdrawn, but the zona pellucida remains as the *vitelline membrane* which supports and protects the egg.

Retention in the Ovary

The oocyte reaches its full growth toward the end of the second season, but it is not immediately laid. The third phase, therefore, is one of quiescent retention in the ovary until the next breeding season. Then, sensitized by hormones of the pituitary and stimulated by the clasp of the male, the ova erupt through the peritoneum and enter the body cavity. This is *ovulation*. It is accompanied by the dissolution of the germinal vesicle and the spreading of its liquid contents across the animal hemisphere. The first meiotic spindle forms, and the first polar body is given off. As the egg enters the oviduct, a second meiotic spindle forms, and the second meiotic division progresses as far as the metaphase. At this stage it pauses, and no further progress is made until and unless the ovum is fertilized.

Laying is a reflex act which is brought about in part by the clasping of the male. Ovulation and then laying can be stimulated artificially in the frog any time from autumn until spring by injecting a pituitary substance into the abdomen of the female and then, two days later applying gentle pressure to her body.

THE NEWLY LAID EGG

Each amphibian egg, as it comes from the cloaca of the female, is a single cell ¾ to 3 mm in diameter. It is surrounded by the vitelline

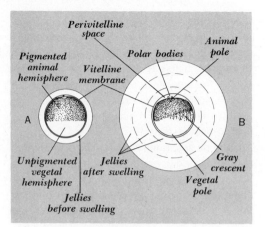

FIGURE 5–2 An amphibian egg before and after laying. *A.* Before laying. *B.* After laying and fertilization. Seen from the left side. Approximately × 3½.

membrane and by several layers of adhesive jelly which were added as it passed through the oviduct (Fig. 5–2). In some species the eggs are laid singly, but in most cases they adhere together in clusters. On reaching water, the jelly layers swell and lose much of their stickiness. The outer layer of jelly becomes firmer, especially in salamanders, while the inner layer remains more fluid.

The eggs of amphibians are fertilized at the time they are laid. In frogs, the male clasps the female (amplexus) and sheds a suspension of sperm over the eggs as they leave her cloaca. In some tailed amphibians, packets of sperm (spermatophores) are deposited by the male on the bottom of a pond and are picked up by the female and placed in her cloaca. In other tailed amphibians, the male deposits the spermatophores directly in the cloaca of the female. In either case, the eggs are fertilized as they pass through the cloaca. Following fer-

tilization, the vitelline membrane (now called the fertilization membrane) rises off the surface of the egg (page 67), and the egg becomes free to rotate within its membranes.

The eggs of amphibians are spherical cells, quite abundantly supplied with yolk. Some of them are so soft that they would flatten under the influence of gravity if it were not for the support they receive from the vitelline membrane.

Polarity

The amphibian egg shows polarity in several ways: (1) The yolk granules are fewer and smaller in the animal hemisphere and are larger and more packed together in the vegetal region. (2) The nucleus is located nearer the animal pole, and it is here that the polar bodies are given off. (3) A thin cortical layer containing dark pigment granules covers the animal hemisphere and extends downward toward the vegetal pole. (4) The vegetal hemisphere, being heavier than the animal hemisphere owing to the greater density of its yolk, rotates downward, and the darker hemisphere comes to be upward. This "rotation of orientation" serves as a sign that fertilization has been accomplished. (5) Gradients of a chemical nature have also been demonstrated in the newly laid amphibian egg, including gradients in the distribution of ribosomes.

Bilaterality

Soon after fertilization (in some species even before fertilization) the amphibian egg acquires bilateral symmetry. The pigmented part of the cortex glides to one side like a cap slipping sidewise on a head. Most commonly it glides to the side on which the sperm entered.

By so doing it uncovers an area of yolk-rich cytoplasm on the opposite side of the egg. Because of its color in the frog's egg, this area is known as the *gray crescent* (Fig. 5-2, *B*). It appears, however, as a clear or gleaming crescent in tailed amphibians. Vertical streaks of pigment, torn from the edge of the retreating dark cortex, are sometimes seen strewn across it.

This shift of the pigmented cortex has been called the "rotation of symmetrization," although the inner cytoplasm does not rotate at all. The gray crescent is important because it is the first visible indication of the side which will become the dorsal side of the embryo. The meridian of the egg, which passes through the animal and vegetal poles and the center of the gray crescent, is destined to become the mid-dorsal line of the embryo.

Other influences besides the entrance of the sperm are capable of affecting the localization of the dorsal side of the embryo, but they are effective only if they act before the eight-cell stage. For example, tilting the egg or compressing it may cause the heavier yolk to become rearranged within. In effect, any region, even the ventral side, where white yolk becomes uncovered by the retreat of the pigmented cortex so that it comes to be close to the surface of the egg, can become functionally a gray crescent and give rise to the dorsal side of an embryo. After the eight-cell stage, however, the dorsal side is stabilized and is no longer subject to environmental influences. The ventral side, on the other hand, is still labile. Curtis has found that if a bit of the cortex of the region of the gray crescent (after it is determined) is transplanted to the ventral side of another egg, it will cause dorsal organs to develop on the ventral side.

Numerous experiments have been performed in which amphibian eggs have been pinched in two, or parts have been removed. In general, it has been found that any major part of the egg which contains material of the gray crescent has the capacity to develop into a whole embryo of reduced size. One of the first to demonstrate this was Hans Spemann. In 1901 he separated the first two blastomeres of a salamander egg from each other with a hair noose between them. If, as usually happens, the first cleavage plane passed through the gray crescent, then both blastomeres became whole embryos of half size. It is of interest that in this case the embryo which is derived from the right blastomere often showed reversed asymmetry of its heart and viscera. If the first cleavage plane did not pass through the gray crescent, then only the blastomere which received gray crescent material became a whole embryo. The other blastomere became a "belly piece" without dorsal parts of any sort —no neural tube, no notochord, no somites. Figure 5-3 illustrates a similar experiment of Spemann's (1928) in which the noose partially divided an uncleaved egg into halves. The half which at first lacked a nucleus was later supplied when a descendent nucleus migrated across the narrow isthmus of cytoplasm and underwent cell division. The fates of the halves depended upon the materials of the cytoplasm which they possessed, and not upon the source of their nuclei.

The opposite experiment was performed by Mangold and Seidel (1927). They took eggs at the two-cell stage and, after removing the vitelline membrane, placed them on top of each other crosswise. The eggs united and developed as one (Fig. 5-4). If the gray crescents of the two eggs were joined together, then a giant

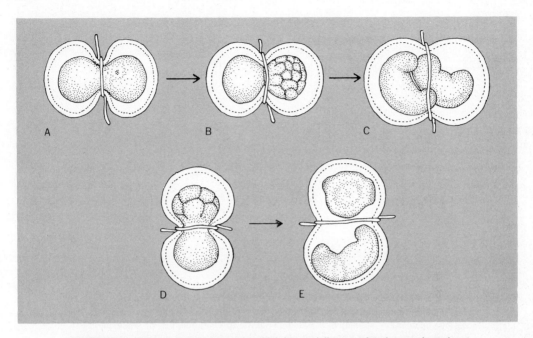

FIGURE 5–3 Spemann's experiment in which he partially constricted an uncleaved egg with a hair noose. One of the descendent nuclei crossed the isthmus of cytoplasm, and cleavage became complete on both sides. *A to C.* When the hair noose passed through gray crescent material, both sides developed as whole embryos of half size. *D to E.* When one side lacked gray crescent material, it developed into a "belly piece." *(Modified from H. Spemann, 1928, Zeitschr. Wiss. Zool., 132:105–134.)*

embryo of twice the normal size developed. If, however, the gray crescent materials were not adjacent, then double or even triple monsters resulted.

The cytoplasm of an amphibian egg contains DNA—enough, so it is said, to supply 5,000 or more nuclei. Apparently, this DNA is inactive during cleavage and is not mobilized until the yolk platelets are digested. Presumably, it has no genetic significance but is nutrient in function, serving as a source of precursors for the DNA and RNA needed later in development.

CLEAVAGE

In amphibians, the first plane of cleavage is meridional; it passes through the animal and vegetal poles (Fig. 5–5). Usually, it passes through the center of the gray crescent. The early experimentalists were misled, however, when they supposed that the first cleavage was a means of separating the potentialities of the right side from the left, for it does not always pass through the center of the crescent.

The first cleavage begins as a shallow furrow at the animal pole and progresses gradu-

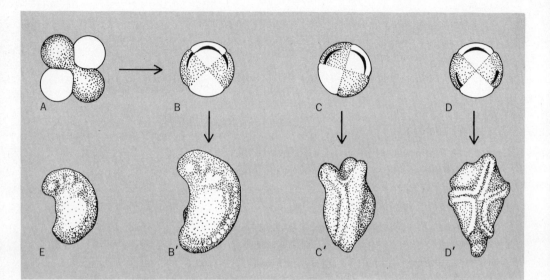

FIGURE 5–4 Mangold and Seidel's experiment in which two eggs at the two-cell stage (the vitelline membranes having been removed) were laid crosswise of each other. One egg had been vitally stained with Nile blue sulfate to distinguish it. *A.* The experiment. *B* and *B′.* Giant embryos were produced when the gray crescent material was together. *C* and *C′.* Double monsters resulted when the gray crescent material was separated and two blastopores formed. *D* and *D′.* Triple monsters were produced when there were three invaginations. *E.* A normal embryo drawn for comparison. *(After O. Mangold and F. Seidel, 1927, Arch. Entwicklungsmech., 111:393–665, Abb. 2, 14, 17, and 20.)*

ally through and around the egg until it reaches the vegetal pole. Normally, it divides the egg into two blastomeres of approximately equal size.

The second cleavage, like the first, is meridional. It is approximately at right angles to the first. It begins at the animal pole even before the first cleavage has completed itself at the vegetal pole. The result is four nearly equal blastomeres.

The third cleavage is typically at right angles to the first two cleavages, and hence cuts latitudinally, that is, horizontally, when the egg's axis is vertical. It passes well above the equa-

tor, so that the eight-cell stage commonly consists of four smaller animal blastomeres (micromeres) above and four larger, yolk-laden vegetal blastomeres (macromeres) below. There is, however, much variation in the direction of the cleavage planes. As a consequence, the sizes of the cleavage cells vary. It frequently happens that the third cleavage planes are vertical, that is parallel, to the first cleavage planes and at right angles to the second planes. It is evident, therefore, that the particular pattern made by the blastomeres has no significance with respect to the normality of development. Rather, it is the distribution of the materials and the localization of the chemical processes which determine development.

During the early phases of cleavage the cell divisions are synchronous and rhythmic (pp. 71, 72). But toward the close of cleavage they become less rapid and more irregular. Instead of 60 percent or so of the cells being in division at one time, the mitotic index sinks to 5 percent or less. The great majority of cells are in interphase.

THE BLASTULA

A cavity is present at the center of the group of blastomeres beginning at the eight-cell stage. It increases in size as cleavage progresses. This is owing, at least in part, to the fact that cleavage planes cut more often at right angles to the outer surface than they do parallel to it. The egg thus becomes a hollow ball of cells, i.e., a blastula (Fig. 5–5, *E* and *F*). Its cavity, the blastocoel, is roughly hemispherical. It has a dome or roof of smaller animal hemisphere cells and a floor of larger yolk-rich vegetal cells.

As cleavage progresses, the roof of the blastocoel expands and actually becomes thinner. Its margins push downward on all sides, while at the same time the floor of the yolk-laden vegetal cells bulges upward from below. Cells from the center of the vegetal pole area fountain upward internally until some of them reach the floor of the blastocoel (Fig. 5–5, *F*).

FIGURE 5–5 Cleavage and blastulation in amphibians, seen from the left. *A* to *D*. Harrison's stages 2, 4, 6, and 8 for *Ambystoma*. (Compare stages 3, 5, 7, and 8 for *Rana pipiens*.) *E* and *F*. Median sections of stages 6 and 8. The arrows indicate the mass movements of the cells of the vegetal hemisphere according to Schechtman. Approximately × 8.5.

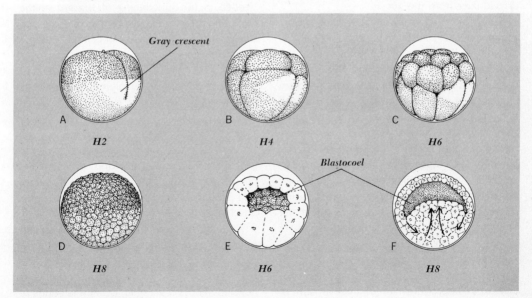

Most of them retain connection with the external surface. As a result of these shifts, the blastocoel comes to have the shape of an inverted bowl. These changes are preliminary to the mass movements of gastrulation which follow.

The Role of the Nucleus and Cytoplasm during Cleavage

The chemical activity of nuclei during cleavage was discussed in Chapter 4 (pp. 71, 72). In amphibians the replication of DNA increases sharply beginning about one-half hour after fertilization. But the synthesis of rRNA and tRNA is not resumed until near the end of cleavage. Nucleoli are absent during cleavage stages, and new ribosomes are not produced. The proteins needed for the manufacture of chromosomes and mitotic apparatus are synthesized with the aid of old RNA and ribosomes carried over from the oocyte.

Indeed, it would seem that the store of ribosomes derived from the oocyte is sufficient to provide for the needs of the embryo until the early tadpole stage. Brown and Gurdon have described a mutant form of the South African clawed toad, *Xenopus,* which lacks nucleoli and presumably is unable to produce ribosomes. Only the heterozygotes (mutant X normal) have a nucleolus and are viable. The homozygotes (those without nucleoli) develop as far as the swimming stage and then die. The normal embryo, on the other hand, begins to synthesize new rRNA and tRNA about the time that gastrulation gets under way. It is then that nucleoli reappear.

Messenger RNA, the form of RNA by which the genes transmit information, behaves in a somewhat different fashion. A small amount of mRNA is said to be synthesized during cleavage, but apparently it is masked and takes no part in the protein syntheses of cleavage. It is not until gastrulation begins that mRNA becomes active and new mRNA is produced. Migrating from the nucleus into the cytoplasm, the chains of mRNA join with ribosomes to form polyribosomes, and new proteins are synthesized. The process is selective and results in different regions of the gastrula giving rise to different tissues.

What controls the changes in nuclear activity? Experiments performed by Gurdon and Brown on the eggs of *Xenopus* indicate clearly that cytoplasm plays the leading role. These investigators took nuclei from late stages of development and substituted them for the nuclei of unfertilized eggs. They even used nuclei from differentiated cells of the epithelium of the intestine of swimming tadpoles. Now, if these nuclei had been left in place, only those few genes would have been active which were needed in the functioning of the gut epithelium. The rest would have remained repressed. But when one of these nuclei is implanted into an uncleaved egg, it quickly (within an hour) takes on the character of a zygote nucleus! It increases 30 times in volume; its nucleolus disappears; it ceases to synthesize RNA. Instead, it resumes the replication of DNA. In short, it actually becomes a zygote nucleus. Normal cleavage follows. Then, in due time, gastrulation commences, and the expected chemical changes follow. In a few of Gurdon and Brown's experiments, mature, fertile male and female frogs resulted.

What do we conclude from these experiments as to the roles of the nucleus and cytoplasm? We conclude that (1) any differentiation of nuclei which takes place during early development is reversible; (2) all the genes remain inviolate within the differentiated cells, even though only a few genes are called upon to function; and (3) in some unknown manner the cytoplasm which surrounds a nucleus controls its activity. At the beginning of cleavage, the cytoplasm turns DNA synthesis on and RNA synthesis off. As the time of gastrulation approaches, the cytoplasm turns DNA synthesis off and RNA synthesis on.

Why is it that in so many nuclear transplantation experiments some of the operated eggs failed to develop normally? Gurdon and Brown suggest that possibly the cytoplasm of the uncleaved egg, acting on the nucleus implanted from a differentiated cell, forces the implanted nucleus to divide before it is fully ready. It has not had time to wholly regain the character of a zygote nucleus.

Fate Map of the Blastula at the Beginning of Gastrulation

It will be of great help in describing the mass movements which follow blastulation if we consider the fates of different regions of the blastula. If one could mark an individual cell, or group of cells, on the surface of the blastula and then follow the mark through the movements of gastrulation and neurulation, one would be able to say that one region is prospective brain, that another is prospective epidermis, or muscle segments, etc. Thus, by reasoning backward, one could construct a "fate map" of the blastula, in which each region would be designated, not according to any present difference which it may or may not have, but according to its *prospective significance,* given normal development.

From 1925 to 1929 Walther Vogt did this very thing (Fig. 5–6). He took small chips of agar, which he had stained with a vital dye,

and placed them for a few minutes in close contact with the surface of the egg. Brightly stained spots remained on the egg's surface. The dyes he used were Nile blue sulfate and neutral red. They colored granules in the cytoplasm, but had little effect on the course of development. Vogt observed the changing positions of the marks and was able to draw his now classic fate map of the surface of the blastula or beginning gastrula. Only minor changes in the map have been made since.

It is very important not to confuse the prospective significance of a cell (given normal development) as recorded on a fate map with its actual capacities for development. If we interfere experimentally with the mass movements of embryo formation so that a cell ends up in a different location from that which it normally would have occupied, or if we transplant a group of cells to a new location in another egg, or, again, if we isolate (explant) a group of cells and so free them from the influence of adjacent cells, then the cell, or group of cells, may become something quite different from what it would normally have become.

A fate map of an amphibian blastula at the beginning of gastrulation, such as drawn by Vogt (Fig. 5–7), shows three primary zones:

1 The zone of *prospective ectoderm* (epidermis and neural plate) corresponds roughly to the animal hemisphere of the egg. It might be called the "zone of expansion," for during gastrulation it expands downward and then converges toward the vegetal pole. It gives rise to the entire outer germ layer of the gastrula.

2 The zone of *prospective notochord and mesoderm,* often called the *marginal zone,* is roughly a subequatorial belt around the blastula, although it is broader and extends above the equator on the dorsal side. Its dorsal part is approximately the original gray crescent.

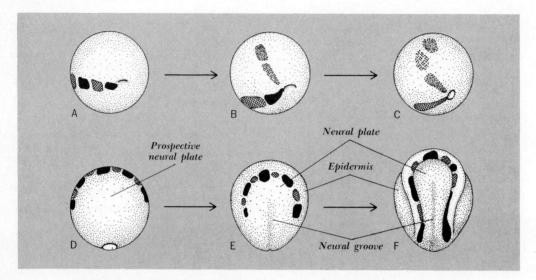

FIGURE 5–6 Vogt's method of marking the surface of amphibian eggs with spots of vital die, namely Nile blue sulfate or neutral red. In *B* and *C* some of the spots have entered through the blastopore and are seen through the outer layer of cells. *(After K. Goerttler, 1925, Arch. Entwicklungsmech., 106:503–541, Abb. 12 and 3.)*

The prospective chorda-mesoderm might be called the "zone of involution," for its fate, as we shall soon see, is to roll around and beneath the lips of the blastopore. Then, after having moved upward and dorsally on the inside of the egg, it becomes the notochord and mesoderm (chorda-mesoderm).

3 The zone of *prospective endoderm* is composed, in large part, of the large yolk-rich cells of the vegetal hemisphere; but it includes also some smaller and more active cells which border the prospective chorda-mesoderm. This zone could be designated the "zone of invagination," for during gastrulation it sinks and glides into the interior of the egg and becomes endoderm. In effect, it is overgrown by the descent of the lips of the blastopore. It is within this zone, a little below its dorsal margin, that

the first slitlike blastopore appears. The position of this first invagination is indicated by a small arrow on the fate map.

Each of the three primary zones may be further subdivided:

1 During the neurulation which follows gastrulation, approximately the dorsal half of the prospective ectoderm folds inward and becomes the brain and spinal cord. This area is the *prospective neural plate*. The ventral half of the prospective ectoderm is the *prospective epidermis*.

2 During neurulation, also, the cells of the prospective chorda-mesoderm become divided into a central notochordal rod and several regions of mesoderm, namely, prechordal mesoderm, axial mesoderm (epimere), intermediate mesoderm (mesomere), and lateral mesoderm (hypomere). These subareas are shown on the fate map and will be defined later (page 116).

3 That part of the prospective endoderm which first enters the egg during gastrulation gives rise to the lining of the foregut, mainly to the pharynx (as distinguished from the ectodermal stomodaeum). From it, the mouth and gill clefts form. The rest of the prospective endoderm becomes the yolk-laden midgut from which the intestine is formed. In fact, the large yolk-filled cells of the vegetal pole come to lie on the floor of the intestine and contribute to the nutrition of the embryo.

GASTRULATION

The mass movements of gastrulation actually begin in the late blastula when the roof of the blastocoel thins, expands, and presses downward around the equator. The descending margins of the roof appear to force the yolk cells of the floor of the blastocoel upward into the blastocoel. But this certainly is not the mechanism which is involved, for these mass movements take place even in isolated pieces (explants). They are therefore autonomous.

FIGURE 5–7 "Fate map" of an amphibian blastula (salamander) at the beginning of gastrulation. *A*. View from the left side. *B*. View from the dorsal side. The short arrow (short dark line in *B*) marks the location of the beginning invagination of endoderm. The lower heavy line indicates the initial lateral and ventral lips of the mesodermal blastopore. It is along this line that prospective mesoderm splits away from prospective endoderm. The upper line marks the lips of the final blastopore after involution is complete. *(Mainly based on W. Vogt, 1929, Arch. Entwicklungsmech., 120:385–706, Abb. 1 b.)*

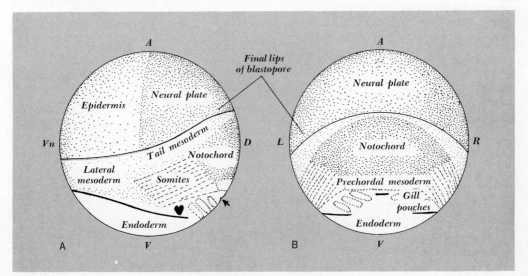

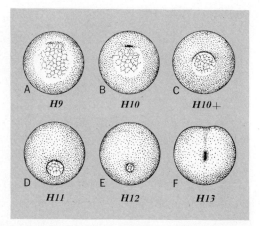

FIGURE 5–8 Gastrulation in the amphibian egg
as seen from the vicinity of the vegetal pole.
A and *B*. Gastrulation begins by an invagination of prospective endoderm on the dorsal
side about halfway between the vegetal pole
and the equator. *C* to *E*. Gastrulation continues by the involution of prospective chorda-
mesoderm at the dorsal, lateral, and finally at
the ventral lips of the blastopore. *F*. Late
gastrula, showing the closure of the blastopore by the apposition of its lateral lips. Harrison's stages 9, 10, 10 +, 11, 12, and 13.
(Compare Shumway's stages 10—, 10, 11, 12,
13—, and 13.) Approximately × 7.

Invagination of the Endoderm

The account which follows applies primarily
to the eggs of salamanders. Frogs eggs perform in much the same manner, although they
differ in certain details.

The superficial aspects of gastrulation are
readily described. A shallow, transverse, and
slightly pigmented groove appears on the dorsal surface of the egg (Fig. 5–8). It is located
in the region of prospective endoderm, about
halfway from the equator to the vegetal pole.

The groove narrows, deepens (invaginates),
and then widens laterally into a crescent concave toward the vegetal pole. This crescent is
the initial or *endodermal blastopore.*

How is this initial invagination accomplished? Stained sections through the early
blastopore show that the cells invaginate by
changing their shapes. They become flask-
shaped by the contraction of the cortex at their
outer surface and the withdrawal of most of
their substance inward (Fig. 5–9). They will do
this in explants even when the surrounding tissue has been cut away. The exposed tips of
the flask-shaped cells are pigmented. Hence
the furrow is dark.

The initial furrow of invagination deepens as
the flask-shaped cells sink farther inside the
egg. At the same time, the prospective endoderm cells which border the blastopore move
toward the furrow and follow the flask-shaped
cells in. Once inside the egg, the cells regain
their cuboidal shape and expand upward toward the animal pole (Fig. 5–10). Thus is

FIGURE 5–9 Median section through the beginning blastopore of an amphibian egg. *(After
W. Vogt, 1929, Arch. Entwicklungsmech.,
120:385–706. Abb. 39.)*

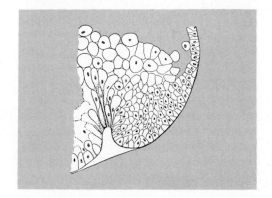

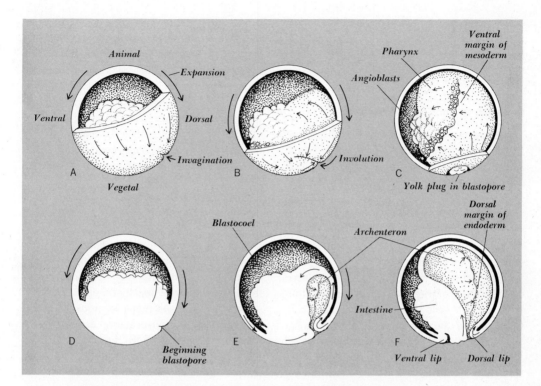

FIGURE 5–10   Diagrams of the mass movements of gastrulation. In *A, B,* and *C* the prospective epidermis and neural plate of the left side have been removed in order to expose the blastocoel, endoderm, and mesoderm to view. Approximately Harrison's stages 10, 11, and 12. *D* to *F.* Median views of the same three stages. Approximately × 11.

formed the beginning of the *archenteron* (primitive gut).

For a brief time, the early gastrula is a two-layered structure consisting of an outer and an inner layer. The inner layer is definitely endoderm; but the outer layer is prospective chorda-mesoderm and endoderm, as well as ectoderm. The terms *epiblast* and *hypoblast* are sometimes applied to these two transitional layers.

The invagination of endoderm continues during the involution of the chorda-mesoderm,

which is to be described next. The prospective endoderm of the vegetal area glides beneath the descending lips of the blastopore (Fig. 5–10). Most of it disappears from view beneath the dorsal and dorsolateral lips and becomes the floor of the archenteron. The lateral margins of the endoderm split away from the mesoderm and glide upward toward the animal pole. They also move dorsally toward the midline. Thus the endoderm, after invagination, forms a sort of trough which is open dorsally.

The early splitting away of the endoderm

from the mesoderm is less obvious in the frog than it is in the salamander, for the cells of the two layers are difficult to distinguish. Later they separate, apparently by delamination.

Involution of the Chorda-Mesoderm

The crescent-shaped blastopore widens to a semicircle, then to the shape of a horseshoe, and finally it closes to a complete circle. As it does so, its lips migrate downward toward the vegetal pole or rather to a point near it. Thus the blastopore comes to be a small circle, with a *yolk plug* of endoderm in its center. Toward the close of gastrulation, the yolk plug is withdrawn into the interior of the egg.

Careful observations of spots of dye placed on the surface of the blastula clearly show that the superficial cells stream downward and converge toward the lips of the blastopore (Fig. 5–10, *A* and *B*). They move especially toward the dorsal and dorsolateral lips. When they reach the lips, they roll around them and so disappear from sight. Once inside the egg, they migrate upward (toward the animal pole), keeping in close contact with the inner surface of the ectoderm. Their fate is to become notochord and mesoderm. Note carefully that the chorda-mesoderm cells enter the interior of the egg by *involution*. Instead of becoming flask-shaped with their broad ends inward, as did the invaginating endoderm, they become temporarily wedge-shaped with their broad ends directed outward.

The actual beginning of the movements of the prospective chorda-mesoderm is internal and is not visible from the outside. Certain cells which are situated around the borders of the blastocoel (labeled *angioblasts* in Fig. 5–10) split away from the adjacent endoderm cells and wander upward toward the animal pole. As they do so, they keep in close contact with the under surface of the roof of the blastocoel. The splitting away begins on the dorsal side and spreads ventrally, although it does not involve the immediate middorsal area. The angioblasts are followed by an internal ring (incomplete dorsally) of upward-wandering mesoderm cells. The involution of the chorda-mesoderm, which is seen from the outside, follows directly in the wake of these inner cells.

In normal development the involution of chorda-mesoderm appears to be the direct successor of endodermal invagination. Observations on abnormal development, however, show that this is not the case. The two processes are independent. When, for some reason, the invagination of endoderm is retarded, it sometimes happens that two grooves appear: an endodermal groove below, and a chorda-mesodermal groove above it (nearer the equator) (Fig. 5–13). Each groove has its own dorsal lip. The lower endodermal groove develops on the dorsal side only. It does not spread laterally. The mesodermal groove, on the other hand, although it begins on the dorsal side, spreads laterally until it forms a complete circular blastopore.

Having now entered the egg by involution, the chorda-mesoderm streams upward, mainly on the dorsal side. For a time, it forms the roof of the archenteron. It would form the side walls of the archenteron, also, if it were not for the fact that the upward cupping of the sides of the endoderm separates it from that cavity.

Closure of the Blastopore

The yolk plug, which has been protruding between the lips of the small circular blastopore, finally withdraws into the gastrula (Fig. 5–8, *F*). Then the blastopore closes. Its lateral lips come together in the midline, and as a result the blastopore becomes a vertical slit. The closed blastopore of an amphibian is comparable to the last remnant of the "primitive streak" of bird and mammal embryos, and it is often referred to by that name. The closure marks the end of the invagination of the endoderm, but the expansion of ectoderm and the involution of chorda-mesoderm continue for a time.

As a consequence of the shift of the yolk-laden endoderm toward the ventral side, the center of gravity of the gastrula is changed. The ventral side is now heaviest. As a result, the egg rotates until its dorsal side is uppermost. The animal pole area is now anterior, and the region of the blastopore is posterior.

Transplantation Experiments on Early Gastrulas

In 1918 Spemann began an important series of experiments in which he transplanted small discs of cells from one region of an early gastrula to another region of another gastrula. In his early experiments, he used different species of salamanders as donors and hosts in order that he might later distinguish donor cells from host cells. Later, the method of staining the donor embryo with Nile blue sulfate was adopted. What happened depended on where the discs came from and where they were planted.

In Spemann's original work, he found that a disc transplanted from one part of the prospective ectoderm to another part of the same zone developed according to its new location (Fig. 5–11). Thus, a disc of prospective epidermis which was transplanted into the area of the future neural plate developed as an integral part of the neural plate. Conversely, a disc of prospective neural plate, when planted into the region of prospective epidermis, became epidermis. A disc of prospective ectoderm, when transplanted to a region of prospective chorda-mesoderm, may become notochord and mesoderm.

We conclude, therefore, that at the start of gastrulation the cells of the animal hemisphere (prospective ectoderm) are either undiffer-

FIGURE 5–11   Spemann's classic experiment, 1921, in which he exchanged small discs of cells between two species of salamanders. *A* to *C.* Discs of prospective epidermis of *Triton cristatus* were implanted in the region of the prospective neural plate of *Triton taeniatus.* They developed according to their new location into brain and retina. *D* to *F.* The converse experiment in which discs of prospective neural plate were implanted in the region of prospective epidermis. The discs became epidermis. *(From H. Spemann, 1921, Arch. Entwicklungsmech., 48:533–570, Abb. 1 to 6.)*

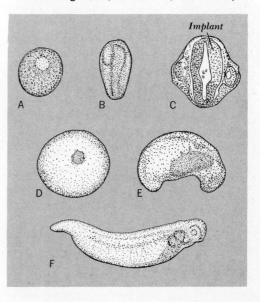

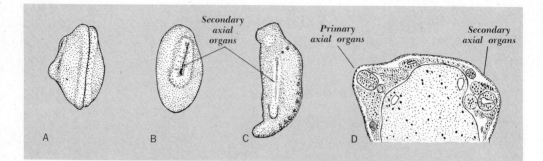

FIGURE 5–12   The "organizer" experiment of Spemann and Hilda Mangold, 1924. A disc
of prospective chorda-mesoderm from the dorsal lip of the blastopore was implanted
to the flank of a beginning gastrula. The implant organized itself and the tissue of the
host which surrounded it into a second embryonic axis. (From H. Spemann and H.
Mangold, 1924, Arch. Entwicklungsmech., 100:599–638, Abb. 19, 20, 21, and 24.)

entiated and undetermined as to their fate
(uncommitted) or else they are labile and capable of reversing their differentiation when
transplanted to changed surroundings.

What determines whether prospective ectoderm will become epidermis or neural tissue?
Transplantation experiments reported in 1924
by Spemann and Hilda Mangold leave no doubt
as to the answer. The fate is decided by the
relation of the ectoderm to adjacent chorda-
mesoderm. If a disc of dorsal chorda-mesoderm is transplanted to the ventral side of
an egg, or if it is placed in the blastocoel so that
it comes into contact with prospective ectoderm, it brings about the differentiation of
axial structures of an embryo (Fig. 5–12). The
disc itself becomes notochord and axial mesoderm, and it induces the adjacent ectoderm
to become a neural plate. The lines are not
sharp, however, between the implant and the
host tissues. Some cells of the implant may
become neural tissue, and some cells of the
host may form mesoderm and even notochord.
The important point is that host and implant
tissues together constitute one self-regulating

"embryonic field" (in the sense in which Spemann used the word), and they differentiate
as an organized whole.

So important is this material of the dorsal
chorda-mesoderm in organizing the structures
of an embryo that Spemann tentatively called
it "the organizer." It is, roughly speaking, the
substance which was the gray crescent. Its
action in stimulating the adjacent ectoderm
and mesoderm to become the dorsal structures
of an embryo is the now-classic example of
embryonic induction. Some of the work which
has been done to discover the chemical nature
and mode of action of the organizer is reviewed
later in this chapter (pp. 112–114).

Exogastrulation

The capacities of gastrula cells to move and
differentiate are shown in the abnormal type
of gastrulation known as exogastrulation (Fig.
5–13). Holtfreter removed blastulas and beginning gastrulas from their membranes and
placed them in a slightly hypertonic solution.
The prospective endoderm at first formed an

endodermal blastopore. Then the prospective chorda-mesoderm, instead of rolling inward (involution), rolled outward. The endoderm, which had begun to sink inward, everted. Thus, both the endoderm and the chorda-mesoderm turned inside out. They retained no contact with the prospective ectoderm.

Much has been learned from these exogastrulas:

1 The prospective endoderm, which glided outward instead of inward, differentiated into the same histological structures (glands and epithelia) which it would have formed if it had invaginated normally.

2 The prospective chorda-mesoderm went through the same mass movements in exogastrulation that it would normally have gone through, except that in this case the movements were outward instead of inward. First, the cells converged toward the blastoporal lips, especially the dorsal lip, in a quite normal fashion. Then they narrowed at the blastopore, as though they were passing through a ring. Following this they stretched and spread outwardly, much as they would do inside the egg. But they were outside the egg, and they became enfolded within the endoderm instead of folding around the endoderm. The chorda-

FIGURE 5–13 Exogastrulation in an amphibian (schematic diagram). Note that the mesoderm and endoderm differentiate "inside out" and that the prospective ectoderm remains undifferentiated. *(Modified from J. Holtfreter, 1933, Arch. Entwicklungsmech., 129:669–793, Abb. 7, with endodermal blastopore added from Abb. 2.)*

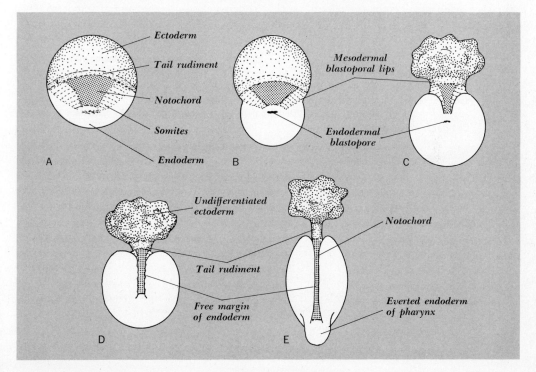

mesoderm then differentiated into the notochord and mesodermal structures.

3 The prospective ectoderm expanded, but
it did not differentiate. Instead, it remained as
a primitive epithelium.

There is much to comment on: the independence of the endodermal and the chorda-
mesodermal blastopores; the apparent self-
organizing capacity of the endoderm and
chorda-mesoderm and the interrelations between them (see page 110); the complete
dependence of the ectoderm on the other germ
layers for its differentiation.

THE GASTRULA

At the time the blastopore closes to a slit, the
amphibian gastrula consists of three layers
of cells: ectoderm, chorda-mesoderm, and

endoderm. It is a triploblastic structure (Fig.
5–14). The outer layer is ectoderm except at
the posterior end where some one-fifth of the
apparent ectoderm is still destined to undergo
involution and become notochord and mesoderm. The chorda-mesoderm forms the roof
of the archenteron. Its free lateral margins
have pushed ventrally at the sides, between
the ectoderm and endoderm. The endoderm
underlaps the lateral chorda-mesoderm internally and forms the sidewalls and floor of
the cavity of the archenteron. This description
applies to salamanders. In the frog, the relations of the chorda-mesoderm and endoderm
are not easy to demonstrate.

The anterior portion of the archenteron is
broad and thin-walled. It consists of the endodermal cells which first invaginated. It becomes the foregut. Near its anterior end it
adheres closely to the ventral ectoderm. This

**FIGURE 5–14 Late gastrula (Harrison's stage 13). The yolk plug has been with-
drawn, and the blastopore has closed to a vertical slit. The location of certain organ
rudiments is indicated. A. View from the left. B. Same view after removal of the prospective epidermis and neural plate of the left side. Approximately × 12. (B is
adapted from E. Witschi, 1956, "Development of Vertebrates," W. B. Saunders
Company, Philadelphia, Fig. 57.)**

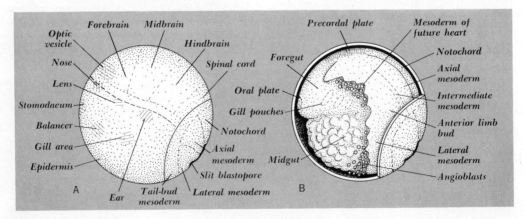

is the location of the oral plate which later breaks through and becomes the mouth opening. Elsewhere, except at the gill clefts and the anus, the ectoderm and the endoderm abhor contact. The major portion of the endoderm, however, consists of the large yolk-laden cells of the *midgut,* that is, of the intestine. Later the cells of the ventral blastoporal lip give rise to the *hindgut,* from which the cloaca is formed.

Transplantation Experiments on Late Gastrulas

We noted that when Spemann transplanted discs of prospective ectoderm of early gastrulas to new locations they differentiated *according to their new location*. When he did the same thing with discs of ectoderm of late gastrulas they differentiated *in accord with their origin*. Something had taken place between the beginning and end of gastrulation. The cells had undergone *determination*.

What is determination? It is a change of some sort which a cell or region of cells undergoes, by which it becomes specified to develop in a certain way. Previous to determination, the region is "indifferent" to its fate and subject to the inductive influences of its environment. Following determination, it is committed to its fate. It is no longer subject to its surroundings. Since the change is not at first visible, it has been called invisible or chemical differentiation. After determination, the gastrula is a mosaic of self-differentiating regions.

It must not be supposed that determination is a sudden process which takes place all at once. On the contrary, it may take place by stages. First, the cells acquire a general predisposition ("bias"), or competence, to respond to further stimulation. This has been called "labile determination." Then later, as

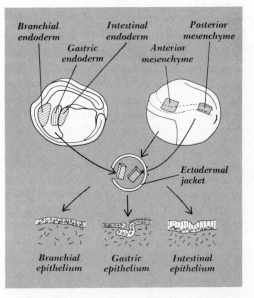

FIGURE 5–15 Okada's experiment in which he took fragments of endoderm, either branchial, gastric, or intestinal endoderm, from an early neurula and placed them in ectodermal jackets along with mesenchyme from adjacent to the neural folds. What the endoderm became was shifted forward or backward according to the source of the mesenchyme. It became more anterior tissue if the mesenchyme came from adjacent to the anterior neural fold, more posterior tissue if it came from near the posterior neural fold. *(Based on T. S. Okada, 1957, J. Embryol. Exptl. Morphol., 5:438–448; and 1960, Arch. Entwicklungsmech., 152:1–21.)*

a result of continuing stimulation or of a second inductive influence, the determination becomes irreversible.

This may be illustrated by some experiments of Okada on the determination of the endoderm (Fig. 5–15). If a fragment of prospective endoderm from any part of a late gastrula (or early neurula) is removed and enfolded in an envelop of ectoderm (of an early gastrula), the endoderm becomes nothing more than a shriveled mass of yolk-laden cells. But if fragments of mesoderm are included with the endoderm, the latter differentiates into recognizable tissues of the gut. What it becomes depends upon where the endoderm came from and also where the mesoderm came from. The endoderm has a predisposition of its own, but it requires the influence of the mesoderm in order to realize its predisposition; moreover, the mesoderm may modify its fate.

In Okada's experiments, the fate of the endoderm was altered according to whether the mesoderm came from a more anterior or more posterior location. If the mesoderm was head mesenchyme, the fate of the endoderm was shifted to a more forward organ. What normally would have been stomach endoderm now became endoderm of the gill region. If, on the contrary, the mesoderm was from the flank, the fate of the endoderm was shifted backward. Prospective gill endoderm became intestinal endoderm. The influence of mesoderm and endoderm is reciprocal, for the fate of the mesoderm is also shifted. The result is that mesoderm and endoderm, differentiate harmoniously into the several organs of the alimentary tract.

NEURULATION

Neurulation is the process by which a neural plate forms and then folds or otherwise transforms into a neural tube (Fig. 5–16). During neurulation, also, the chorda-mesoderm subdivides into the notochord and several regions of mesoderm. The mesoderm pushes anteriorly and laterally between the ectoderm and endoderm until it almost completely surrounds the embryo. At the same time, the free dorsolateral margins of the endoderm move dorsally and close in beneath the notochord. (In the frog, the margins of the endoderm reach the notochord earlier, about midgastrulation.) Other changes also take place during neurulation, so that by the time the process nears completion a "young embryo," recognizable as such, may be said to have taken form.

Neural Plate and Neural Tube

The *neural plate* becomes visible about the time the blastopore closes to a slit. It is a racket-shaped area of dorsal ectoderm, broadest at its anterior end, and narrowing posteriorly toward the blastopore (Fig. 5–6). It is somewhat thicker than the surrounding epidermis and is usually pigmented.

The neural plate shows a shallow, central longitudinal groove, the *neural groove*. This extends from the center of the broad forward area of the plate back to the dorsal lip of the blastopore. The plate thickens and narrows because its cells become columnar. At the same time the adjoining epidermis becomes thinner and spreads. Soon the margins of the neural plate rise up as the *neural folds,* and move toward the middorsal line. They come together, fuse, and so form the *neural tube*. Hence it is that the middorsal lines of the epidermis and neural tube are lines of suture. The cavity of the neural tube is known as the *neurocoel*.

The fusion of the neural folds begins in the region of the future midbrain and progresses forward and backward. An opening, the *anterior neuropore,* remains for a brief time at the anterior end. At the posterior end, the neural folds terminate at the lateral lips of the blastopore. When ultimately they come

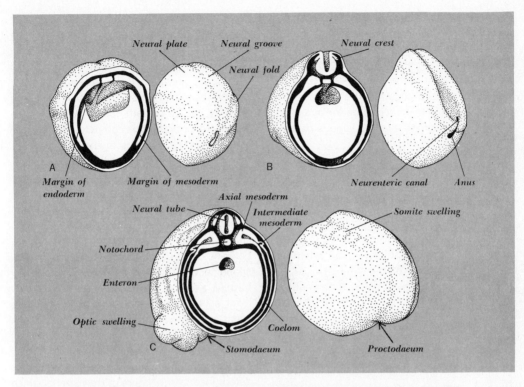

FIGURE 5-16 Diagram of neurulation in salamander development (highly schematic).
The germ layers are in reality closely packed together.

together, they enclose the dorsal opening of the slitlike blastopore. Thus, for a brief time, there is potentially a passageway between the cavity of the neural tube (neurocoel) and the cavity of the gut (enteric cavity), which is known as the *neurenteric canal* (Fig. 6–23).

Neural Crest

Particular note must be taken of the cells along the margins of the neural plate. They form what is known as the *neural crest* (Fig. 5–16). Unlike the cells of the neural plate proper, they early lose their epithelial char-

acter, become loose, and migrate individually. They give rise to various tissues, notably nerve ganglia, pigment cells, and certain cartilages of the head. We shall have occasion to refer to them again in the next chapter.

The Chorda-Mesoderm

The subdivision of the chorda-mesoderm into notochord, axial mesoderm, intermediate mesoderm, and lateral mesoderm will be referred to when we describe the neurula (page 116). Differences between these tissues are not easy to recognize in cross sections of young frog embryos, for the cells are closely packed together and are full of yolk granules. Yet, although they look alike, they are different, for they engage in different morphogenic movements. By the time the neural folds come together, they have acquired the capacities to differentiate and to induce ectoderm and endoderm to differentiate into the various parts of the embryo.

The involution of chorda-mesoderm continues throughout the period of neurulation. At the beginning of neurulation roughly one-fourth of the open neural plate is still destined to roll around the lips of the slit-shaped blastopore and then move forward on the inside of the egg as a part of the archenteric roof (Fig. 5–14). The material of the future tail bud is a band across the neural plate, somewhat forward of its posterior end (Fig. 5–14, A). The posterior end of the neural plate becomes the underside of the tail bud just posterior to the anus.

The Endoderm

We described the endoderm as a trough which is open dorsally. During neurulation, its free lateral margins move upward to the midline beneath the notochord. Here they fuse and form a new or secondary roof over the gut. Hence the gut is no longer an archenteron (primitive gut) but an enteron. In the frog, the upward pushing of the endoderm margins is precocious so that, as seen in sections, the endoderm seems to split away (delaminate) from the mesoderm.

ANALYSIS OF PRIMARY INDUCTION

The determination of the neural plate is the result of influences arising in the underlying chorda-mesoderm. It is a classic example of *embryonic induction*. How is this accomplished? What is its mechanism? A great deal of experimentation has been carried on in seeking an answer to this question.

It has been found that induction is a process which is limited in both space and time. Only prospective ectoderm is competent to respond to the inducing stimulus of the chorda-mesoderm and to give rise to neural tissue. Moreover, it is only for a definite period during gastrulation that it will so respond. As gastrulation progresses, the competence wanes. Then, soon after neurulation has begun, it ceases entirely. Neurula ectoderm has new competencies; for example, its epidermis will respond to the stimulus of an optic vesicle by forming a lens.

For a time it was thought that the inducing agent acts only as a releasing stimulus, much as pulling the trigger of a gun releases energies pent up in the cartridge. According to this interpretation, the nature of what is induced is determined by the nature of the material which responds. This view was based on the discovery that in salamanders many different materials are able to induce, not only chorda-mesoderm, but also brain tissue, muscle segments, and even adult tissues, both living and dead. Furthermore, various chemical substances were found which bring about the differentiation of ectoderm into nervous tissue. But this effect could be explained in many cases as an indirect effect: the chemicals injure the living tissue so that the tissue releases substances with inductive powers.

This view that the inducing agent acts as a stimulus only has in recent years been changed by the discovery by Spemann, Yamada, and others that different agents have different inductive effects. What is induced is determined, at least in part, by the nature of the inducing agent. (1) Thus the dorsal lip of the blastopore, at the beginning of gastrulation, induces ectoderm to form forebrain, eyes, and nose. It has been called an *archencephalic inductor*. (2) The dorsal lip at a somewhat later stage (the cells of the initial dorsal lip having entered within the gastrula) induces the production of midbrain, hindbrain, and inner ears (otocysts). It is termed a *deuterencephalic inductor*. (3) Still later, when gastrulation is nearing completion, the now dorsal lip stimulates the ectoderm to become spinal cord and even notochord and somites of the tail. It is spoken of as a *spinocaudal inductor*. (4) Finally, some inducing agents have been found which cause early gastrula ectoderm to become mesodermal and endodermal tissue.

It will be noted that in each of these cases the inductor determines some *region* of the embryo, not particular organs. Thus, the same agent will cause a variety of organs to be produced—ectodermal, mesodermal, and even endodermal; but they all belong to the same region of the embryo. (There are exceptions, and certain unnatural agents may induce the production of tissues which are chaotically disorganized.)

For a long time, it seemed that the inducing influence passes from the inductor to the induced only when there is actual contact between the two. But it was found by Twitty and Niu that in salamanders if a bit of embryonic tissue which is capable of acting as an inducing agent is explanted into a small drop of suitable saline solution (on a hanging-drop slide) and grown there for a week or more, the solution becomes "conditioned." It acquires the capacity to induce. If now a few cells of ectoderm are placed in the solution for 24 hours or more (the original tissue having first been removed), they will transform into nerve cells, meso-

dermal tissues, etc. Thus it seems evident that the inducing substance is soluble.

Much of this work on induction has been facilitated by the technique of inserting the material to be tested into a sandwich of ectoderm taken from the animal hemisphere of a beginning gastrula. The cut margins of the ectoderm heal together and form a vesicle with the test object inside. If the test object lacks inducing power, then the ectoderm becomes a wrinkled epithelium which soon degenerates. If, on the other hand, it has the capacity to induce, then the ectoderm responds by differentiating into recognizable tissues of an embryo.

This sandwich method has been employed in studying the inductive action on salamander ectoderm of adult tissues of guinea pigs and rats. Of course, tissues of an adult mammal are not the normal inductors of amphibian development, but they have served as models to point the way to principles which possibly apply to normal development. Hayashi found, for example, that liver tissue has predominantly archencephalic induction properties (forebrain, eyes, nose), while kidney tissue induces, for the most part, deuterencephalic and spinocaudal structures (midbrain, hindbrain, ears, spinal cord, notochord, and somites of the tail). Similar experiments with bone marrow have shown that it possesses the capacity to induce gastrula ectoderm to become mesodermal structures (limb buds, notochord, pronephric tubules, and mesenchyme), and even endodermal structures (pharynx, esophagus, lung buds, stomach, and intestines) but not neural structures.

As a result of observations such as these, Nieuwkoop has proposed the view that two substances are involved in induction. Toivonen refers to them as (1) a "neuralizing factor" which is concentrated toward the dorsal side and favors the development of the organs of the dorsal side, and (2) a "mesodermalizing factor" which increases in concentration toward the posterior end and which tends toward the production of mesoderm. According to this interpretation, the neuralizing factor, if present alone (as in liver), induces archencephalic structures. The neuralizing factor, along with a fair amount of mesodermalizing factor, favors deuterencephalic structures. If the mesodermalizing factor is present in still greater proportion (as in kidney), the result is spinocaudal organs. If, finally, the mesodermalizing factor strongly predominates (as it does in bone marrow), only mesodermal and endodermal structures are produced.

Of the two factors, the neuralizing factor is the more stable. It is not destroyed by heat, alcohol, and various chemical treatments. The mesodermalizing factor, on the other hand, is thermolabile. Toivonen found, for example, that when kidney tissue is boiled, its deuterencephalic and spinocaudal induction capacities are reduced, but its archencephalic induction powers largely remain. This two-factor hypothesis has been put to the test by placing both a pellet of guinea-pig liver and a pellet of guinea-pig bone marrow in the blastocoel of an early amphibian gastrula. Now, a pellet of bone marrow by itself will induce mesodermal structures only. But the two tissues placed together within the blastocoel induce structures belonging to all levels of the embryo. What is induced depends upon the relative amounts of the two agents which are present.

The work reported above was performed on salamander (urodele) material. The ectoderm of frog and toad gastrulas is not responsive to the same wide variety of inducing agents which affect salamander ectoderm. It indeed responds to living chorda-mesoderm cells of frog or salamander eggs by forming neural tissue; but, generally speaking, it is not influenced by dead or unnatural agents. An explanation for this discrepancy has not yet been found.

THE NEURULA

During neurulation, the three germ layers reach their final positions with respect to one another, and their fates begin to be determined. These are matters of importance, for the development of unified organs depends upon the normal interaction between germ layer and germ layer.

Figure 5–17 indicates the approximate location of certain organ rudiments in an early neurula of a salamander. Note that the neural plate is divisible into forebrain, midbrain, and hindbrain. The forebrain and midbrain overlie that part of the chorda-mesoderm which is known as *prechordal mesoderm*. The central portion of the latter is still a part of the roof of the archenteron and is termed the *prechordal plate*. Its lateral portion, here labeled "mandibular mesoderm," is underlaid by endoderm. It receives contributions of cells from the neural crest and gives rise to the upper and lower jaws. The prechordal plate becomes separated from the roof of the pharynx by the closing in of endoderm beneath it. It then separates to the right and left and gives rise to the extrinsic muscles of the eyeballs.

The most anterior region of the neural plate is the *optic area*. As a result of inductive action by the prechordal plate which lies beneath it, it acquires the capacity to become retinas and optic nerves. The fact that it forms two eyes depends upon the prechordal plate separating to the right and left. If, by reason of retarded development, this separation fails to occur, a single median eye is formed, a condition known as *cyclopia.*

Anterior and lateral to the optic areas are the *olfactory areas* which give rise to the nasal pits. In cyclopian embryos, the nasal pit is single and dorsal to the median eye.

The central axis of the hindbrain overlies the notochord. Its flanks overlie the head portion of the axial mesoderm. Inductive interactions control the differentiations of this part of the head. The *auditory area* of the epidermis, for example, is related both to underlying axial mesoderm and to the forward part of the hindbrain. It gives rise to the auditory vesicles (otocysts), from which the labyrinth of the inner ear is derived.

The epidermis ventral to the forward tip of the neural plate is free of mesoderm at the neural plate stage and has immediate contact with the forward wall of the foregut (pharynx). The epidermis responds by developing the *stomodaeum,* the ectodermal-lined part of the oral cavity. Here is formed the *oral plate,* which later breaks through.

Note that the mesoderm has not yet entered the region immediately ventral to the mandibular mesoderm. This is where the first pharyngeal pouch (endodermal) comes into contact with the epidermis at the side of the head. It forms a plate, but it does not break through in amphibians. The other pharyngeal pouches push through the mesoderm, make contact with the epidermis, and give rise to gill openings.

Loose mesenchymal cells are shown in Fig. 5–17 as having pushed forward from the margin of the lateral mesoderm. These are *angioblasts,* which are destined to become the endo-

FIGURE 5-17 Diagram of an early neurula of a salamander. *(Based on H. B. Adelmann, 1932, J. Morphol., 54:1–67, Fig. 2.)*

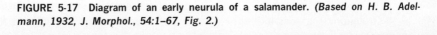
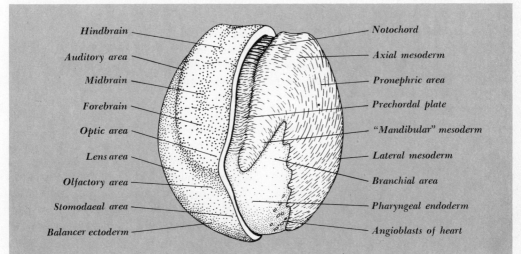

thelial linings of the heart (endocardium) and blood vessels. The muscles of the heart (myocardium) will come from lateral mesoderm which moves ventrally and enfolds the endocardium (Fig. 6–21).

Posterior to the head region the chordamesoderm is divisible into notochord, axial mesoderm (from which the body segments, or somites, are derived), intermediate mesoderm, and lateral mesoderm. The forward portion of the intermediate mesoderm gives rise to the pronephric tubules, the first excretory organs of the embryo. The lateral mesoderm becomes the mesoderm of the visceral and body walls, with the body cavity between. The anterior limb buds (not labeled in the figure) come from the lateral mesoderm of the body wall just ventral to the pronephric area.

CHANGES IN CELL PROPERTIES DURING GASTRULATION AND NEURULATION

The onset of gastrulation is associated with important changes in the chemistry of the cells: (1) basic proteins increase within the nucleus; (2) the synthesis of ribosomal RNA, absent during cleavage, is resumed; (3) nucleoli, not present during cleavage, reappear within the nuclei; (4) messenger RNA, entering the cytoplasm or becoming unmasked, joins with the ribosomes to form polyribosomes; and (5) the ribosomes become increasingly attached to the endoplasmic reticulum.

These events are presumably related. The proteins and the ribosomal RNA unite in the production of the substance of the nucleoli. The nucleoli in turn are the stuff of which ribosomes are made. The polyribosomes engage in the synthesis of the new proteins characteristic of the regions of the differentiating embryo. Their attachment to the endoplasmic reticulum signifies that proteins are about to be secreted.

The evidence for these statements comes from biochemical studies of several sorts. In some studies the cells are labeled with radioactive precursors of RNA and protein, and then the location of the label is ascertained by means of radioautographs. In other studies, the new proteins are detected by immunological methods. A third technique involves treating the cells with substances which inhibit synthetic processes. It will be recalled that actinomycin D is an antibiotic which specifically inhibits the production of new RNA at gene (DNA) templates; that is, it prevents transcription (page 72). The old ribosomes and mRNA carried over from the oocyte are sufficient for the needs of cleavage and blastula formation. But new mRNA and new types of protein are required for gastrulation, neurulation, and the differentiations which then get under way.

Another observation which points in the same direction concerns interspecies hybrids (pp. 32, 33). These develop quite normally through cleavage but fail to gastrulate. Again the explanation seems to be that the old mRNA already present in the egg cytoplasm is sufficient for the needs of cleavage, but the unfamiliar genes which are brought in by the foreign sperm are unable to cooperate with the egg's cytoplasm in the production of the new mRNA needed for differentiation. As a result, the proteins needed for further development are not produced, and death results. Strangely enough, it has been found that if the blocked tissue is grafted into a normal host, "revitalization" occurs. Possibly the new RNA which is lacking in the graft is supplied by the neighboring host cells.

The ribosomes are not distributed evenly

throughout the cytoplasm of the egg. They are most abundant around the animal pole and on the dorsal side. Any treatment which modifies the arrangement of the ribosomes by increasing or decreasing the gradient of their distribution produces characteristic abnormalities. For example, a strong centrifugal force can produce an accumulation of ribosomes on one side of the egg and so give rise to a secondary embryonic axis. Chemical treatments which reduce the steepness of the gradient can result in microcephaly (underdevelopment of the head).

During the course of neurulation, the processes of differentiation enter a new phase. We have already noted that gastrulas and early neurulas are sensitive to actinomycin D. But later neurulas and the tail-bud stages which follow are less sensitive. Why is this so? The probable explanation is that at these more advanced stages sufficient new RNA has already been produced to provide for the production of the proteins needed in differentiation. On the other hand, those drugs, such as puromycin, which interfere with translation, that is, with the new production of protein by RNA, bring differentiation to an immediate halt. It is during the period of translation when the cells have lost much of their sensitivity to actinomycin that differentiations first become visible.

The mass movements and differentiations which take place at the time of gastrulation and neurulation no doubt require energy. A progressive increase in respiration (in the utilization of oxygen) does indeed accompany these processes. But just how much of this increase is due to a general speeding up of life processes, and how much is related specifically to morphogenetic activities, cannot be determined with certainty. The increased demand for energy is accompanied, as might be expected, by an increase in the number of mitochondria, the "powerhouses of the cell." Also, the outer layer of the yolk granules is dissolved, glycogen is consumed, and water enters the cells. As a result the volume of protoplasm is increased. These processes take place first and most notably in those cells which are engaged, or are about to engage, in morphogenetic movements, namely, in the cells of the dorsal blastoporal lip and later in those of the neural plate. But this fact in itself does not prove that the increased energy is being used to support morphogenetic movements.

THE BEHAVIOR OF FRAGMENTS OF EMBRYOS AND OF DISSOCIATED CELLS

Blastula cells are loosely bound together. When a fragment of a blastula is placed in a calcium- and magnesium-free salt solution, the bonds between the cells weaken and the cells fall apart. If cells which have been dissociated in this manner are then transferred to an abundance of a suitable physiological salt solution, they may live for days nourished by the yolk granules in their cytoplasms. But, generally speaking, they do not differentiate. Some adhere to the bottom of the dish or move about in amoeboid fashion; but this is as far as they go.

When the dissociated cells of a blastula come into contact with one another, they adhere and draw together into tight balls. After a day or two, some of the balls may develop cilia and roll around. But only if the mass is of sufficient size do some of the cells in the center of the balls show a tendency to differentiate.

The behavioral properties of cells change

during gastrulation and neurulation. Cells which form epithelial layers adhere together with ever-increasing strength, as any one who dissects living embryos will discover. To bring about dissociation it is now necessary to use powerful methods, such as exposing the fragments to an alkaline solution or to a weak solution containing trypsin. If, after dissociating, the cells are promptly returned to a suitable physiological salt solution, they appear to be uninjured. After sinking to the bottom of the dish, some types undergo differentiation. For example, future nerve cells produce one or more pseudopod-like outgrowths, the forerunners of axons; future notochordal cells develop a fringe of clear protoplasm around their peripheries; and future muscle cells produce lobes at opposite poles and become spindle-shaped.

When gastrula or neurula cells of different sorts come into contact with one another, they tend to adhere. Then they crawl around on each other and rearrange themselves in characteristic manners. Holtfreter has described the results and has noted that they are similar to normal processes of development. First he devised a salt solution—which now bears his name—in which cells or fragments will live for days and behave quite normally. To bring about dissociation, he added KOH until the pH became about 9.8. Then he returned the cells to his physiological salt solution at pH 8.0. The cells did not attract one another. But when they came into contact they adhered like soap bubbles. Holtfreter called the principle "affinity," although "adhesiveness" would seem to be an adequately descriptive term. A few of Holtfreter's results are as follows:

When dissociated prospective ectoderm cells of a gastrula reaggregate, they form a cyst. The wall of the cyst thins and becomes wrinkled. This process may be compared to the expansion by which epidermis spreads over the outside of an embryo. Prospective endoderm cells round up into a compact mass. Then, about the time that the margins of the endoderm of a gastrula are migrating dorsally and closing in beneath the notochord, some of the endoderm cells spread out on the glass bottom of the containing dish and form a layer one cell thick. The exposed surface corresponds to the exposed surface of an intestinal epithelium.

When prospective ectoderm and endoderm cells are intermingled, the endoderm draws toward the interior, and the ectoderm spreads out on the surface as a cap. This indicates, according to Steinberg's interpretation (page 54), that the adhesion of endoderm for endoderm is stronger than of ectoderm for ectoderm and that the sum of their tendency to adhere to their own kind is greater than the tendency of endoderm to adhere to ectoderm. About the time that gastrulation takes place, the endoderm and ectoderm separate, and the exposed surfaces of both become nonadhesive. This may explain why the lips of the blastopore do not adhere to the yolk plug.

The most adhesive and at the same time the most independent of the three germ layers is the mesoderm. Fragments of the prospective chorda-mesoderm from an early gastrula tend to go through the mass movements of gastrulation. Those cells which undergo the motions of involution differentiate as notochord and mesoderm. Those which fail to do so, even though they are prospective chorda-mesoderm, become ectodermal tissues, i.e., epidermis and neural tissue.

Ectodermal and mesodermal cells which

are mingled together draw into a compact cluster. Then they segregate, and the mesoderm migrates inward. Soon, however, the mesodermal cells pinch away from the ectodermal cells, much as they do during normal gastrulation (compare exogastrulation, Fig. 5–13), but they remain connected by a narrow waist. Later they come together again, the ectoderm on the surface, the mesoderm as a compact mass within. The mesoderm then self-differentiates into notochord, somites, occasional pronephric tubules, and some free mesenchyme. The ectoderm, under the inductive influence of the mesoderm, differentiates into neural plate and epidermis. Mesoderm plus endoderm behaves somewhat like mesoderm plus ectoderm. First the mesodermal cells move to the interior of the mass, where the mesoderm differentiates into notochord and somites. The endoderm differentiates with its epithelial surface facing outward (again compare exogastrulation, Fig. 5–13).

What happens when ectoderm, endoderm, and mesoderm are explanted together? Both ectoderm and endoderm tend to enclose mesoderm, but ectoderm tends to enclose endoderm. The problem is solved, as in the embryo, by ectoderm spreading on the outer surface, mesoderm becoming an intermediate layer, and endoderm forming a hollow tube within. The epithelial surfaces of the ectoderm and endoderm become nonadhesive. Their surfaces in contact with mesoderm, however, are strongly adhesive. Indeed, the adhesion of ectoderm and endoderm with mesoderm may be stronger than the adhesion of mesoderm with mesoderm, for cavities appear in the mesoderm which may be compared to the coelom of the embryo.

New factors appear during neurulation. For example, neural tissue becomes distinct from epidermis. When the ectoderm of a neurula is disaggregated and then reassembled, the future neural cells move inward and become a compact internal mass. The epidermal cells spread on the surface. The boundary between the two types is clean cut. Indeed, a space forms between them. Then a cavity appears in the neural mass by cavitation. Regional differentiation of the neural tissue does not take place, presumably because the inductive influence of the chorda-mesoderm is lacking. But if mesenchyme is present, or even if mesectoderm from the neural folds is present, the neural tissue may undergo some differentiation. For example, an optic vesicle may form and cave in to form an optic cup.

SELECTED READINGS

Descriptive accounts of the development of the frog will be found in the textbooks of McEwen, Rugh, Huettner, and Witschi. See Appendix.

Normal stages in the development of frogs and salamanders are figured in: Rugh, R., 1962. 2d ed. Experimental Embryology. Burgess Publishing Company, Minneapolis.

The Development of the Oocyte

Witschi, E., 1956. Development of Vertebrates. W. B. Saunders Company, Philadelphia. Pp. 24–33.

Nace, G. W., and L. H. Lavin, 1963. Heterosynthesis and autosynthesis in the early stages of anuran development. Am. Zool., 3:193–207.

Wischnitzer, S., 1966. The ultrastructure of the cytoplasm of the developing amphibian egg. Advan. Morphogenesis, 5:131–179.

Symmetrization and the Role of the Cortex

Newport, G., 1854. Researches on the impregnation of the ovum in the Amphibia; and on the early stages of the development of the embryo. Phil. Trans. Roy. Soc. London, 1854, 229–244. Reprinted in Gabriel and Fogel (eds.), 1955. Great Experiments in Biology. Prentice-Hall, Inc., Englewood Cliffs, N.J.

Clavert, J., 1962. Symmetrization of the egg of vertebrates. Advan. Morphogenesis, 2:27–60.

Pasteels, J. J., 1964. The morphogenetic role of the cortex in the amphibian egg. Advan. Morphogenesis, 3:363–388.

Curtis, A. S. G., 1963. The cell cortex. Endeavour, 22:134–137.

Cleavage and the Blastula

Dan, K., 1960. Cyto-embryology of echinoderms and amphibia. Intern. Rev. Cyto., 9:321–368.

Dettleff, T. A., 1964. Cell division, duration of the interkinetic states, and differentiation in early stages of embryonic development. Advan. Morphogenesis, 3:323–362.

Fate Maps of Amphibians

Vogt, W., 1929. Gestaltungsanalyse am Amphibienkeim mit örtlicher Vitalfärbung. Arch. Entwicklungsmech., 120:384–706.

Pasteels, J., 1942. New observations concerning the maps of presumptive areas of the young amphibian gastrula. J. Exptl. Zool., 89:255–281.

Carpenter, Esther, 1937. The head pattern in Amblystoma studied by vital staining and transplantation methods. J. Explt. Zool., 75:103–129.

Adelmann, H. B., 1932. The development of the pre-chordal plate and mesoderm of Amblystoma punctatum. J. Morphol., 54:1–67.

Primary Induction

Spemann, H., 1938. Embryonic Development and Induction. Yale University Press, New Haven, Conn.

Gray, G. W., 1957. The organizer. Sci. Am., 197 (November):79–88.

Holtfreter, J., and V. Hamburger, 1955. Embryogenesis: Amphibians. In Willier, Weiss, and Hamburger (eds.), Analysis of Development. W. B. Saunders Company, Philadelphia. Pp. 230–298.

Yamada, T., 1961. A chemical approach to the problem of the organizer. Advan. Morphogenesis, 1:1–54.

Saxen, L. and S. Toivonen, 1962. Primary Embryonic Induction. Prentice-Hall, Inc., Englewood Cliffs, N.J.

Tiedemann, H., 1966. The molecular basis of differentiation in early development of amphibian embryos. In Moscona and Monroy (eds.), Current Topics in Developmental Biology, Academic Press Inc., New York. Pp. 85–112.

Chemical Activity During Cleavage

Brown, D. D., 1964. RNA synthesis during amphibian development. J. Exptl. Zool., 157:101–114.

———— and J. B. Gurdon, 1964. Absence of ribosomal RNA synthesis in the anucleolate mutant of Xenopus laevis. Proc. Natl. Acad.

Sci., **51**:139–146. Reprinted in Bell (ed.), 1965. Molecular and Cellular Aspects of Development. Harper and Row, Publishers, Incorporated, New York. Pp. 333–340.

Gurdon, J. B., and D. D. Brown, 1965. Cytoplasmic regulation of RNA synthesis and nucleolus formation in developing embryos of Xenopus laevis. J. Mol. Biol., **12**:27–35.

Behavior of Dissociated Cells and Fragments

Holtfreter, J., 1939. Tissue affinity, a means of embryonic morphogenesis. Translated and reprinted in Willier and Oppenheimer (eds.), 1964. Foundations of Experimental Embryology. Prentice-Hall, Inc., Englewood Cliffs, N.J. Pp. 186–225.

Townes, P. L., and J. Holtfreter, 1955. Directed movements and selective adhesion of embryonic amphibian cells. J. Exptl. Zool., **128**:53–120. Reprinted in Bell (ed.), 1965. Molecular and Cellular Aspects of Development. Harper and Row, Publishers, Incorporated, New York. Pp. 3–40.

Elsdale, T., and K. Jones, 1963. The independence and interdependence of cells in the amphibian embryo. In Society for Experimental Biology, Cell Differentiation (17th Symposium). Academic Press Inc., New York. Pp. 257–273.

chapter **6**

THE
YOUNG
AMPHIBIAN
EMBRYO

This chapter is written with a young amphibian embryo in mind, yet much of what is stated applies in a general way to the embryos of other air-breathing vertebrates. The "young amphibian embryo" begins at approximately the late neurula or tail-bud stage of a frog or salamander. It may be compared in many respects to a 13- to 18-somite chick (33 to 38 hours of incubation), or to a 2.5-mm mammal (15-day pig or 22- to 26-day human). Although the earliest stages of vertebrate development are vastly modified by the amount of yolk which may be present in the egg and by the presence or absence of fetal membranes, yet, by the time neurulation gets under way, the embryos of all vertebrates, especially air-breathing vertebrates, are astonishingly similar. This is true whether the embryo's outer habitat is water (as in amphibians), air (as in reptiles and birds), or the maternal uterus (as in mammals). The late neurula stage has been the most conservative stage in the evolution of the vertebrates.

Of course any stage in embryonic development is an embryo. We are using the word in a narrow, and probably improper, sense to refer to those stages in which the form of the new individual is becoming apparent. As thus defined, the young embryo is quite straight and, for the most part, symmetrical. We visualize it in a standard position, namely, with the head forward and the back upward. Anterior (cephalad), therefore, means toward the head, and posterior (caudad) means toward the tail. Dorsal (dorsad) is upward, and ventral (ventrad) is downward. (Unfortunately, the terms anterior and posterior are used otherwise in human anatomy. They are used as synonyms of ventral and dorsal.) Right and left always refer to the embryo's right and left.

FIGURE 6–1 Young embryos of the leopard frog *Rana pipiens. A* to *E.* Shumway's stages 15, 16, 17, 18, and 18 +. Approximately × 12.

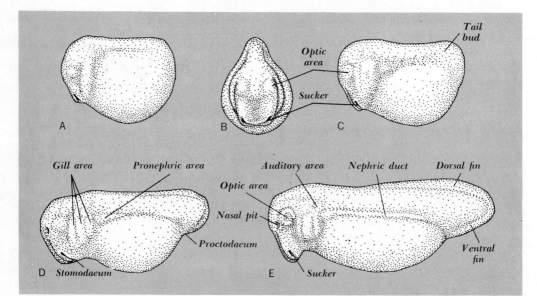

TABLE 6–1. "NORMAL STAGES" IN THE DEVELOPMENT OF RANA PIPIENS

Based on Waldo Shumway, 1940. Stages in the normal development of Rana pipiens. I. External form. Anat. Record, 78:139–147.

Stage 1. Unfertilized egg.

Stage 2. Fertilized egg showing gray crescent. (1 hour).

Stage 3. Two-cell stage. (3.5 hours.)

Stage 4. Four-cell stage. (4.5 hours.)

Stage 5. Eight-cell stage. (5.7 hours.)

Stage 6. Sixteen-cell stage. (6.5 hours.)

Stage 7. Thirty-two cell stage. (7.5 hours.)

Stage 8. Midcleavage. (16 hours.)

Stage 9. Late cleavage. (21 hours.)

Stage 10. Dorsal lip of blastopore appearing. (26 hours.)

Stage 11. Midgastrula; semicircular blastopore. (34 hours.)

Stage 12. Late gastrula; circular blastopore, one-fifth the diameter of the egg. (42 hours.)

Stage 13. Early neurula; broad neural plate; shallow neural groove. The blastopore is in the process of closing. (50 hours.)

Stage 14. Midneurula. The neural folds have moved about halfway toward the midline. (62 hours.)

Stage 15. Late neurula. The neural folds are in contact one-half to three-fourths of their length. The rotation of the embryo (a result of the development of cilia by the epidermis) begins. (Fig. 6–1, *A* illustrates a somewhat older stage. 67 hours.)

Stage 16. 2.5 mm long. The neural folds have closed. The swellings of the optic vesicle, mandibular arch, and gill areas are seen. (Fig. 6-1, *B*. 72 hours.)

Stage 17. 3 mm. The tail bud is beginning to grow out. (Fig. 6–1, *C*. 84 hours.)

Stage 18. 4 mm. Lateral bending, due to unilateral muscle contraction, is seen for the first time. At this time the tail bud is about one-fifth the length of the embryo. (Fig. 6–1, *D*. 96 hours.)

Stage 19. 5 mm. The heart begins to beat. The tail is approximately one-third the length of the whole embryo. (Fig. 6–1, *E* is somewhat younger than stage 19. 118 hours.)

Stage 20. 6 mm. Circulation of blood can be seen in the external gills as soon as they appear. The tail is approaching one-half the total length. During this stage the embryo hatches unless, by reason of crowding or temperature, it has already done so. (Fig. 6–2, *A*. 140 hours.)

Stage 21. 7 mm. The external gills have now fully grown out; the mouth is open. The corneas of the eyes are becoming transparent. Occasional spontaneous swimming takes place. (162 hours.)

Stage 22. 8 mm. The opercular folds are clearly seen. The first circulatory loops are visible in the tail fins. (Fig. 6–2, *B*. 192 hours.)

Stage 23. 9 mm. Teeth beginning to appear. (216 hours.)

Stage 24. 10 mm. The operculum has covered the gill on the right. (240 hours.)

Stage 25. 11 mm. The opercular folds are complete (except for the spiracle on the left side). Soon the hind limb buds will appear. (Fig. 6–2, *C*. 284 hours.)

Medial means near the midplane, and lateral means away from it. Proximal (referring to appendages) is toward the body, and distal is away from it. When an embryo, in the case of birds and mammals, becomes flexed like a hibernating squirrel, it may come about that the anterior and posterior ends come close together. This need cause no difficulty in terminology if the embryo is thought of as straightened out.

"Stages" in the Development of the Frog

For those who use the egg of the leopard frog, *Rana pipiens,* in the laboratory, Table 6–1, based on a paper by Shumway, will prove useful. The approximate age in hours (at 18°C) is indicated in parentheses. However, the stage in development varies, not only with temperature, but with crowding and the amount of oxygen in the water.

The young embryo may be described as tubular and consisting of six longitudinal elements (Fig. 6-3):

1 The outer epidermal covering
2 The dorsal neural tube
3 The notochordal rod
4 The endodermal canal
5 The right mesodermal plate
6 The left mesodermal plate

Between these elements there are loose wandering cells known as mesenchyme. Although the mesenchyme is mostly of mesodermal origin, yet the other germ layers, especially the ectoderm, contribute to it. It gives rise to

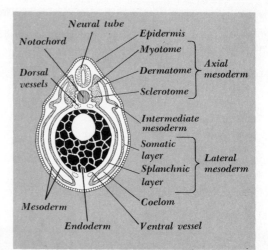

FIGURE 6-3 Diagrammatic cross section of a young embryo. The labels indicate the longitudinal elements and the subdivisions of the mesoderm.

supportive tissues, to muscles, to the blood vessels, and to the blood.

THE EPIDERMIS

The ectoderm of the outer surface of the young embryo is the epidermis. The midline of the epidermis of the head, the dorsal sagittal ridge, and the tail are lines of suture (seams) where the neural folds and lateral lips of the blastopore came together. Since elongation is more rapid along these lines of suture than elsewhere, it comes about that the head fold rolls forward and the tail bud grows backward. In salamanders the back arches upward (Fig. 6-4).

Local swellings of the epidermis reveal the positions of internal structures, such as eyes, brain vesicles, somites, gill area, kidneys (pro-

FIGURE 6-2 Tadpoles of the leopard frog. A to C. Shumway's stages 20, 22, and 25. × 9, × 6½, and × 5½, respectively.

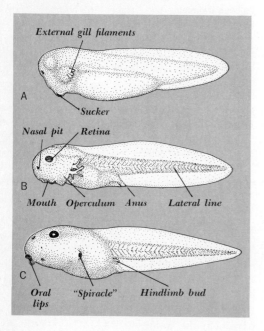

nephroi), and limb buds. The epidermis also develops placodes, or local thickened areas, which sink inward and form sensory vesicles. It shares in the formation of closing plates, such as the oral plate and the gill plates. These break through and become openings into the gut.

Frogs and toads are exceptional in that the epidermis is divided into two layers and only the inner layer, the so-called nervous layer, develops placodes. The outer layer consists of cells bearing dark pigment and, interspersed among them, glandular cells which secrete mucus.

Sensory Placodes

The most anterior of the sensory placodes are the *olfactory placodes* near the anterior tip of the embryo close to the forebrain (Fig. 6–6, *A*). These become the olfactory pits and

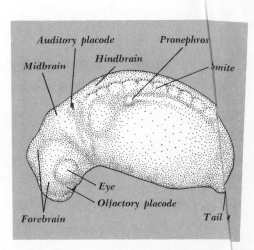

FIGURE 6–4 Salamander embryo (*Amb_stoma*) from the left, showing the regions of the epidermis. Harrison's stage 28. Approximately × 9½.

FIGURE 6–5 Lateral dissection of a 6-mm leopard frog embryo. Shumway's stage 20. The levels of the sections shown in Figs. 6–6 to 6–8 are indicated. Note the three pronephric tubules and the primary nephric duct. × 15.

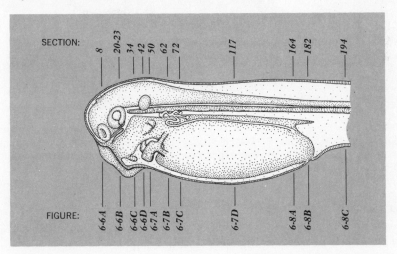

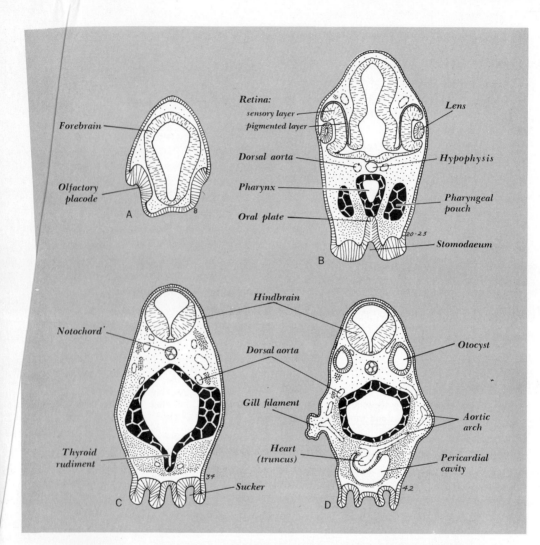

FIGURE 6–6 Cross sections of a 6-mm embryo of the leopard frog. *A.* Section through the olfactory pit. *B.* Section through the eyes, hypophysis, and tip of the pharynx. *C.* Section through the second pharyngeal pouches and thyroid rudiment. *D.* Section through the otocysts, third aortic arches, and truncus. × 40.

ultimately give rise to the olfactory membranes of the nasal cavities.

The *lens placodes* are next in order. They form where the developing optic vesicles come into contact with the epidermis. Soon they cup inward, and after pinching away from the epidermis, become the lenses of the eyes (Fig. 6–6, *B*).

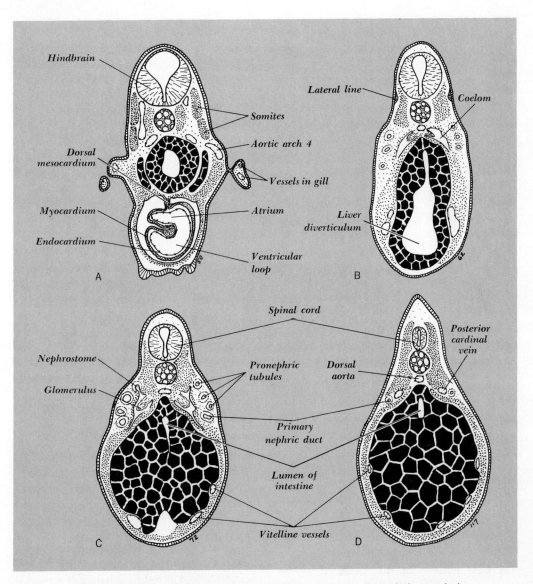

FIGURE 6–7 Cross sections of a 6-mm frog embryo. *A.* Section through the ventricular loop and gill filaments. *B.* Section through the liver diverticulum and the anterior tip of the coelom. *C.* Section through pronephric tubules and glomeruli. *D.* Section through the intestine. × 40.

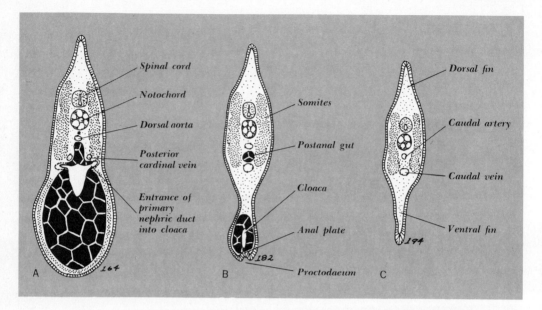

FIGURE 6–8 Cross sections of a 6-mm frog embryo. A. Section through the entrance
of the primary nephric duct into the cloaca. B. Section through the anal membrane and
the postanal gut. C. Section through the tail. × 40.

The *auditory placodes* develop at the sides
of the hindbrain (Fig. 6–6, *D*). They sink inward
as the auditory pits and give rise to the
otocysts (auditory vesicles), from which the
membranous labyrinths of the inner ear (for
equilibrium and hearing) are derived. In the
frog, only the inner layer of the epidermis
sinks inward and is involved in producing the
lens and otocyst.

The Determination of the Sensory Placodes

What determines the location where these
several placodes will arise? At the beginning
of gastrulation, this has not been decided.
The prospective epidermis is still undifferentiated. If the region of a placode is transplanted
to a new location, it will develop in harmony
with its new location. When, at the late gastrula
or early neurula stage, it comes under the
inductive influence of endoderm or chorda-
mesoderm, it acquires a capacity which it did
not have before, namely, to respond to a further stimulus or combination of stimuli in a
specific manner and become the organ which
is appropriate to its location.

The determination of the lens of the eye
has been the most thoroughly studied. Spemann found that head epidermis of a neurula
is competent to form a lens, for it responds
to an underlying optic vesicle by becoming a
lens. If an optic vesicle is removed, a lens, or
at least a normal lens, is not formed. In some
species even flank epidermis, if grafted over

an optic vesicle, will produce a lens. Or, if an optic vesicle in transplanted to the flank, a lens may form from the epidermis of the flank. Note, however, that in all cases it is the epidermis *of a neurula* which is competent to respond. Prospective epidermis of a gastrula responds to a similar contact with an optic vesicle by producing neural tissue.

Similar principles apply to the otocyst. During neurulation, the epidermis of the side of the head responds to an induction arising in head mesenchyme by becoming competent to form an otocyst. But this is not its final determination. Later, in the tail-bud stage, the location of the otocyst is made definite and precise as a response to a second inductive stimulus which originates in the hindbrain. A similar double induction seems to account for the determination of the olfactory placodes. First, the epidermis of the neurula is influenced by an induction arising in endoderm of the forward wall of the pharynx. (Mesoderm has not yet entered this area, Fig. 5–17.) Then later, it is acted upon by an induction of forebrain origin.

Jacobson has restudied the determination of the sensory placodes. After extensive explantation and transplantation experiments he concludes that determination is a continuing process which begins in gastrulation when the epidermis first comes into contact with endoderm and mesoderm. It continues as mesoderm pushes out from beneath the neural plate. Finally, it becomes complete through the inductive action of the neural plate itself. Although any one of these several influences may call forth the production of placodes, yet it takes all of them, acting in proper sequence, to bring about the normal differentiation and positioning of the nose, lens, and ear. Strangely enough, the placodes themselves do not influence each other. They may develop in contact with one another, even out of sequence.

Stomodaeum, Hypophysis, and Branchial Furrows

The *stomodaeum* is a shallow ectodermal depression on the underside of the head fold where the endoderm of the floor of the foregut comes into contact with epidermis (Fig. 6–6, *B*). The *oral plate* which is thus formed breaks through and becomes the oral opening of the embryo. The ectodermal stomodaeum, at first shallow, becomes deeper as a result of the thickening of the mesoderm on each side of it.

Balinsky and others have shown that the formation of the stomodaeum is a response of the epidermis on the underside of the head to contact with the endoderm of the foregut. During later gastrulation, much of the epidermis of the head acquires the competence to so respond. Other epidermis does not, or acquires it to a less extent. If no contact is made between epidermis and endoderm, no stomodaeum is formed. But once contact is made, a stomodaeum will form even though the endoderm is afterward removed.

Epidermal cells of the forward wall of the stomodaeum form a median thickening which sinks inward. This is the *hypophyseal rudiment*. In birds and mammals it is hollow, but in amphibians it is a solid cord of cells (Fig. 6–12). It migrates upward, through the head mesenchyme, until it comes into contact with the infundibulum on the underside of the forebrain. Ultimately it becomes the anterior lobe of the hypophysis (pituitary body). (It gives rise to a part of the posterior lobe as well. See legend of Fig. 10–1.)

A series of ectodermal furrows develops on

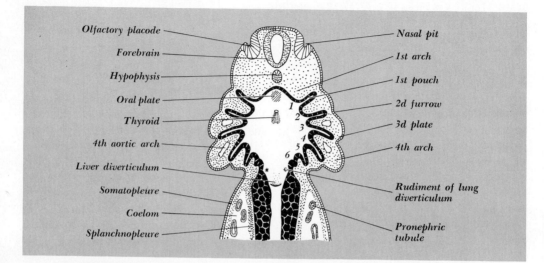

Olfactory placode
Forebrain
Hypophysis
Oral plate
Thyroid
4th aortic arch
Liver diverticulum
Somatopleure
Coelom
Splanchnopleure

Nasal pit
1st arch
1st pouch
2d furrow
3d plate
4th arch
Rudiment of lung diverticulum
Pronephric tubule

FIGURE 6–9 Schematic frontal section through the head of a frog embryo. The oral plate, thyroid, and liver diverticulum are shown beyond the plane of the section. The asterisks mark the location of the future lung evaginations. *(Adapted from several sources.)*

each side of the head, lateral to the foregut (Fig. 6–9). We shall call them *branchial furrows*. They originate in a manner similar to that by which the stomodaeum is formed; that is, the endoderm of a pharyngeal pouch (see page 115) pushes laterally through the head mesenchyme and comes into contact with epidermis at the side of the head. The *branchial plates* (gill plates) which are thus formed consist of epidermis and endoderm with a very little mesoderm between. In the frog, the gill plates break through (except the first pair) and become open *branchial clefts* (gill slits).

The tissue at the sides of the head, anterior and posterior to each branchial plate or cleft, is known as a *branchial arch*. The first pair of branchial arches are also termed the *mandibular arches*. They are lateral to the stomodaeum and later become the upper and lower

jaws. Posterior to the mandibular arches are the first branchial plates. These plates close the *hyomandibular clefts*. In the shark, they break through and the resulting openings are known as the spiracles. In amphibians and other air-breathing vertebrates they do not break through (or if they do they close again), but instead they become the ear drums (tympanic membranes).

The second branchial arches are the *hyoid arches*. These support the tongue. Then come the second branchial plates, and so on, until finally we come to the sixth branchial arches.

The several branchial furrows do not develop simultaneously. In the frog, the first (or hyomandibular) furrows appear at about stage 15 or 16. They separate the mandibular arches anteriorly from a swelling on each side of the head which we shall call the *gill areas*

(Fig. 6-1). The fifth branchial furrows also appear at about the same time at the posterior margin of the gill areas. They mark off the gill areas from the trunk. Later, at about stage 18, two vertical grooves are seen crossing each gill area. These are the second and third branchial furrows. The fourth furrow develops later. (Utimately, there is a fifth furrow and, so it is said, a rudimentary sixth.)

(*Note:* Many authors use the term "visceral" for the several structures which we have termed "branchial." They reserve the latter word for the clefts and arches which actually develop gills. Primitively all the arches bore gills. It is customary to refer to the homologous structures of mammal embryos as "gill arches and clefts" in spite of the fact that the mammal embryos have no gills and the clefts do not break through to the outside.)

Processes of one sort or another grow out from the gill arches. In salamanders, the first branchial arches give rise to tentacle-like outgrowths known as *balancers* (Fig. 6-10, *A*), which serve to support the larva as it rests on the substrate. After the limbs appear, they degenerate. In the frog, the ventral surface of each mandibular arch produces a sucker (Fig. 6-1). These also are temporary structures. By them the newly hatched tadpole attaches to objects in the pond. Balancers and suckers may possibly be homologous organs, for in a remarkable experiment performed by Schotté prospective ectoderm of a frog gastrula was substituted for the prospective head epidermis of a salamander gastrula (Fig. 10, *B*). The larva which resulted was a salamander larva with a frog mask over its face. The head mesenchyme of the salamander had induced the epidermis of a frog to produce frog structures in the manner in which normally the mesenchyme induces its own epidermis to

FIGURE 6-10 *A.* Head of a salamander larva. Harrison's stage 40. *B.* Salamander larva with a frog mask, the result of grafting a patch of prospective epidermis from a beginning frog gastrula to the region of the prospective mouth of a beginning gastrula of a salamander. *(After Hans Spemann, 1938, "Embryonic Development and Induction," Yale University Press, New Haven, Conn., p. 364. From the work of O. Schotté.)*

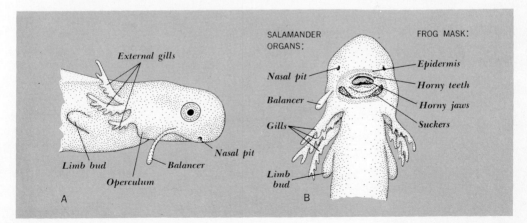

form the facial features of a salamander. Suckers were induced by mesenchyme of the mandibular arches of a salamander.

This experiment points up the important distinction which must be made between *species characters* and *organ characters*. The ectoderm of the frog responded to the inductive stimulus of the mesoderm of the salamander by forming *organs* of a character appropriate to their location in the host. But the *species* character of the organs, indeed their individual character, was determined by the protoplasm of the donor. The epigenetic factors which determine organ characteristics have remained remarkably stable during the course of evolution, while the genetic factors which control species and individual characteristics have evolved.

The hyoid arches of amphibians give rise to opercular folds which grow backward and cover the open gill clefts. The third, fourth, and fifth branchial arches of salamanders produce external gill filaments which serve as respiratory organs. In frogs, the third and fourth branchial arches produce external gills which are soon overgrown by opercular folds. Finally only a small opening remains to the outside,

namely, the "spiracle," situated on the left side of the body (Fig. 6–2, *C*). By this time the external gill filaments have been replaced by internal gill filaments derived from the endoderm of the pharyngeal pouches.

THE NEURAL TUBE

Almost from its beginning, the neural tube is divisible into four regions: forebrain, midbrain, hindbrain, and spinal cord (Fig. 6–11). It is conventional to further subdivide the forebrain and hindbrain as shown in the outline at the foot of this page.

Telencephalon

The first division of the forebrain is the *telencephalon* (Fig. 6–12). Its anterior extremity is close to the olfactory epithelia. In relation to this, indeed partly in response to it, the forebrain develops nerve centers dominated by the nerves of smell. The sidewalls of the telencephalon give rise to lateral swellings which become centers of higher sensory correlation, the paired cerebral hemispheres.

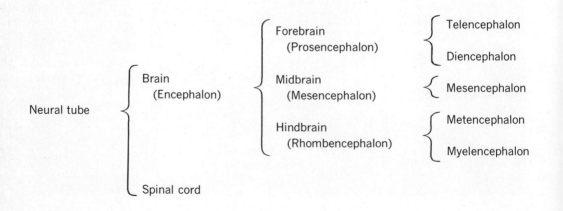

Diencephalon

The second part of the forebrain is the *diencephalon*. Early in the development of the neural tube, its sidewalls bulge laterally and form the *optic vesicles*. These then cave in and become optic cups, from which the retinas of the eyes are developed (Fig. 6–6, *B*). Beneath, and somewhat posterior to the optic vesicles, the floor of the diencephalon pushes downward as the so-called *embryonic infundibulum*. It comes into contact with the hypophyseal rudiment, and together they produce the pituitary body, or *hypophysis*. The hypophyseal rudiment becomes the embryonic anterior lobe. The embryonic infundibulum becomes most of the posterior lobe. It also becomes the stalk (infundibulum proper) which attaches the pituitary body to the brain. The sidewalls of the diencephalon thicken and develop the important nerve centers of the *thalamus* and *hypothalamus*. From the roof of the diencephalon, a small outgrowth pushes upward known as the *epiphysis*.

Mesencephalon

The midbrain, or *mesencephalon,* located posterior to the eyes, is dominated by the centers of reflexes based on vision. In those animals in which reflexes based on sight are highly organized—this is true of amphibians and birds—the dorsal side of the midbrain becomes large and forms two *optic lobes*. The ventral side of the mesencephalon develops the nerve centers of two pairs of nerves which supply muscles that move the eyeballs.

The neural tube bends downward in the midbrain region. This is the *cranial flexure*. Posterior to the midbrain, the neural tube narrows to the *isthmus* (Fig. 6–12). This marks the boundary between midbrain and hindbrain.

FIGURE 6–11   Neural tube and gut of a 3½-mm embryo of the leopard frog *Rana pipiens* Shumway's stage 18. × 24.

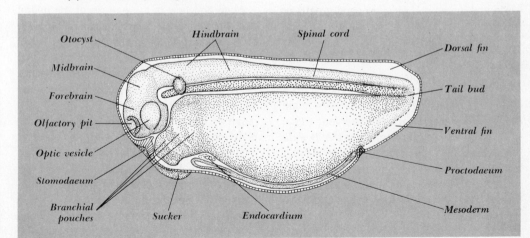

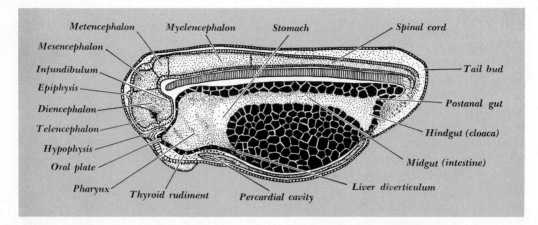

FIGURE 6–12   Median view of the right half of a 3½-mm leopard frog embryo. × 24.

Metencephalon

The anterior part of the hindbrain, namely, the *metencephalon,* is situated just anterior to the otocysts, the sensory organs of equilibrium. Hence it is appropriate that the metencephalon give rise to the *cerebellum* in which are located the nerve centers for balance. The cerebellum is small in amphibians but large in birds and mammals, which depend greatly on balance.

Myelencephalon

Posterior to the metencephalon is the *myelencephalon.* Most of its roof becomes broad, thin, and completely lacking in nervous tissue. Its thick sidewalls and floor form the *medulla oblongata.* Since the entire hindbrain lies dorsal to the pharynx, it is understandable that it contains the nerve centers for the sensory and motor nerves which supply the branchial arches.

Spinal Cord

The hindbrain tapers gradually to the spinal cord. The nerves to the body wall, limb buds, and most of the nerves to the viscera originate from the cord.

Experiments on the Neural Plate

When and how is the pattern of the neural plate determined? When is it decided what each of its regions will become? At the beginning of gastrulation, as we have already seen, the prospective neural plate and the prospective epidermis are not yet fully committed in their capacities for development. But when, as a result of gastrulation, the prospective neural plate comes into contact with the chordamesoderm of the archenteric roof, it undergoes its initial determination; that is, it acquires the capacity, if transplanted or explanted, to become nervous tissue. But it is still labile and dependent on underlying notochord and mesoderm for the pattern of its further differentiation.

Many years ago, Spemann discovered that if a square is cut from the anterior part of the neural plate, rotated 180° and replaced, the material which would have become midbrain, becomes forebrain and eyes, while the material which would have become forebrain becomes midbrain (Fig. 6–13). The result is entirely different if the underlying archenteric roof (chorda-mesoderm) is taken up along with the square of neural tissue and both tissues are rotated and reimplanted together. In this case, eyes form posterior to the otocyst, or if the eye-inducing mesoderm has been divided in the operation, four eyes may form, two in front and two behind. The same results are obtained also when only the chorda-mesoderm is rotated and the neural tissue is not rotated.

It is clear from these experiments that the differentiation of the pattern of the neural plate takes place progressively during neurulation and is dependent on inductive influences emanating from the chorda-mesoderm beneath it. The eye, for example, depends on the prechordal mesoderm. If this is removed before the open-neural-plate stage, no eye is formed. If, as a result of chemical or other inhibition, the prechordal mesoderm remains in its original position anterior to the notochord and fails to divide to the right and left, a single median eye will result. This condition is *cyclopia* (page 293). If, however, as normally takes place, the prechordal mesoderm separates half to one side and half to the other, then two eyes are produced.

The differentiation of the cross section of the neural tube is dependent on the notochord and mesoderm. Holtfreter found that an isolated piece of neural plate (with no mesoderm) will round up and differentiate nerve cells, but it does so inside-out; that is, the nuclei of the nerve cells are now crowded next to the outside of the piece, and solid white matter (axons) forms within (Fig. 6–14). However, if the isolated piece is surrounded by mesenchyme, the arrangement of nervous tissue is more normal. White matter is now on the outside, as it

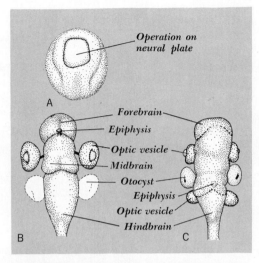

FIGURE 6–13 Determination of the brain by the roof of the archenteron. *A.* Spemann's experiment, 1912, in which he lifted and revolved a rectangular piece of the neural plate of a salamander neurula. He reimplanted it in reversed orientation. *B.* If the underlying roof of the archenteron was not revolved, normal development of the embryo took place. *C.* If, however, the neural tissue and the archenteric roof were lifted and revolved together, then displacement and duplication of the brain parts resulted. *(Modified from J. S. Huxley and G. R. deBeer, 1934, "Elements of Experimental Embryology," Cambridge University Press, London, p. 246. After H. Spemann, 1912, Zool. Jahrb., Suppl., 15:1–48.)*

138

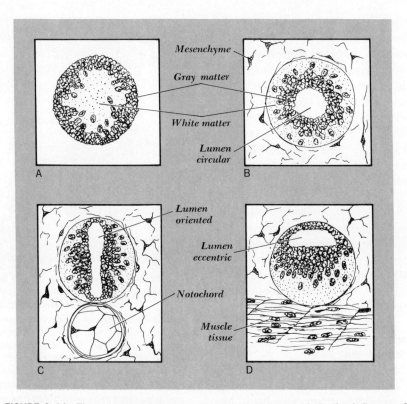

FIGURE 6–14 The structure of the neural tube arises under the inductive influence of surrounding notochord and mesoderm. *(From J. Holtfreter, 1934, Arch. Exptl. Zellforsch., 15:281–301.)*

should be, and the nuclei crowd around a central lumen. But even so, the central lumen is circular. If, however, the piece is in contact on one side with the notochord, then the lumen is slitlike in cross section, as it is in normal development, with one end of the slit directed toward the notochord. If the piece is in contact with muscle cells, there is an increase in the white matter on the side next to the muscle cells. Many details of the patterning of the nervous system, however, remain to be discovered.

THE NEURAL CREST

When the neural folds rise up and close inward toward the middorsal line, certain cells at the margins of the neural plate known as mesectoderm break loose from adjacent epidermis and form *neural crests* along the top of the neural tube. These cells have varied and important fates. Instead of remaining astride the midline, they migrate in loose streams to the sides of the embryo (Fig. 6–15, *A*). One stream passes in front of the eye and ultimately gives rise to some of the cartilages of the skull. A second stream descends behind each eye and becomes the cartilage of the first (mandibular) arch. Other streams farther back enter the remaining arches and give rise to branchial cartilages (Fig. 6–15, *B*).

The cranial ganglia of the head are formed in part from neural-crest cells; but they also receive contributions from adjacent epidermal placodes related to the branchial arches. Furthermore, much of the mesenchyme of the head, from which the dermis and subcutaneous tissue is formed, is of neural-crest origin. This includes the papillae of the teeth.

In the trunk, the neural-crest cells wander in amoeboid fashion to the sides of the embryo and form loose longitudinal strands. The strands which lie between the neural tube and the somites become subdivided into a series of segments, the *spinal ganglia*. This segmentation, however, is secondary to the segmentation of the mesodermal somites, for if a somite is removed, a discrete ganglion is not formed. Other strands of crest cells form along the aorta and in tissue ventral to it. Here they give rise to the ganglia of the autonomic nervous system and to the medulla of the adrenal glands. Still other neural-crest cells migrate individually, multiply, and become pigment cells (melanophores). They appear to repel one

FIGURE 6–15 Differentiation of the neural crest in the salamander. A. Migration of the neural crest (shown by stippling) of the head. (*After L. S. Stone, 1926, J. Exptl. Zool., 44:95–131, Fig. 2.*) B. Contribution of the neural crest to the cartilages of the head. Cartilages derived from the neural crest are shown by stippling, whereas those which originate from head mesenchyme are not stippled. (*Adapted from S. Hörstadius, 1950, "The Neural Crest," Oxford University Press, Fair Lawn, N. J.*) C. Distribution of pigment cells (melanophores) derived from the neural crest of the trunk (Drawn from a living *Ambystoma gracile* larva.) × 7.

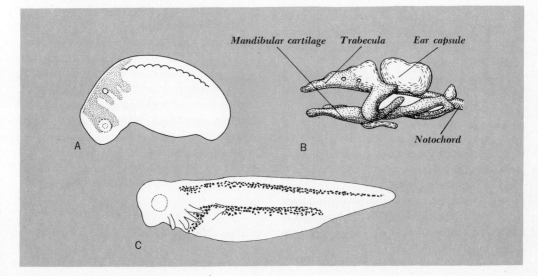

another, but finally they settle in certain regions as though they were attracted by these regions (Fig. 6-15, C).

A few neural-crest cells remain in the mid-dorsal line where they give rise to the mesenchymal core of the dorsal fin fold. If neural-crest cells are transplanted to the flank of an embryo, they often induce the formation of a fin fold in their new location. A fin fold on the underside of the tail bud arises in a similar way, induced in this case by neural-crest-like cells which are derived from the most posterior part of the neural folds. The initial determination of the neural-crest cells, like that of the neural plate, is the result of embryonic induction by underlying chorda-mesoderm.

THE NOTOCHORD

The notochord extends along the midline of the embryo beneath the hindbrain and spinal cord. Because of the cranial flexure and also because of the elongation of the notochord itself the anterior end of the notochord comes close to the infundibulum on the underside of the forebrain. In amphibian larvae and in some of the lower fish the notochord functions as a skeletal rod. As such, it consists of a core of pithlike cells containing gelatinous vacuoles. A sheath and membrane develop around the cord, apparently having been produced by the notochord itself. In most vertebrates, however, the notochord is surrounded and then more or less completely replaced by skeletal elements derived from mesoderm.

If the notochord is experimentally removed or prevented from forming, the embryo fails to elongate properly and the somites of opposite sides may be united across the midline.

THE ENDODERMAL CANAL

The endodermal canal, or gut, begins at the oral plate and ends posteriorly at the anal plate (Figs. 6-11 and 6-12). Actually it begins a bit forward of the oral plate in a shallow temporary recess termed Seessel's pocket. The gut ends behind the anal plate in a temporary extension of the gut known as the postanal or *tail gut*. In the amphibian, the endodermal canal is commonly described as consisting of three regions: foregut, midgut, and hindgut. (These terms must not be confused with the same terms used in invertebrate zoology, where "foregut" refers to the stomodaeum, and "hindgut" to the proctodaeum.)

The *foregut* is that portion of the endodermal canal which is anterior to the main yolk mass. It gives rise to the pharynx, esophagus, stomach, liver, pancreas, and the forepart of the duodenum. Its walls are at first one cell thick. The *midgut* is the future intestine. Its sides and floor are the yolk-rich cells of what was the vegetal area. In salamanders the center of its roof is a suture (seam) formed during neurulation at the place where the lateral margins of endoderm closed in beneath the notochord (page 112). The *hindgut* is that portion of the endodermal canal which is posterior to the main yolk mass and ventral to the tail bud. It gives rise to the cloaca and the temporary postanal gut. These definitions apply to amphibians. They must be modified when applied to birds and mammals, which have a yolk sac.

The Pharynx

The oral plate (pharyngeal membrane) forms on the underside of the pharynx near its anterior end where a downpocket of endoderm comes into contact with the ectoderm of the stomodaeum. Later the plate breaks down and disappears so that no clear evidence remains as to where ectoderm ends and endoderm begins.

Behind the oral plate the pharynx is broad, but it narrows toward its posterior end (Fig. 6–9). We have already described how, from each side of the pharynx, a series of five pharyngeal pouches push through the mesenchyme until they make contact with the epidermis at the sides of the head. The gill plates which are thus formed later break through (except the first pair) and become open gill clefts (branchial clefts).

A median ventral downpocketing from the floor of the pharynx between the second pair of pharyngeal pouches is the primordium of the *thyroid gland* (Fig. 6–6, *C*). Later, after the embryo has hatched, the floor of the pharynx posterior to the thyroid forms a midventral longitudinal groove known as the *laryngotracheal groove*. From its posterior end *lung buds* push laterally and grow backward. It has long been debated whether or not the lung buds represent a posterior pair of pharyngeal pouches (Fig. 6–9).

Esophagus, Stomach, and Liver Diverticulum

In addition to the pharynx, the foregut gives rise to the esophagus, stomach, liver, and parts of the pancreas and duodenum. At first these organs are represented by narrow transverse bands of endoderm posterior to the pharynx; but later, as the embryo grows in length, the bands constrict, elongate, and take the form of the several organs.

The *liver diverticulum* is formed very early

in development as a deep downpocketing of the foregut just anterior to the yolk of the midgut (Fig. 6–12). Later, its anterior wall forms cords of cells which grow forward and ramify among the endothelial cells of the future ventral blood vessels. Together, the endodermal cords and the endothelia of the blood vessels become the liver. Farther back, the posterior extremity of the liver diverticulum becomes the *gall bladder*. (A strange feature of frog development is that the liver diverticulum perforates the ventral endoderm and unites with the remnant of the blastocoel.)

The Intestine

The yolk-rich cells of the amphibian midgut become the intestine. As the digestion of the store of yolk proceeds, these cells organize as an intestinal epithelium. From the beginning its lumen is narrow. In some amphibians it becomes completely closed, and the final lumen is a new cleft which opens through the center of the mass of yolk cells.

The Cloaca

The hindgut lies between the yolk mass of the midgut and the lips of the closed blastopore. It forms the *cloaca* (Fig. 6–23). As the tail bud grows backward, the dorsal wall of the cloaca is drawn out with it. This is the temporary tailgut (postanal gut). It has been said to represent the neurenteric canal, but actually it is more dorsal in position. The primary nephric ducts from the pronephros grow backward at the sides of the embryo until they reach and then open into the cloaca. Thus, by definition, the cloaca is the common passageway for the wastes from both the alimentary canal and the excretory organs. Later it serves the reproductive organs as well.

The anus appears to be a new formation. The ventral opening of the slitlike blastopore closes, but the ectoderm and endoderm remain in contact and constitute an *anal plate* (cloacal membrane). The narrow groove which appears in the ectoderm is the *proctodaeum*. After hatching, the anal plate breaks down, so that the anus (like the mouth) is a new opening to the outside.

Far along in the course of development, namely, about the time of metamorphosis, a pocket pushes forward from the anterior wall of the cloaca and becomes the urinary bladder.

Experimental Observations

When and how does the endodermal canal become organized into its several regions? What are the steps in its determination? It will be recalled that if the prospective chorda-mesoderm of the dorsal lip (organizer) of a gastrula is implanted to the ventral side of another gastrula, a secondary embryonic axis results. This axis may include a pharynx with pouches and even a secondary intestinal lumen. The endoderm, therefore, like the ectoderm, is dependent on chorda-mesoderm for its initial organization. By the time gastrulation is complete, the gut has already undergone anteroposterior determination, although it is still subject to inductive influences of the chorda-mesoderm (pp. 109, 110). For a time it remains dorsoventrally undetermined. Thus if the entire endoderm of an early neurula is removed and replaced upside down (i.e., inverted dorsoventrally) within the outer ectodermal-mesodermal shell, a fairly normal embryo may develop; but, if it is reversed end for end (anteroposteriorly), abnormality results.

After neurulation is complete, the parts of the endoderm, if transplanted or explanted, develop according to their original prospective fates and are no longer dependent on their surroundings. The determination of the gut has been accomplished.

THE MESODERM

When discussing neurulation it was noted that the right and left sheets of mesoderm are divisible into several regions (Fig. 5-16):

1 The mesenchyme of the head
2 The dorsal or axial mesoderm (epimere)
3 The intermediate mesoderm (mesomere)
4 The lateral mesoderm (hypomere)

In addition to these there are loose cells (angioblasts) which, during gastrulation, migrate ahead of the mesoderm proper. They ultimately give rise to blood islands (essentially a single island in amphibians), blood vessels, blood cells, and possibly to the germ cells as well.

Mesenchyme of the Head

The mesoderm anterior to the notochord (prechordal mesoderm) and the axial mesoderm anterior to the otocysts break down into loose cells (mesenchyme) which wander forward in the head and fill the spaces between the other tissues. This mesenchyme receives important contributions from the neural crest (mesectoderm) and from the epidermal placodes of the cranial nerves. As we have seen, the mesoderm has much to do with the determination of the structures of the head. Ultimately it gives rise to the head skeleton (Fig. 6-15, *B*) and to the muscles of the eye.

The Àxial Mesoderm: the Somites

The axial mesoderm (epimeres) consists of longitudinal bands, one on each side of the notochord and neural tube (Fig. 6–3). Beginning in the region of the hindbrain and progressing posteriorly, the bands become divided by transverse clefts into a series of blocklike segments, the *somites* (Fig. 6–4). The last somites which form are always the most posterior. Each somite may develop an inner cavity, the *myocoel,* but these are transitory.

The outer wall of each somite is called the *dermatome* (Figs. 6–3 and 6–16). It is adjacent to the epidermis and ultimately forms much of the inner or connective tissue layer of the skin

(the dermis) and the subcutaneous tissues which lie beneath the skin. The dorsal edge of the inner wall of each somite is the *myotome.* At first it is small; but it grows tremendously, and its cells elongate in a longitudinal direction to become the axial muscles of the body. The dermatomes and myotomes push dorsally and ventrally beneath the epidermis and, having received major contributions from the lateral mesoderm, completely surround the embryo. The ventral part of the inner wall of each somite is the *sclerotome.* It breaks down into loose cells which migrate around the notochord and spinal cord and give rise to the supporting tissues (connective tissues, cartilage, and later bone) of the vertebral column.

FIGURE 6–16 Diagrammatic cross section of a 6-mm frog embryo illustrating the differentiation of the mesoderm.

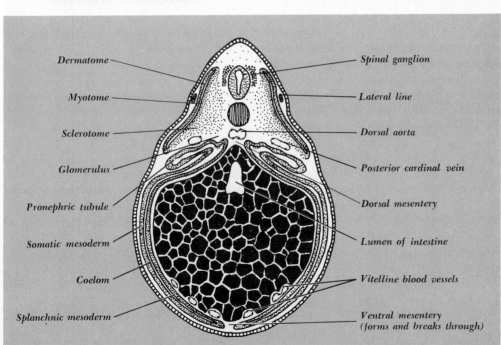

The Intermediate Mesoderm

The intermediate mesoderm (mesomeres) consists of narrow bands of cells which border the axial mesoderm laterally. It gives rise to the excretory and reproductive organs. In most vertebrates the anterior portion of each band divides into a series of segments known as *nephrotomes*. At first each nephrotome is connected to a somite by a narrow neck, but this connection is transitory (Fig. 18–3). It is usual for the lateral wall (outer layer) of each nephrotome, beginning with the most anterior, to produce a tubule, a *pronephric tubule*. This grows outwardly toward the epidermis and then, turning backward, it joins the tubule immediately behind it. The result is a longitudinal duct, the *primary nephric duct*. The duct, with some possible addition from the intermediate mesoderm farther back, grows posteriorly along the lateral border of the intermediate mesoderm until it comes to the region of the hindgut. It then turns ventrally and enters the hindgut (Figs. 6–5 and 6–8, *A*).

The medial end of each pronephric tubule opens into the body cavity by a ciliated funnel known as a *nephrostome*. The median wall of the body cavity adjacent to each nephrostome develops a tuft of blood capillaries known as an *external glomerulus* (Fig. 6–7, *C*). Each glomerulus is fed by a short nephric artery directly from the dorsal aorta (see Fig. 18–3, *C*).

In the case of the frog, the development of the pronephros is somewhat more direct than has just been described. The outer layer of the intermediate mesoderm ventral to somites 2, 3, and 4 thickens and projects ventrally as a "shelf" between the epidermis and the lateral mesoderm (Fig. 6–16). At first the shelf appears to be solid, but then cavities appear which, uniting, become the primary nephric duct and three pronephric tubules. (In salamanders there are generally two pronephric tubules.) The duct grows posteriorly in the manner which already has been described, and the tubules develop ciliated nephrostomes (Figs. 6–5 and 6–7, *C*). The tubules become convoluted and are bathed by blood of the posterior cardinal vein. The glomerular tufts in frogs are united on each side into a single body commonly referred to as a *glomus*.

References to some of the experiments which have been performed on the development of the pronephros of amphibians will be made in the last chapter of this book.

The Lateral Mesoderm: the Coelom

The lateral mesoderm, or hypomeres, consists of sheets of cells which cover the flanks of the embryo. The lateral mesoderm never divides into segments. Rather, each sheet splits into two layers: an outer layer next to the ectoderm, known as the *somatic layer;* and an inner layer next to the endoderm, termed the *splanchnic layer* (Figs. 6–3 and 6–16). The cleft between the two layers is the body cavity, or *coelom*. Note that it is a cavity *within* the mesoderm, not a cavity between the germ layers. The ectoderm and somatic layer of the hypomere, together with cells derived from the dermatome and myotome, constitute the body wall or *somatopleure*. The splanchnic layer and the endoderm form the visceral wall or *splanchopleure*. Actually, the split between the somatic mesoderm and splanchnic mesoderm is not everywhere evident in young amphibian embryos. Only in the region of the heart and pronephric tubules is it at first easily recognized.

Figure 6–17 depicts in a highly schematic fashion the regions of the coelom and the location of mesenteries in a generalized vertebrate larva. Observe that a mesentery is a double sheet of mesoderm which suspends an organ within the body cavity. It is formed when the splanchnic mesoderm pushes medially both above and below the endoderm of the gut until it almost completely surrounds the gut. The *dorsal mesentery* suspends the esophagus, stomach, and intestine in the body cavity. It carries the blood vessels and nerves which supply these organs. The *ventral mesentery* of the pharynx (it has no dorsal mesentery) wraps itself around the heart tube (the endocardium, presently to be described) and forms the surface and muscular layers of the heart (epimyocardium). Mesenteries are formed above and below the heart; but they disappear, so that the heart becomes free (except at its anterior and posterior ends)

within the anterior portion of the body cavity, the portion known as the *pericardial coelom*. The ventral mesentery of the stomach is invaded by the endodermal cords of the hepatic diverticulum, as has already been described. In principle, it is divided by the liver into a dorsal mesentery between the stomach and the liver (gastrohepatic ligament) and a ventral mesentery between the liver and the ventral body wall (ventral hepatic ligament). Across the posterior border of the gastrohepatic ligament, the bile duct (derived from the hepatic diverticulum) travels from the liver to the intestine. The ventral mesentery of the intestine breaks down. The result is that, in the abdominal region, the body cavity of one side opens to the body cavity of the other side beneath the gut.

Figure 6–17 also indicates schematically that there is an adhesion on each side, just anterior to the liver, between the splanchnic mesoderm of the heart and the somatic mesoderm of the body wall. These adhesions unite with the ventral mesentery to make a partition,

FIGURE 6–17 Schematic diagram to illustrate the development of the body cavity and mesenteries of a generalized vertebrate.

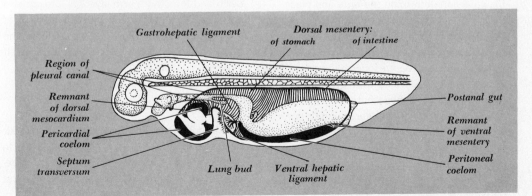

the *septum transversum,* which partially separates the pericardial coelom in front from the *peritoneal coelom* behind. The transverse septum, however, is only a partial partition, for above it, on each side, communications remain between the pericardial and peritoneal cavities. The lung buds push laterally into these communications so that they (the communications) may be called the *pleural canals,* or recesses. They become cut off from the pericardial cavity in front by a pleuropericardial membrane, but in amphibians they remain widely open to the peritoneal cavity behind.

Experiments on the Mesoderm

The evidence is overwhelming that the chorda-mesoderm takes the lead in the organization of the embryo. Both ectoderm and endoderm obtain their cues from inductive influences emanating from the chorda-mesoderm adjacent to them. But how is the organization of the chorda-mesoderm itself accomplished? For an answer we are thrown back on some form of the gradient theory. The genes of the nucleus can supply the possibilities. But only some spatial factor beyond the gene can determine the time and place.

Experiments were described in the previous chapter (pp. 113, 114) which have been interpreted as indicating that at the close of gastrulation the chorda-mesoderm presents two gradients: (1) a dorsoventral gradient in the distribution of a hypothetical "neuralizing factor" (strongest on the dorsal side); and (2) an anteroposterior gradient of a supposed "mesodermalizing factor" (strongest at the posterior end). The biochemical nature of these factors has not certainly been identified. Apparently the head end is relatively free from the mesodermalizing factor. It is possible that

the neuralizing factor is a carry-over from the cortical material of the gray crescent and that the mesodermalizing factor has some relation to the time which has elapsed since the material in question gastrulated. Note that the head mesenchyme was the first mesoderm to enter in gastrulation and that the tail mesoderm was the last. However this may be, the initial determination (that is, the commitment of the regions of the chorda-mesoderm to their several fates) may be visualized as taking place in a two-dimensional grid. By the time neurulation has begun, the main territorial subdivisions or "embryonic fields" have been pretty well staked out, and the chorda-mesoderm has become a mosaic of self-differentiating regions. In this respect a field resembles the original state of the egg itself.

At first the several fields are not sharply delimited. On the contrary, they broadly overlap one another, and each constitutes within itself a "harmonious equipotential system" which is capable of self-regulation. The classic example of such a field is the region of the prospective limb bud, to which topic we next turn our attention. It must not be thought, however, that the relation between the mesoderm and the other germ layers is altogether a one-way relation. The continued patterning of the mesoderm requires the presence of the other tissues whose differentiation it has itself induced.

The Limb Buds

In amphibians, the limb buds first appear as moundlike swellings of the body wall. In salamanders, the buds of the forelimbs become visible at about the tail-bud stage just ventral to the pronephric swelling. This in turn is ventral to the 3d, 4th, and 5th somites (Fig.

6-18, *A*). Hindlimb buds arise later in the region of the anus. In frogs, the limb buds are slower to develop. The forelimb buds become enclosed in the gill cavities which are formed when the operculum grows backward over the external gills. At metamorphosis, the right forelimb erupts through the operculum, while the left forelimb emerges through the opercular opening (spiracle). The hindlimbs (the rudiments of which are shown in Fig. 6-2, *C*) grow rapidly during metamorphosis.

The forelimb buds of salamanders grow out-ward and backward, and then twist (pronate) so that the original ventral or flexor surface is turned toward the body (Fig. 6-18, *E* to *G*). They become paddle-shaped by the flattening of their distal ends, with the future palm of the hand or sole of the foot facing inward. The radial border (preaxial border) becomes ventral; the ulnar border (postaxial) becomes dorsal. Soon flexion is seen at the elbow joint.

The region of the prospective forelimb mesoderm has been identified as early as the beginning gastrula in the future lateral meso-

FIGURE 6-18 The development of the forelimb of the salamander *Ambystoma punctatum*. A to D. Harrison's stages 29, 35, 37, and 45. Approximately × 7. E to G. The fate of the quadrants of the limb disc *(as described by F. H. Swett, 1923, J. Exptl. Zool., 37:207–218, Figs. 8 to 11).*

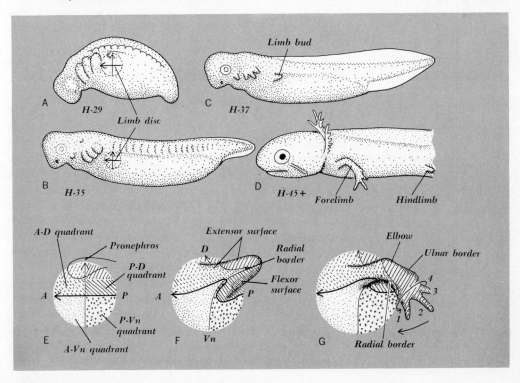

derm. Later, namely, at the tail-bud stage, loose mesenchyme cells split away from the outer surface of the somatic layer of mesoderm. They multiply to form the swelling which is the limb bud. In shark embryos, the fin buds receive contributions from the nearby axial mesoderm, but this process has not been observed in amphibians.

Experiments on Limb Buds

How are the limbs determined? Harrison (1918) reported that in salamander embryos the capacity to form a forelimb resides in a disc of somatic mesoderm situated ventral to the 3d to 5th somites (Fig. 6–18, A). Wherever this disc may be transplanted, whether beneath its own ectoderm, or covered over by strange ectoderm, it will form a limb. Actually, the capacity to form a limb extends beyond Harrison's disc, for if the disc is removed, cells from the bordering somatic mesoderm move in, the wound heals, and a limb is reorganized. Indeed, it is necessary to remove a disc of mesoderm about five somites in diameter to avoid such reorganization. But this is not the entire story: A large part of the somatic mesoderm of the flank is competent to form a limb bud, provided it is sufficiently stimulated. Balinsky found that if an otocyst or an olfactory vesicle is implanted in the lateral mesoderm of a neurula anywhere between the fore- and hindlimb buds, it may stimulate the development of a supernumerary appendage. In short, the limb area is a field in which three degrees of capacity to form a limb grade into another: (1) The smaller disc (three somites in diameter, or at least a part of it) is the prospective limb. It becomes a limb, given normal development. (2) A larger disc (five somites

in diameter) is potentially a limb and will form a limb if the smaller area is removed. (3) The entire flank mesoderm has the competence to form a limb, but it does so only if it is sufficiently stimulated.

Other experiments performed by Harrison on the amphibian limb bud have further emphasized its field character. It is an "harmonious equipotential system" in Driesch's sense (pp. 6, 7). Any half of the limb disc, whether dorsal, ventral, anterior, or posterior, may be removed, and the half which remains will form a whole limb. The mesoderm of two limb buds may be superposed, one on top the other, and a single limb may result. Remarkably enough, a considerable regulation toward normal size takes place.

The nature and orientation of the limb is not determined all at once. Whether it will be a forelimb or a hindlimb is decided during gastrulation. The anterior-posterior polarity of the limb (AP axis) seems to be carried over from the polarity of the egg. It is not reversible. If a limb disc is removed and then implanted either in the same place or another place, but with reversed polarity, the limb which develops points anteriorly rather than posteriorly. This is true whether the limb disc comes from the same side (homopleural) (Fig. 6–19, B) or from the opposite side (heteropleural) (Fig. 6–19, C). Yet the surrounding tissue is not without its influence, for more often than not, a duplicate limb develops which is oriented harmoniously with the host (see dotted outline in Fig. 6–19, B and C).

Swett has showed that dorsal-ventral polarity (DV axis) is decided during the period that the tail bud is growing out. Anytime before Harrison's stage 32, a limb disc which is transplanted with its ventral side up will develop a limb having its dorsal side up, as it should be. The first digit will be ventral, and flexion will be downward. But after stage 35 (Fig. 6–18, *B*), if a limb disc is inverted, the limb which develops is upside down. The first digit then points upward, and flexion is upward. Swett found that the mediolateral axis (ML axis) is not fixed until stage 37 (Fig. 6–18, *C*). If before this stage the mesoderm of a limb bud is removed and reimplanted with its inner surface outward, a normal limb may nevertheless result. Not until the limb has actually begun to bulge outward is the ML axis settled. (Further discussion concerning the develop-

FIGURE 6–19 Determination of the axes of the limb according to the work of R. G. Harrison, 1921 *(J. Exptl. Zool., 32:1–136)* and F. H. Swett, 1927 *(ibid., 47:385–432).* Limb discs of the left (L) or right (R) side were grafted to the left flank of salamander embryos at the tail-bud stage in four different orientations. When they first grew outward, the limb buds "pointed" posteriorly according to the anteroposterior axis (AP axis) of the bud; but they "pointed" dorsally according to the dorsoventral axis (DV axis) of the host. Duplications frequently occurred when the AP axis was reversed (*B* and *C*).

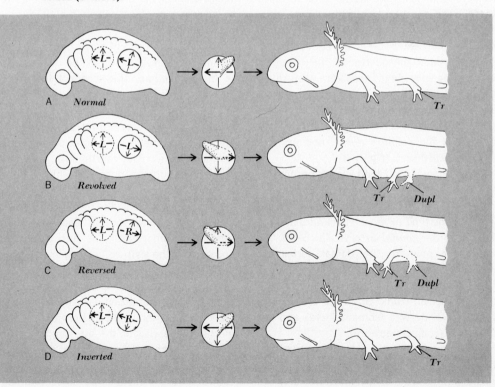

ment of limbs, including the important role of the ectoderm, is presented in Chapter 11.)

THE HEART AND BLOOD VESSELS

The Primary Dorsal and Ventral Blood Vessels

In Chapter 5 we noted that during gastrulation certain cells of the prospective chordamesoderm, which border the blastocoel, break away from the adjacent endoderm and wander upward within the blastocoel in front of the involuting mesoderm (see page 105, Fig. 5–10). These are the *angioblasts,* the forerunners of the linings (endothelia) of the blood vessels and the blood cells. In the neurula, angioblasts are found beneath the pharynx, in the gill arches, and between the endoderm and the splanchnic mesoderm of the intestinal wall. As the splanchnic mesoderm wraps itself around the endoderm of the gut, strands of angioblasts migrate before it, both above the gut and below it. These strands are the precursors of the *primary dorsal* and *primary ventral blood vessels* (Fig. 6–3).

The primary dorsal blood vessels become (1) the *dorsal aortas;* their extensions forward into the head are (2) the *internal carotid arteries,* and their extension backward into the tail is (3) the *caudal artery* (Fig. 6–20, *A*). Above the pharynx, the dorsal aortas are paired, right and left; but from the pharynx backward they unite in the midline and form a single median trunk. Branches of the aorta go to the body wall (somatopleure) and visceral wall (splanchnopleure).

The primary ventral vessels, which are located beneath the gut, have a more varied fate. In the ventral wall of the intestine, they are associated with a more or less unified *blood*

island, from which also the first blood cells originate. The island becomes channeled and forms a pair of lateral *vitelline veins* which course forward to the liver. These correspond to the veins of the yolk sac of amniotes (reptiles, birds, and mammals). We have already noted that ventral to the stomach the primitive ventral vessels are invaded by endodermal cords of the liver diverticulum. The combined tissues give rise to the liver. Beneath the posterior part of the pharynx, the ventral vessels fuse together in the midline and become the inner endothelial tube of the heart, the *endocardium.* Farther forward they produce the *ventral aortas.* The splanchnic mesoderm which enfolds the endocardium becomes the *epimyocardium,* from which the muscles of the heart are derived.

A series of six pairs of *aortic arches* of endothelial origin develop within the mesenchyme of the gill arches. They connect the ventral aortas with the dorsal aortas above. The first two pairs (those of the mandibular and hyoid arches), however, are transitory. The third aortic arches remain as the roots of the internal carotid arteries. The fourth, fifth, and sixth arches retain their connections with the dorsal aortas. The third and fourth arches (and in tailed amphibians also the fifth arches) sprout capillary loops into the gill filaments. At the time of metamorphosis the fifth arches disappear, and the sixth arches acquire branches to the lungs, namely, the *pulmonary arteries.*

Thus a complete visceral, or splanchnic, circulation is established early in the development of the embryo. Blood from the capillary plexus in the wall of the intestine is carried forward by the vitelline veins to the liver. It then is pumped through the heart and into the ventral aortas. Next it flows dorsally through the aortic arches and then backward in the dorsal aorta. It finally is distributed by the vitelline branches of the dorsal aorta to the capillaries of the intestine from which it started.

The Cardinal Veins

How does the blood which goes forward to the head by way of the internal carotid arteries return to the heart? How does the blood which enters the body wall by way of the dorsal and lateral branches of the aorta return? Or again, how does the blood which goes to the tail by the caudal artery get back to the heart? To complete these somatic circulations of the body, the embryo develops within the somatopleure a system of veins known as the cardinal system (Fig. 6–20, *B*). A pair of *anterior cardinal veins* (precardinals) brings the blood posteriorly from the head. A pair of *posterior cardinal veins* (postcardinals) carries the blood forward in the body wall from the tail and

FIGURE 6–20 Schematic diagrams which illustrate the primary blood vessels of the amphibian embryo. *A.* The primary dorsal and ventral blood vessels and the splanchnic circulatory arc. *B.* The cardinal system of veins and the somatic circulatory arcs.

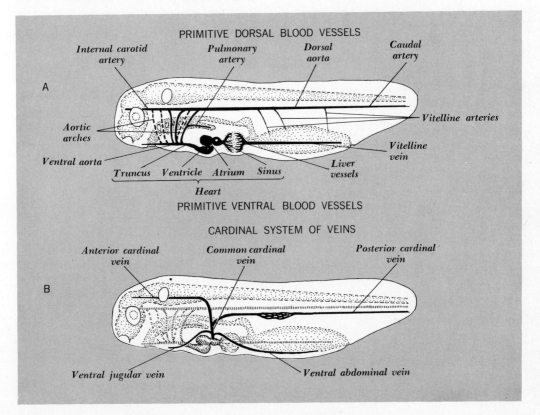

PRIMITIVE DORSAL BLOOD VESSELS

Internal carotid artery   *Pulmonary artery*   *Dorsal aorta*   *Caudal artery*

A

Aortic arches   *Vitelline arteries*

Ventral aorta   *Vitelline vein*

Truncus Ventricle Atrium Sinus   *Liver vessels*

Heart

PRIMITIVE VENTRAL BLOOD VESSELS

CARDINAL SYSTEM OF VEINS

Anterior cardinal vein   *Common cardinal vein*   *Posterior cardinal vein*

B

Ventral jugular vein   *Ventral abdominal vein*

trunk. These join together lateral to the heart and form the right and left *common cardinal veins* (ducts of Cuvier). The latter cross from the body wall to the sinus venosus of the heart by way of the transverse septum, which thus serves as a bridge between the somatopleure and splanchnopleure. Just before each common cardinal vein crosses its bridge, it is joined by a vein from the ventral side of the gill area (ventral jugular vein) and a vein from the ventral abdominal wall (ventral abdominal vein).

The Heart

The heart (endocardium and epimyocardium) grows in length faster than the pericardial cavity in which it lies. It becomes a counterclockwise spiral (Fig. 6–22, *A*). Its chambers, named from behind forward (in the direction of blood flow), are (1) the *sinus venosus,* which receives blood from the liver and from the two common cardinal veins; (2) the *atrium,* located just anterior to the sinus venosus; (3) the *ventricle,* which is bent into a loop beneath the sinus and atrium; and (4) the *truncus,* which discharges forward into the ventral aortas.

By careful observation, the first twitching of the frog heart can be seen at about stage 19, soon after the tail bud has begun to grow out. The circulation of the blood can be seen in the gill filaments as soon as these appear (stage 20).

Experiments on the Heart

Staining experiments have made it possible to identify prospective heart tissue (epimyocardium) as early as the beginning gastrula. It is situated internally in the dorsolateral marginal zone (prospective mesoderm) not far from the endoderm of the future pharynx. (The location is marked by a heart symbol in Fig. 5–7, *A*.)

The heart rudiments move anteriorly (upward) as the leading edge of the lateral mesoderm. They are preceded by the loose cells (angioblasts which are destined to form the endothelial lining (endocardium) of the heart (Fig. 5–14, *B*). At the stage of the open neural plate, the heart rudiments lie ventral to the lateral neural folds and beneath the epidermis of the future auditory placodes. During neurulation, the rudiments of the two sides move ventrally between the epidermis and pharyngeal endoderm until at an advanced neurula stage they come together ventral to the posterior part of the pharynx (Fig. 6–21). Here they surround the endothelial cells; and, together, mesoderm and endothelium give rise to the heart.

When and how does the heart-forming material acquire the capacity to become a heart? Is the heart a self-differentiating system, or is it induced by adjacent tissues? Apparently the answer is not simple. Prospective heart substance, in its migration, comes into relation to both endoderm and epidermis. Do these have an inductive influence? Initially, they possibly do not. The work of Ekman, Copenhaver, and Bacon demonstrates that beginning about the stage of the horseshoe-shaped blastopore, the material of the prospective heart, if isolated, will self-differentiate and give rise to pulsating tissue. But it does not form an organized heart. The presence of pharyngeal endoderm is necessary if the latter is to occur. The capacity to self-differentiate histologically as cardiac muscle, therefore, is not the same as the ability to differentiate morphologically into a heart.

At first the primordium of the heart is a field in the same sense that a limb bud is a field. Its boundaries to begin with are not sharply defined. If the prospective heart mesoderm is removed, adjacent mesoderm will move in and produce a heart. Mesoderm of the gill area as late as a tail-bud stage, if substituted for heart mesoderm, is able to produce a heart. Not until a late tail-bud stage is this capacity restricted to the heart material itself. Also, it is not until then that the heart material is independent of pharyngeal endoderm in its surroundings. Jacobson has uncovered a strange relation. If the pharyngeal endoderm of a neurula is removed, no heart forms. But if brain tissue is also removed, a heart does form! It seems that there is an inhibitory influence which emanates from the brain and which confines heart differentiation to the ventral side of the body.

The heart field of an early neurula is an harmonious equipotential system. If one of the lateral rudiments of a heart is removed, the remaining rudiment will form an entire heart. In fact, any half—anterior, posterior, dorsal, or ventral—will similarly organize itself into a whole. Or, again, two heart rudiments, if placed together, will organize into a single whole organ. It is also during neurulation that the axiation of the heart rudiment takes place. The story is similar to the axiation of the limb buds.

FIGURE 6–21 The development of the salamander heart. A to C. The location of prospective heart materials at the stages of the horseshoe-shaped blastopore, neural folds, and tail bud. Harrison's stages 11, 15, and 28. In A the prospective heart material (epimyocardium) has not yet undergone involution; in B and C it is seen through overlying epidermis. D to F. Diagrammatic sections through the developing heart. D is an early neurula (approximately stage 13). E is an early tail-bud stage (stage 27). F is a schematic section through a young embryo. (Adapted from papers by Bacon, Jacobson, Wilens, Vogt, and Copenhaver.)

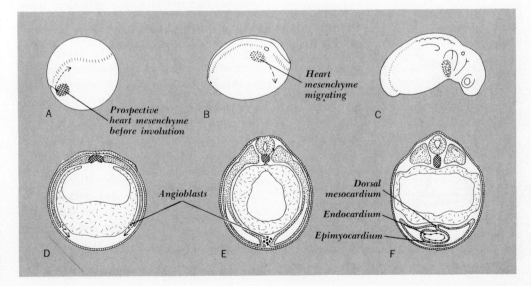

First, the anteroposterior axis is determined. If a heart rudiment of an early neurula is removed, revolved 180°, and reimplanted, it develops normally; but later, when the tail bud has appeared, the same operation results in a reversed heart. Yet, even at the early tail-bud stage, if the heart rudiment is inverted and implanted with the ventral side up, a normal heart may be produced. Still later, the dorsoventral and mediolateral axes become fixed.

The asymmetry of the heart is a matter of considerable interest. If the two sides of a heart are prevented from uniting in the midline, two hearts will form (Fig. 6-22, *B*). Under these circumstances, the left heart is always of normal asymmetry. It spirals counterclockwise like a left-handed screw. The right heart frequently shows reversed asymmetry. It spirals clockwise. How is this *situs inversus cardis* brought about? It has been observed that a defect inflicted on the mesoderm of the left side of a gastrula may result in reversed asymmetry. In some way, the mesoderm of the left side seems to dominate the right side. But when the right side is freed from this dominance, or when the left side is weakened, a reversal of asymmetry may take place. It has been suggested that asymmetry is implicit in the cortex of the gray crescent as early as the one-cell stage of development.

FIGURE 6–22 **A. Ventral view of the heart of a 6-mm frog embryo. B. Twinning of the heart results when the union of the right and left heart rudiments is blocked by a strip of branchial mesoderm having been placed between them. Note that the right heart develops reversed asymmetry.** *(Adapted from G. Ekman, 1929, Arch. Entwicklungsmech., 116:327–347, Abb. 1.)*

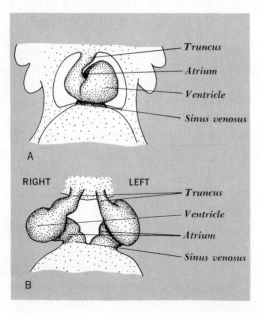

A
RIGHT   LEFT

Truncus
Atrium
Ventricle
Sinus venosus

Truncus
Ventricle
Atrium
Sinus venosus

B

THE TAIL BUD

The tail bud merits special attention. It will be recalled (page 112) that at the beginning of neurulation the material which is to be the tail bud is a band across the neural plate, somewhat forward of its posterior end (Fig. 5–14). The posterior fifth of the neural plate is prospective chorda-mesoderm.

Soon after the end of neurulation, when the neural folds have come together, the tail bud begins to push backward (Fig. 6–23). The result is that the material of the former neural plate becomes bent upon itself. Its dorsal limb becomes the neural tube and notochord. Its ventral limb, extending from the tail bud to the anus, becomes mesoderm and endoderm. It has been shown, by vital staining, that it gives rise to muscles of the tail.

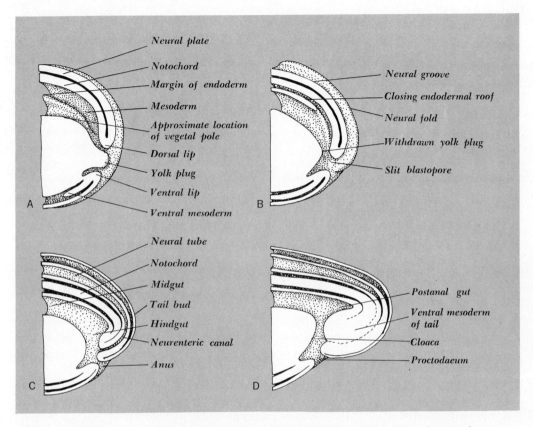

FIGURE 6–23 An interpretation of the development of the salamander cloaca and tail bud.

As the tail bud grows backward, the endoderm also is drawn backward as the postanal gut, or *tail gut*. Note that the tail gut is not the same as the neurenteric canal. It is more dorsal in position. Both are temporary structures.

The dorsal and ventral sides of the tail bud are sutures where the neural folds come together. They are also regions of active cell proliferation. Under the influence of neural-crest cells, they give rise to the caudal fins of the tadpole.

METAMORPHOSIS

There is much of interest in the further development of amphibians, but space permits only a brief reference to the remarkable phenomenon of metamorphosis and to the role which hormones play in the process.

Gudernatsch (1912) noted that when tadpoles are fed desiccated thyroid tissue, they promptly transform into little frogs. Their limbs grow rapidly, their tails are resorbed, and the gill structures disappear. The intestine shortens in adaptation to a more completely carnivorous diet, the skin becomes cornified, and the lungs—already in use—enlarge. The liver begins to produce urea as the principal nitrogenous waste.

B. M. Allen (1918) made the reverse observation; namely, that if the thyroid gland of a frog embryo (tail-bud stage) is surgically removed, the tadpole does not transform into a frog. It feeds and grows until it is many times the size of a normally metamorphosing tadpole, but it does not become a frog. However, if it is fed thyroid, or if it is put in a solution containing thyroxin (the hormone of the thyroid gland, an amino acid containing iodine in its composition), it begins to transform at once.

By similar surgical procedure, it has been found that the embryonic hypophysis also plays a part in metamorphosis. If it is removed, the tadpole remains small, becomes light in color, and does not transform into a frog. Here, then, we have two glands, both of which are necessary if metamorphosis is to take place. But their roles are different. If, for example, both the thyroid and hypophysis are removed, and if then thyroid material is fed, metamorphosis promptly takes place. But if both glands are removed and anterior lobe material (derived from the hypophysis) is grafted into the tadpole, metamorphosis does not take place. Growth is stimulated, but metamorphosis fails. (If pars intermedia, also derived from the hypophysis, is implanted, the tadpole becomes dark owing to the expansion of its melanophores.)

The well-authenticated explanation of these results is that the anterior lobe acts as a "master gland" which produces several hormones, one of which, the so-called thyrotropic hormone, stimulates the thyroid gland to produce and release thyroxin.

Similar results have been obtained with salamanders. In their case, metamorphosis involves a change in texture of the skin, loss of gills and tail fins, and the development of lungs for respiration.

Of special interest is the case of the Mexican salamander known as axolotl, which inhabits lakes near Mexico City. It does not normally metamorphose. Instead it grows, matures, and reproduces while it is yet a larva. (Reproduction by a larva is termed neoteny or paedogenesis.) Why does it not transform? One might guess that the failure is due to a lack of iodine in the water since thyroxin contains iodine in its molecule; but this is not the case. A second guess would be that the thyroid glands of axolotl are deficient in the production of thyroxin. But if the thyroid gland of axolotl is substituted for the thyroid gland of a tiger salamander (which does metamorphose), it proves competent to bring about the metamorphosis of the tiger salamander. The answer to the problem is that axolotl does not metamorphose because its tissues are less responsive to thyroxin than are the tissues of other salamanders. It has been found, however, that if a sufficient amount of thyroid tissue is grafted into an axolotl, it does metamorphose. In this case we see a creature not normally found in nature, namely, a land-adapted axolotl.

It will be noted that during metamorphosis the tissues of a tadpole respond to thyroxin in different ways. Some tissues are stimulated, others are depressed. Eye and limb tissues are stimulated to grow, and will do so even if they have been implanted into the tail and are surrounded by degenerating cells. Tail muscle degenerates even if it is implanted into the trunk next to muscle which does not degenerate. Similar principles apply to the gills. It appears, therefore, that the response of the tissues to the hormone is a direct effect, not one mediated through the surroundings.

SELECTED READINGS

Accounts of experiments on amphibians will be found in:

Huxley, J. S., and G. R. deBeer, 1934. The Elements of Experimental Embryology. Cambridge University Press, London.

Spemann, H., 1938. Embryonic Development and Induction. Yale University Press, New Haven, Conn.

Willier, B. H., P. A. Weiss, and V. Hamburger (eds.), 1955. Analysis of Development. W. B. Saunders Company, Philadelphia. Especially J. Holtfreter, and V. Hamburger, Section VI, Chap. 1. Amphibians, pp. 230–296.

Sensory Placodes; the Balancers

Twitty, V., 1955. Organogenesis: Eye. In Willier, Weiss, and Hamburger (eds.), Analysis of Development. W. B. Saunders Company, Philadelphia. Pp. 402–414.

Yntema, C. L., 1955. Organogenesis: Ear and Nose. In Willier, Weiss, and Hamburger (eds.), Analysis of Development. W. B. Saunders Company, Philadelphia. Pp. 415–428.

Jacobson, A. G., 1963. The determination and positioning of the nose, lens, and ear. I, II, and III. J. Exptl. Zool., **154:**273–284, 285–292, 293–304.

Mangold, O., 1931. Experimental analysis of the development of the balancer in urodeles; an example of interspecific induction of organs. Naturwissenschaften, **19:**905–911.

Translated and reprinted in Gabriel and Fogel (eds.), 1955. Great Experiments in Biology. Prentice-Hall, Inc., Englewood Cliffs, N.J. Pp. 219–224.

Neural Plate and Neural Crest

Kingsbury, B. F., 1922. The fundamental plan of the vertebrate brain. J. Comp. Neurol., **34:**461–491.

Weiss, P. A., 1955. Organogenesis: Nervous system. In Willier, Weiss, and Hamburger (eds.), Analysis of Development. W. B. Saunders Company, Philadelphia. Pp. 346–401.

Horstadius, S., 1950. The Neural Crest. Oxford University Press, Fair Lawn, N.J.

Limb Buds

Harrison, R. G., 1918. Experiments on the development of the forelimb of Amblystoma, a self-differentiating, equipotential system. J. Exptl. Zool., **25:**413–462.

———, 1921. On relations of symmetry in transplanted limbs. J. Exptl. Zool., **32:**1–136.

Swett, F. H., 1923. The prospective significance of the cells contained in the four quadrants of the primitive limb disc of Amblystoma. J. Exptl. Zool., **37:**207–218.

Nicholas, J. S., 1955. Organogenesis: Limb and girdle. In Willier, Weiss, and Hamburger (eds.), Analysis of Development. W. B. Saunders Company, Philadelphia. Pp. 429–439.

Balinsky, B. I., 1965. An Introduction to Embryology. W. B. Saunders Company, Philadelphia. 2d ed. Pp. 404–415.

The Heart

Copenhaver, W. M., 1955. Organogenesis: Heart, blood vessels, blood, and entodermal derivatives. In Willier, Weiss, and Hamburger (eds.). Analysis of Development. W. B. Saunders Company, Philadelphia, Pp. 440–461.

Bacon, R. L., 1945. Self-differentiation and induction in the heart of Amblystoma punctatum. J. Exptl. Zool., **98**:87–125.

Wilens, Sally, 1955. The migration of heart mesoderm and associated areas in Amblystoma punctatum. J. Exptl. Zool., **129**:576–606.

Jacobson, A. G., 1961. Heart determination in the newt. J. Exptl. Zool., **146**:139–151.

Metamorphosis

Etkin, W., 1955. Metamorphosis. In Willier, Weiss, and Hamburger (eds.), Analysis of Development. W. B. Saunders Company, Philadelphia. Pp. 631–663.

Kollros, J. J., 1961. Mechanisms of amphibian metamorphosis: hormones. Am. Zool., **1**: 107–114.

Frieden, E., 1963. The chemistry of amphibian metamorphosis. Sci. Am., **209** (November): 110–118.

Balinsky, B. I., 1965. An Introduction to Embryology. W. B. Saunders Company, Philadelphia. 2d ed. Chap. 18, Metamorphosis, pp. 559–586.

Etkin, W., 1966. How a tadpole becomes a frog. Sci. Am., **214** (May):76–88.

chapter **7**

THE
EARLY
DEVELOPMENT
OF THE
CHICK

The evolution of land-laid eggs marked a great advance in the history of life on earth. It is likely that the first land-laid vertebrate eggs were those of amphibious ancestors of the reptiles, which lived in water and fed on fish. There were advantages in thus placing the eggs on land, for the embryos, unlike water larvae, were not subjected to the fouling of the water and a lack of oxygen. Neither were they in danger of being consumed by the voracious predators which inhabited the ponds.

But there were also great hazards to be met and problems to be solved in laying eggs on land. The newly hatched young had to be well enough developed to crawl on land and fend for themselves; and this required a large store of nutrient in the egg, in the form of yolk, and an extended embryonic period. The eggs and embryos also had to endure dryness and the harshness of rugged surroundings; and for this purpose they were supplied with a watery envelope of albumen, tough outer membranes, and often a hard shell. Furthermore the embryos needed special adaptations to protect their delicate outer surfaces and to carry on respiratory exchange of gases with the surrounding air. They needed receptacles to retain the wastes which, if they accumulated within the embryo proper, would poison them. To serve these several functions the embryo developed four *fetal membranes,* namely, the chorion, amnion, yolk sac, and allantois. A reorganization of the circulation was also needed, for gills could no longer be used for respiration. Finally there was evolved a method of excretion whereby wastes could be eliminated with the least possible loss of water and useful substances. This was accomplished by substituting a relatively insoluble waste, uric acid, for the highly soluble wastes, ammonia and urea. Eggs which meet these specifications are known as enclosed or *cleidoic eggs* and

are characteristic of reptiles, birds and the most primitive mammals, the monotremes.

The hen's egg has long been the favorite cleidoic egg for embryological study. Its earlier stages are similar to those of mammals, even of man himself. It is easy to procure—or was easy until unfertilized eggs dominated the market. Aristotle's naked-eye observations were made on the developing chick. Malpighi and Buffon formed their opinions of preformation on what they saw—or thought they saw—in opened chick eggs. Wolff and von Baer found in the chick convincing evidence that the preformation doctrine was in error. All these, however, overlooked the earliest stages in development, namely, those which take place within the body of the mother hen before the egg is laid and which therefore can be studied only by killing the hen.

THE FORMATION OF THE HEN'S EGG

The Ovarian Egg or Oocyte

The development of the chick embryo begins in the ovary of the mother when she is herself an unhatched chick. It begins when an oogonium or egg-to-be commences to transform into an oocyte or unripe egg. The microscopic oogonium becomes surrounded by a jacket of cuboidal follicle cells, probably sister cells (or rather cousins) of the oogonium; and with their help it grows, at first gradually, forming within it a central sphere of white yolk. Then, when the time comes for the oocyte to ripen, layer upon layer of yellow yolk alternating with white are added in a daily cycle until the cell—the "yolk" of the egg, as we commonly know it—has become a sphere an inch and a half or so in diameter (Fig. 7–1). In six days it increases its mass as much as two hundredfold. It seems

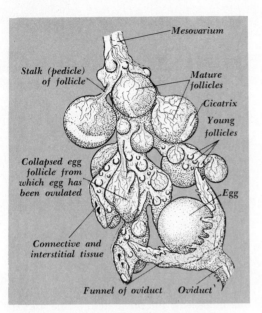

FIGURE 7–1 The development of oocytes in the ovary of the hen. *(From O. E. Nelsen, 1953, "Comparative Embryology of Vertebrates," McGraw-Hill Book Company, New York, Fig. 31.)*

obvious that all this substance could not be synthesized and deposited without the help of other cells, and indeed it is not. Biochemical analyses confirmed by serological studies have revealed that the proteins of the egg yolk are synthesized in the mother's liver, circulated to the ovary, and deposited as yolk fluid and granules. Just what part the follicle cells play is not entirely clear.

The nuclei of the youngest oocytes are at the center of each cell; but as each oocyte grows in size, its nucleus moves from the center, keeping close to one side, namely, the animal pole. The cytoplasm is at first distributed throughout the oocyte; but as growth takes place it becomes more and more limited to a "germinal disc" or *blastodisc* at the ani-

mal pole. At the center of the disc lies the nucleus. For a time a thin layer of cytoplasm remains also at the periphery of the cell, but ultimately it disappears except for the disc. Beneath the disc and extending to the center of the oocyte is a column of white yolk known as the *latebra*.

A connective tissue sheath (theca) develops around the entire follicle and is well supplied with blood vessels except at the pole opposite to the attachment to the ovary. When the follicle has reached its full size, a membrane, the *zona radiata,* forms between the oocyte and surrounding follicle cells. At ovulation the sheath ruptures, and the ovum surrounded by the zona radiata, now known as the *vitelline membrane,* enters the body cavity. Normally the ovary of a hen contains follicles with their oocytes at various stages of development. The youngest are microscopic; the oldest are full-sized yolks ready to ovulate.

The Egg in the Oviduct

Birds have a functional ovary and oviduct on the left side only. Right ovaries and oviducts are present in embryos, but they regress and become mere rudiments. It would seem that a single ovary with its eggs in various stages of growth is about all that a bird of flight can be expected to carry. If the left ovary is removed, the right rudiment may grow, but it is likely to develop into a testis instead of into an ovary.

At ovulation muscle fibers which surround the mature follicle contract, and the egg bursts from its follicular capsule. Active movements of the ostium (mouth) of the oviduct then take place, by which the egg is grasped and swallowed.

As the egg, the so-called yolk, passes down the first or convoluted part of the oviduct (magnum), a thick layer of dense albumen is secreted around it and extends as cords (chalazae) before and behind it (Fig. 7–2). The passage takes about three hours. The egg next spends one hour or so in the short isthmus which follows the convoluted part of the oviduct. While here two shell membranes are secreted around the albumen. Following this the egg with its membranes enters the "uterus," where a watery fluid soaks through the shell membranes and accumulates around and within the investment of denser albumen. This is possibly the result of osmosis, for the dense albumen has a slightly higher osmotic pressure than the more liquid albumen. Thus the egg becomes distended and takes the form of an oval. For 18 or 20 hours it remains in the

uterus, during which time a calcareous shell is formed outside the shell membranes.

The albumen and egg membranes rotate during the period the egg is in the uterus. This is a result of muscle contractions in the uterine wall; but the egg proper, that is, the yolk, now free within the albumen sac, does not rotate. The explanation is that the cytoplasm of the blastodisc is lighter in weight than the rest of the yolk and as a result floats on top. As a consequence of this rotation, the chalazae become spirally twisted, as anyone who opens an egg will observe.

Finally the completed egg as we know it enters the vagina, where it remains for a somewhat variable length of time and is then laid.

The rotation of the egg membranes in the uterus explains the orientation of the axes of the future embryo. According to von Baer's rule (1828), the axis of the embryo is crosswise to the long axis of the egg. The sharp end of the egg is on the embryo's right; the blunt end is on its left (Fig. 7–3). However, there are exceptions. There is a correlation between the direction of coiling of the chalazae and the orientation of the embryo. The head is oriented in the direction toward which the membranes rotated. By removing eggs from the uterus 10 to 12 hours before it is time for them to be laid, and rotating them artificially, Vintemberger and Clavert were able to control the orientation of the embryonic axis at will.

FIGURE 7–2 The egg in the oviduct of the hen. Normally only one egg descends the oviduct at a time.

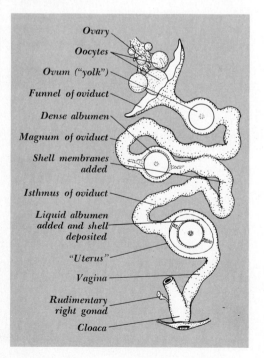

Ovary
Oocytes
Ovum ("yolk")
Funnel of oviduct
Dense albumen
Magnum of oviduct
Shell membranes added
Isthmus of oviduct
Liquid albumen added and shell deposited
"Uterus"
Vagina
Rudimentary right gonad
Cloaca

DEVELOPMENT PRIOR TO LAYING

Maturation and Fertilization

As usual among the vertebrates, the first polar body is given off about the time that the ovum bursts from the ovary. Maturation proceeds as far as the metaphase of the second meiotic division, then pauses and awaits the entrance of the fertilizing sperm.

In the hen, the sperm, if present, will have migrated the full length of the oviduct and will be found in great numbers in the ostium of the oviduct. Since a hen may lay fertile eggs for as long as three weeks after insemination, it is evident that the sperm have long lives. Normally several sperm enter the blastodics as the ovum moves into the ostium. Only one, however, becomes the male pronucleus which combines with the egg pronucleus. The other sperm nuclei undergo abortive attempts at cell division and then disappear. After the sperm has entered the ovum, a second polar body is given off. Then the egg and sperm pronuclei come together and become the zygotic nucleus.

Cleavage

The cleavage of the chick egg is incomplete, i.e., meroblastic. Only the blastodisc at the animal pole subdivides into cells (Fig. 7–4). The first four or five divisions consist of vertical furrows which do not cut very far inward and which do not separate the cells from the underlying yolk. After the fifth cleavage, however, horizontal cleavage planes undercut the central cells. The cytoplasm which remains beneath these cells then disappears so that the disc of cells, now rightly termed the *blastoderm,* is separated from the underlying yolk by a shallow *subgerminal cavity.* Incompletely separated cells and scattered nuclei remain for a time around and beneath the perimeter of the blastoderm, but soon they disappear.

The Blastula

Cleavage continues as long as the egg remains within the body of the hen. For a time

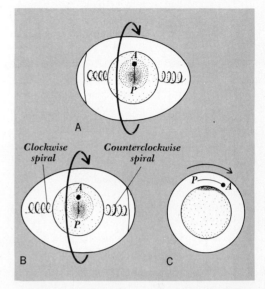

FIGURE 7–3 The rotation of the egg in the hen's uterus determines the orientation of the embryo. The arrows indicate the direction in which the envelopes of the egg revolve. The egg proper ("yolk") does not revolve. As a result the chalazae become coiled, clockwise on the future embryo's left, counterclockwise on its right. The embryo's head is in the direction toward which the egg envelopes revolve. *A.* The orientation is usually according to von Baer's rule: the blunt end of the egg and the clockwise chalaza are on the embryo's left. *B.* Sometimes the sharp end of the egg is on the embryo's left. *C.* View from the future embryo's right side. *(Based on J. Clavert, 1962, Advan. Morphogenesis, 2:27–60.)*

the blastoderm is a delicate layer of rounded and somewhat loose cells a little over 3 mm in diameter and four or five cells thick. The cells of the periphery are somewhat larger and contain more yolk than the central cells. Moreover, they adhere to the underlying yolk. This outer zone is known as the *area opaca* (Fig. 7–5, *A*). The cells of the central area which overlie the

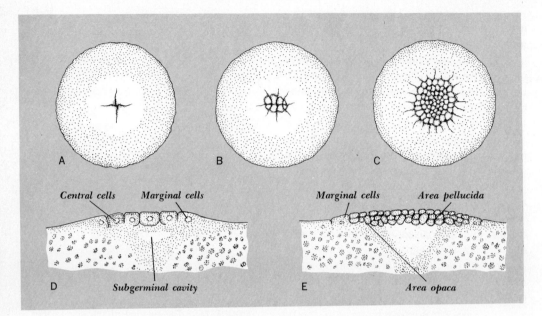

FIGURE 7–4 Meroblastic cleavage in the hen's egg. *A.* Blastodisc following the second cleavage. *B.* Blastodisc following the fourth cleavage. *C.* Blastodisc following the seventh cleavage. *D.* Section through the blastopore after five cleavages. *E.* Section after about eleven cleavages. *(Adapted from J. T. Patterson, 1910, J. Morphol., 21:101–134, Figs. 12, 20, 24, and 26.)*

subgerminal cavity, on the other hand, are smaller and relatively free from yolk. They constitute the *area pellucida.* It has been asserted that at this stage the cells of the pellucid area are of two kinds which differ in size and yolk content and which are intermingled. This interpretation has been challenged.

The nature of the subgerminal cavity also has been a matter of dispute. In the past it has usually been considered to be a blastocoel, homologous with the blastocoel of an amphibian blastula. However, it originates by a disintegration of the cytoplasm and yolk beneath the central cells. At no time is it actually a cleft between cells as it is in the amphibian blastula. Soon, however, a true cleft does ap-

pear in the midst of the cells of the blastoderm. Some authors consider this to be the true blastocoel. The cells of the pellucid area segregate. According to a study of the pigeon egg, the smaller cells which contain less yolk remain at the surface and organize into a thin outer layer known as the *epiblast* (Fig. 7–5, *B*). The larger yolk-laden cells move inward one by one or in loose clusters and form a less organized understratum, the *hypoblast.* The process has been termed *delamination,* and it continues for a time even after incubation has begun. It is noteworthy that this splitting away of the hypoblast from the epiblast takes place first and principally in the posterior quadrant of the pellucid area (Fig. 7–5, *B*).

While delamination is going on, the margins of the blastoderm continue to grow outwardly across the surface of the yolk. This results from active cell proliferation and also from a thinning of the blastoderm. In part it is an actual stretching of the blastoderm due to marginal cells of the area opaca which crawl outward in amoeboid fashion across the surface of the vitelline membrane. Ultimately the epiblast and hypoblast each become only one cell layer thick except in the posterior quadrant. The margins of the expanding blastoderm soon overlap the yolk so that there is a narrow *margin of overgrowth* peripheral to a *zone of adhesion* between the blastoderm and the underlying yolk.

At about this stage the egg is laid, and further development does not take place until incubation has begun. If the egg is not incubated within a few days, regression sets in.

GASTRULATION

Gastrulation in the chick is the process by which the blastoderm gives origin to the three germ layers: ectoderm, chorda-mesoderm, and endoderm.

The Unincubated Egg. Stage 1

Hamburger and Hamilton have arbitrarily defined and pictured "normal stages" in the development of the chick (page 209). They designate the fertile unincubated egg as stage 1 (Fig. 7–6, *A*). As interpreted by Pasteels, it is a late blastula or early gastrula. The actual degree of development varies with the time the egg is retained in the vagina of the hen before it is laid.

At this stage, a morphogenetic movement of

the hypoblast is already under way. It consists of an outward streaming in every direction from a "growth center" located in the posterior quadrant not far from the margin of the pellucid area (Fig. 7–7, *A*). This streaming may be compared to the invagination of endoderm in amphibian development.

Spratt and Haas have demonstrated this streaming in experiments in which they removed blastoderms of unincubated eggs, placed them upside down on clots of an agar medium, and then marked the undersurface (now the top surface) of the hypoblast with bands of carbon or carmine grains. They then recorded the displacement of the particles which followed. Figure 7–7, *B*, is adapted from their paper and illustrates some of their results.

FIGURE 7–5 Schematic sections through the early chick blastoderm. *A.* Before delamination has taken place. *B.* During delamination. *(Adapted from H. L. Hamilton, 1952, "Lillie's Development of the Chick," Holt, Rinehart and Winston, Inc., New York. 3d ed., Figs. 27 and 28. After J. Pasteels, 1945, Anat. Rec., 93:5–21.)*

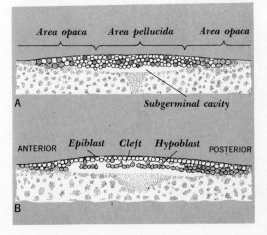

Area opaca   Area pellucida   Area opaca

A   Subgerminal cavity

ANTERIOR   Epiblast   Cleft   Hypoblast   POSTERIOR

B

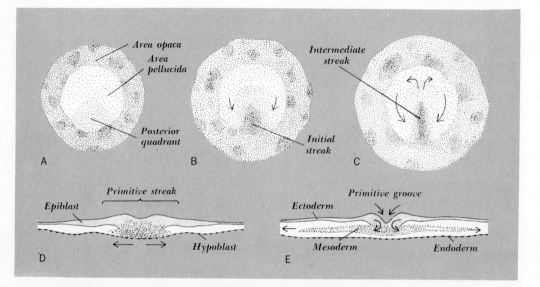

FIGURE 7–6   The development of the primitive streak during the early hours of incubation. *A.* Prestreak stage (stage 1). *B.* Initial streak stage (stage 2). *C.* Intermediate
streak stage (stage 3). The arrows indicate the direction of movement of the epiblast.
*(Adapted from N. T. Spratt, Jr., 1946, J. Exptl. Zool., 103:259–304; and M. E. Malan,
1953, Arch. Biol., 64:149–182.) D.* and *E.* Schematic sections of an early and late
primitive streak.

What is the state of determination of the un-incubated egg? Is it a mosaic of self-differentiating regions, each region committed to its own particular fate? No. Quite the contrary, the unincubated egg possesses great powers of regulation, so great in fact that the term "regulation" (which suggests a revision of a condition already present) hardly seems appropriate. If a blastoderm is divided into halves by a cut in any direction, twin embryos often result. Indeed Spratt and Haas have found that as little as an eighth of the pellucid area—any eighth—may form a whole and complete embryo, provided only that it includes a part of the marginal zone (namely, the zone where the area pellucida joins the area opaca). This ca-

pacity for regulation is retained through the early primitive streak stages.

If the parts of a blastoderm of a newly laid egg are capable of becoming whole embryos, why, then, does only one embryo normally form? Spratt and Haas offer the hypothesis that the hypoblast possesses a gradient pattern with respect to the density or concentration of the cells which compose it. It is thickest at the growth center of the posterior quadrant of the pellucid area. From here its density decreases in every direction, except that it is again somewhat dense in the marginal zone that borders the pellucid area (Fig. 7–7, *A*). It is least dense in the central portion of the anterior half of the pellucid area. Here the hypoblast cells are at first loose and scattered. Now, according to this hypothesis, the active flow of hypoblast cells away from a denser region (normally from the growth center, but away from the marginal zone in experiments with fragments), organizes the axis of an embryo. Bilateral symmetry is therefore a consequence of the eccentric position of the growth center.

We may even have here a hint as to why the axis of the embryo is usually at right angles to the axis of the egg. It seems likely that when the egg rotates in the uterus, the outer cortical layer of the uncleaved blastodisc is dragged somewhat in the direction in which the membranes of the egg rotate. The margin of the blastodisc which trails gives rise to the growth center in the posterior quadrant (Fig. 7–3, C). Recall that in amphibian development, also, the cortex shifts toward the side at which the sperm enters, and then the opposite side (which trails) becomes the gray crescent and later the posterior end of the embryonic axis (pp. 93, 94).

The Initial Primitive Streak. Stage 2

During the first several hours of incubation, the epiblast exhibits no morphogenetic movements other than those involved in the expansion of the blastoderm. Then, at about the sixth or seventh hour, it begins to gather toward the growth center in the posterior quadrant (Fig. 7–6, B). Its cells swing backward in circular orbits in much the fashion that a folding fan might be gathered inward toward one of its radii. At the same time the posterior radius elongates somewhat in a forward direction. By the eighth hour the cells of the posterior radius have begun to heap up and form a short, broad, longitudinal strand which is

thicker than the rest of the epiblast of the pellucid area. This is the *initial primitive streak*.

Formation of the Chorda-Mesoderm

Cells from the undersurface of the epiblast of the primitive streak break away and push forward and laterally between the sheets of the epiblast and hypoblast (Fig. 7–6, D to F). In Spratt's earlier experiments, he placed particles of carbon on the epiblast anterior to the

FIGURE 7–7 Morphogenetic movements of the hypoblast. A. Spratt and Haas's diagram of the undersurface of the chick blastoderm at the beginning of incubation. Note the dense "growth center" in the posterior quadrant from which the hypoblast "fountains" out. B. The pattern which results when the undersurface of the unincubated blastoderm is marked with transverse bands of carbon and carmine particles and then incubated. (After N. T. Spratt, Jr., and H. Haas, 1960, J. Exptl. Zool., 144:139–157. Figs. 3 and 9 b.)

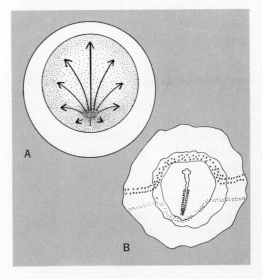

streak and found that they sank inward as soon as the lengthening streak reached them. This labeled epiblast gave rise to pharyngeal endoderm and notochord. (In more recent work Spratt notes that it is mainly the inner layer of the epiblast which contributes to the endoderm and mesoderm.) Epiblast material lateral to the forward end of the streak, moves toward the streak, sinks inward, and becomes axial mesoderm. Farther back it becomes lateral mesoderm. This is a true gastrulation movement, comparable to the involution of chorda-mesoderm which takes place around the amphibian blastopore and later at the tail bud.

We may anticipate a bit by noting that although there is no open blastopore in the chick, yet in ducks, reptiles, and mammals, including man, there is a tongue of cells, the so-called *head process,* which pushes forward from the anterior end of the primitive streak. It acquires a canal, the floor of which fuses with the hypoblast and then breaks down. As a result, the walls of the canal open out, and its most dorsal cells become the roof of the archenteron. Its central axis becomes the notochord. The term head process in chick embryology refers to the notochord.

The Intermediate Primitive Streak. Stage 3

The initial streak grows in length at its anterior end, partly by proliferation, but mainly by the accretion of cells which come to it from the sides. By the 12th hour the intermediate streak is about one-half the length of the circular pellucid area (Fig. 7–6, *C*). It continues to grow in length, but from this stage onward it does so mainly by elongating backward at its posterior end. At the same time the area pellucida becomes oval and then pear-shaped as

though to accommodate the lengthening streak.

Some forward movement within the streak itself also takes place, as a result of which the anterior end of the streak develops a thickening known as the *primitive node,* or Hensen's knot. For a time involution of chorda-mesoderm continues to be especially active just posterior to the node.

The Definitive Primitive Streak. Stage 4

By the 19th hour of incubation the primitive streak has attained its greatest length and its fullest development. It is now referred to as the *definitive streak* (Fig. 7–8, *A*). Just posterior to the primitive node is a pit, the *primitive pit.* Continuing posteriorly from the pit and extending the full length of the streak is a groove, the *primitive groove.* On each side of the groove are ridges, the *primitive ridges.* The streak ends posteriorly in a diffuse area, the *primitive plate.* The node and the ridges are the result of the heaping up of cells which have migrated in from the sides. The pit and the groove are apparently the consequences of the sinking in of cells. By the time the definitive streak stage is attained, the process of involution at the node and pit has ceased. The node has become a region of proliferation comparable to the tail bud of an amphibian. But involution of chorda-mesoderm still continues along the primitive ridges and groove.

At the stage of the definitive streak and immediately following it, the homologies between the chick blastoderm and the late amphibian gastrula (i.e., with a slit blastopore) are clear:

1 The primitive node corresponds to the

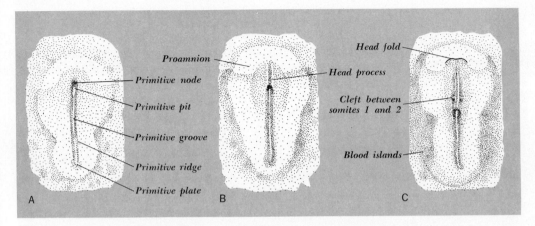

FIGURE 7–8 The embryo begins to take form. A. Definitive streak stage (stage 4). B. Head-process stage (stage 5). C. Head-fold stage (stage 6). *(The stages are defined by V. Hamburger and H. L. Hamilton, 1951, J. Morphol., 88:49–92.)*

dorsal lip of the closed blastopore (future tail bud).

2 The primitive pit represents the dorsal opening of the blastopore (neurenteric canal).

3 The primitive groove and ridges are comparable to the apposed lateral lips of the closed blastopore.

4 The primitive plate may be compared with the ventral region of the blastopore, where later the anus forms.

It is obvious from the above description that the cells of the primitive streak are in continual flux. Epiblast cells come in from the sides, heap up, sink inward, and move away from the streak as chorda-mesoderm.

The Gastrula

The stage of the definitive streak (19th hour of incubation) may be said to mark the end of gastrulation and the beginning of neurulation. The fully formed gastrula consists of three germ layers: ectoderm, chorda-mesoderm, and endoderm. The ectoderm and chorda-mesoderm are in continuity along the axis of the primitive streak, just as they are at the lateral lips of an amphibian blastopore (Fig. 7–9). The endoderm is also united with the mesoderm and ectoderm at the anterior end of the streak and at its posterior end. This is true also of the amphibian blastopore.

Fate Map of the Blastoderm

It should be possible to construct a fate map of the blastoderm of the chick in the manner in which Vogt drew his fate map of the amphibian early gastrula (pp. 99–101). Such a map does not mean that the various "prospective areas" are already differentiated or that they are incapable of becoming something different from their normal fate if they are iso-

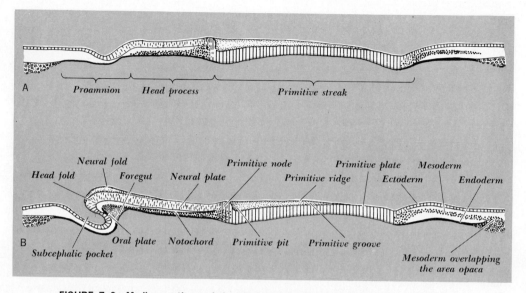

FIGURE 7–9 Median sections of the chick blastoderm. *A.* Head-process stage (stage 5). *B.* Head-fold stage (stage 6). Schematized from longitudinal sections.

lated or transplanted to changed surroundings. It indicates only what the regions will become, given normal development. The construction of such a map is difficult by reason of the great delicacy of the early blastoderm. But it has been attempted with considerable success by several embryologists, including Rudnick, Waddington, Pasteels, Spratt, and Malan. Their findings have not been in complete accord, yet they tend to confirm what one would anticipate on the basis of Vogt's fate map of the amphibian early gastrula.

Let us approach this subject by imagining that an amphibian early gastrula has been expanded tremendously by the addition of yolk until the living tissues are spread out in the form of a flat disc or blastoderm. One might expect that this expansion would take place at the vegetal pole and that the marginal zone of prospective chorda-mesoderm would consti-

tute the boundary of the disc. But this is not the case. The expansion has taken place in the region of prospective epidermis. Figure 7–10, *A,* is a hypothetical fate map of such an expanded gastrula. Each point on the surface is drawn at that same distance from the initial blastopore which it has in the early amphibian gastrula. It will be noted that the amphibian animal and vegetal poles are points on the surface of the disc.

Now the blastoderm of a chick during the early hours of incubation differs from this imaginary, spread-out amphibian gastrula mainly in that most of its endoderm has already invaginated (Fig. 7–10, *B*). Prospective chorda-mesoderm has moved medially from the sides and taken its place. Surrounding the chorda-mesoderm is the prospective ectoderm. However, only the central part of the prospective ectoderm will become a part of the embryo proper. The outlying ectoderm remains outside the embryo and shares in the formation of extra-embryonic membranes, namely, the amnion and chorion.

Each of the embryonic areas can be subdivided as in the amphibian fate map. The prospective ectoderm includes the regions of the prospective neural plate and prospective epidermis. The prospective chorda-mesoderm is divisible into prechordal mesoderm, notochord, axial mesoderm (epimere), intermediate mesoderm (mesomere), and lateral mesoderm (hypomere). Figure 7–11 illustrates the displacement of these prospective areas during the formation of the embryo.

State of the Blastoderm at the Beginning of Neurulation

We noted that when incubation begins the blastoderm is essentially undetermined. What is its state at the definitive streak stage?

Have the fates of its cells been decided? If isolated or transplanted, would its parts self-differentiate as parts, or would they become wholes?

Many experiments have been performed in attempts to answer these questions. Rudnick cut the blastoderm into transverse bands and cultivated them on plasma clots. She found that sections taken from anterior to the primitive node give rise to neural plate, sensory structures of the head, and body wall; that is, they develop according to their expected fates. Sections which include the node produce, in addition, notochord, head mesoderm, and (laterally) heart muscle. These results are most certain if endoderm is included in the explant; less certain if it is absent. Transverse sections behind the node form no axial structures whatever—no neural plate, notochord, or axial mesoderm—whether endoderm is present or not. By the forward migration of streak cells the streak itself lengthens to form a thin, dense

FIGURE 7–10 "Fate maps" of amphibian and chick compared. A. Hypothetical fate map of an amphibian early gastrula which has been expanded around the early blastopore as a center (see text). The short dark line marks the beginning of the endodermal blastopore. B. Fate map of an unincubated chick blastoderm. (B is mainly after M. E. Malan, 1953, Arch. Biol., 64:149–182.)

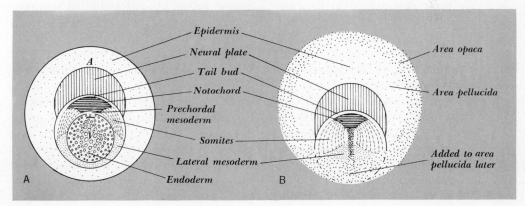

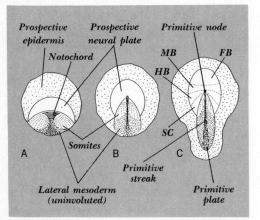

FIGURE 7–11 Movements of the prospective regions of the epiblast during the formation of the embryo. A. The prospective areas at the beginning of incubation. *(Modified from M. E. Malan, 1953, Arch. de Biol., 64:149–182.)* B and C. The same areas at the intermediate and definitive streak stages. *(In part after N. T. Spratt, Jr., 1952, J. Exptl. Zool., 120:109–130.)* Forebrain, midbrain, hindbrain, and spinal cord are indicated.

tongue. Sections taken from the posterior part of the pellucid area produce an abundance of immature blood cells (erythroblasts).

How are these observations to be interpreted? Experiments by Spratt, using various techniques, have established the view that the node is the continuing center of organization. The cells behind the node (corresponding to those cells in the amphibian which have not yet undergone involution) are undetermined. They are incapable, in and of themselves, of developing according to their normal fates. As these posterior cells undergo involution and move forward past the node, their fate becomes decided. Those which move forward through the center of the node become specified as noto-

chord. Those which pass forward at the sides of the node acquire the power to become somites. Those still farther to the side are predestined to produce lateral mesoderm.

NEURULATION

Neurulation in the chick is comparable in almost every way to neurulation in the amphibian. It is the process by which (1) the outer layer (ectoderm) differentiates into neural plate and epidermis, (2) the chorda-mesoderm divides into notochord and the several regions of the mesoderm proper, and (3) the endoderm gives rise to the foregut and later to the hindgut as well. During neurulation, also, the embryo proper begins to separate from the surrounding nonembryonic (extra-embryonic) areas of the blastoderm.

Development of the Head Process. Stage 5

Neurulation may be said to begin immediately following the definitive primitive streak stage (about the 20th hour of incubation) when chorda-mesoderm cells migrate forward from the primitive node and differentiate into the central tongue of cells known as the head process. The axis of the head process becomes the notochord (Fig. 7–8, B).

The notochord elongates by cell multiplication, stretching, and the addition of cells from the primitive streak to its posterior end. If, as Spratt has shown, the notochordal center in the primitive node is removed at an early primitive streak stage, it may regenerate, and further development may take place normally. But if it is removed at the definitive streak stage or later, even though a normal head may form, a new notochord is not produced, and the right and left somites fuse together across the midline.

Axial mesoderm is proliferated from centers on each side of the notochordal center. If one of these centers is removed, no more somites are formed on that side. Thus the primitive node is a processing center through which the material of the primitive streak streams and, as it does so, becomes organized into notochord and axial mesoderm.

Grabowski grafted tissue taken from the primitive node into the region of the lateral mesoderm. He found that mesoderm alone, without accompanying ectoderm, is able to bring about the formation of a head. The ectoderm of the node has no such power. The mesoderm of the donor induces the ectoderm of the host to become neural tissue. Then the neural tissue, in turn, as Fraser has shown, influences the mesoderm to condense into axial mesoderm and to segment into somites.

The Regression of the Primitive Streak

As a result of the streak material moving forward through the centers of the node, the embryonic region in front of the node increases in length, and the primitive streak decreases (compare Figs. 7-8, C, 8-2, 8-4, and 8-12). The process is termed the regression of the streak. In effect, the node migrates backward. Spratt states that the cells of the top surface of the node regress with the node; but for the most part the cells of the streak feed through the regressing node, so that the node is more like a moving wave than a material object.

As the node regresses, material at the posterior end of the streak disperses until finally about all that is left of the streak is the node itself, situated far back at the posterior end of the pellucid area. Its fate is to become the tail

bud of the embryo, from which for many hours the neural tube, notochord, and axial mesoderm (somites of the tail) continue to be proliferated.

The Neural Plate

From about the 20th hour of incubation onward, the neural plate can be recognized in stained whole mounts as a denser region of ectoderm anterior and lateral to the head process (Fig. 7-8, B and C). Posteriorly it fades out opposite the middle of the primitive streak.

During neurulation the central axis of the neural plate, namely, that part which overlies the notochord, sinks inward as the neural groove, while the margins of the neural plate, i.e., the neural folds, rise up and move toward the midline, just as they do in amphibians.

The Chorda-Mesoderm

While the neural plate is folding into the neural tube, the chorda-mesoderm is also differentiating. Its most anterior part, the prechordal mesoderm, gives rise to much of the mesenchyme of the head. Behind it is the notochord. On each side of the notochord are the three longitudinal strands of mesoderm (Fig. 7-12): (1) The thicker medial strands are axial mesoderm, or epimeres, which merge anteriorly into the mesenchyme of the head. (2) Lateral to the axial mesoderm are strands of intermediate mesoderm, or mesomeres. At this early stage they are thin and difficult to recognize. (3) Farthest to the side is the lateral mesoderm, or hypomeres. Only a small part of the lateral mesoderm, however, is destined to become a part of the embryo proper. Its outlying regions remain extra-embryonic and con-

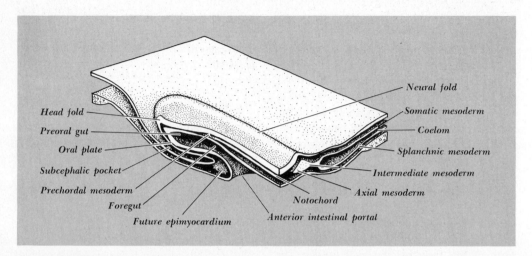

FIGURE 7–12 Stereogram to illustrate neurulation in the chick. Note the head fold, foregut, neural folds, and the derivatives of the chorda-mesoderm.

tribute to the fetal membranes. Early in development the lateral mesoderm splits into somatic and splanchnic layers, with the body cavity (coelom) between.

The Head Fold and Foregut. Stage 6

While the neural plate is folding into the neural tube and the chorda-mesoderm is dividing into the notochord and the several regions of mesoderm, the embryo proper pinches away from the extra-embryonic blastoderm. The process starts anterior to the neural plate when the embryo proper pushes forward. This results in a transverse fold, namely, the *head fold* (Figs. 7–8, 7–9, and 7–12). The pocket of ectoderm beneath the head fold is the *subcephalic pocket.*

From the first the head fold includes endoderm as well as ectoderm. The pocket of endoderm within it is the *foregut.* Its opening pos-

teriorly onto the surface of the underlying yolk is termed the *anterior intestinal portal.*

As the lateral neural folds rise up and join to form the neural tube, they also stretch longitudinally. The result is that the head fold increasingly bulges forward and the subcephalic pocket becomes deeply recessed. The endoderm is not involved in this process. It is true that the foregut lengthens, but it does so in a different manner and in the opposite direction from the lengthening of the head fold. It is added to at its posterior end. The endoderm and accompanying splanchnic mesoderm close in from the sides beneath the foregut, so that the anterior intestinal portal is made to retreat toward the rear.

Meanwhile the blastoderm surrounding the embryo continues to grow outwardly over the surface of the yolk. Its outermost zone consists of ectoderm and endoderm only. Since the cells of the latter are laden with yolk granules, this zone is known as the *area vitellina* (Fig. 8–2). Mesoderm pushes laterally and posteriorly from the embryo a part of the way into the space between the ectoderm and endoderm. This zone containing mesoderm appears mottled in stained whole mounts because of clusters of cells which contain hemoglobin. These clusters are known as *blood islands* and are located between the splanchnic layer of mesoderm and the endoderm. Soon they organize into blood vessels and blood cells, as

will be described in the next chapter. Hence the zone of mesoderm is called the *area vasculosa* (Fig. 7–8, *C*). For a time, mesoderm does not invade a region beneath and anterior to the head, known as the *proamnion* (Fig. 7–8, *B*).

We arbitrarily close this chapter at about the 21st hour of incubation (stage 6), when the head fold has formed and the first somites have just begun to appear (Fig. 7–8, *C*).

SELECTED READINGS

General references on chick embryology include:

Hamilton, H. L., 1952. Lillie's Development of the Chick. Holt, Rinehart and Winston, Inc., New York. 3d ed.

Patten, B. M., 1951. Early Embryology of the Chick. McGraw-Hill Book Company, New York. 4th ed.

Romanoff, A. L., 1960. The Avian Embryo, Structural and Functional Development. The Macmillan Company, New York.

Normal stages in the development of the chick have been defined by:

Hamburger, V., and H. L. Hamilton, 1951. A series of normal stages in the development of the chick embryo. J. Morphol., **88**:49–92. The figures illustrating the stages are reprinted in Hamilton, H. L., 1965. Lillie's Development of the Chick. Holt, Rinehart and Winston, Inc., New York. 3d ed.

Cleavage and Blastulation

Blount, M., 1907. The early development of the pigeon's egg with especial reference to

the supernumerary sperm nuclei, the periblast, and the germ wall. Biol. Bull., **13**:231–250.

Patterson, J. T., 1910. Studies on the early development of the hen's egg. J. Morphol., **21**:101–134.

Pasteels, J., 1945. On the formation of the primary entoderm of the duck (Anas domestica) and on the significance of the bilaminar embryo in birds. Anat. Record, **93**:5–21.

Formative Movements and Fate Maps of the Chick Blastoderm

Rudnick, Dorothea, 1944. Early history of the chick blastoderm. Quart. Rev. Biol., **19**:187–212.

Spratt, N. J., Jr., 1946. Formation of the primitive streak in the explanted chick blastoderm marked with carbon particles. J. Exptl. Zool., **103**:259–304.

————, 1947. Regression and shortening of the primitive streak in the explanted chick blastoderm. J. Exptl. Zool., **104**:69–100.

Malan, M. E., 1953. The elongation of the primitive streak and the localization of the presumptive chorda-mesoderm studied by means of coloured marks of Nile blue sulphate. Arch. Biol., **64:**149–182.

Fraser, R. C., 1954. Studies on the hypoblast of the young chick embryo. J. Exptl. Zool., **126:**349–399.

Spratt, N. J., Jr., and H. Haas, 1960. Morphogenetic movements in the lower surface of the unincubated and early chick blastoderm. J. Exptl. Zool., **144:**139–157.

———, 1965. Germ layer formation and the role of the primitive streak in the chick. I. Basic architecture and morphogenetic tissue movements. J. Exptl. Zool., **158:**9–38.

Regulative Capacities of the Chick Blastoderm

Clavert, J., 1962. Symmetrization of the egg of

vertebrates. Advan. Morphogenesis, **2:**27–60.

Spratt, N. J., Jr., and H. Haas, 1960. Integrative mechanisms in the development of the early chick blastoderm. I. Regulation potentialities of separated parts. J. Exptl. Zool., **145:**97–137.

Eyal, H., and N. J. Spratt, Jr., 1965. The embryo-forming potencies of the young chick blastoderm. J. Embryol. Exptl. Morphol., **13:**267–274.

Spratt, N. J., Jr., 1966. Some problems and principles of development. Am. Zool., **6:**9–19.

Grabowski, C. T., 1962. Neural induction and notochord formation by mesoderm from the node area of the early chick blastoderm. J. Exptl. Zool., **150:**233–245.

chapter **8**

THE
YOUNG
CHICK
EMBRYO

In a broad sense any stage of development from fertilization to hatching is an embryo; but we shall use the word in a narrow sense to refer to those stages during which the main features of the new organism take form.

The "normal stages" in the development of the chick (Table 8–1) which Hamburger and Hamilton defined have served a useful purpose especially to experimental embryologists. Unfortunately their stages do not represent equal intervals in the course of development. Stating the age of the embryo in hours of incubation, as is usually done, also is not exact; for chick eggs vary in the amount of development which takes place before they are laid, and variation in the temperature of incubation makes a very considerable difference. From the 21st hour to the 40th hour it is convenient to indicate the stages of development by the number of pairs of somites. After the 40th hour somites are formed at less frequent intervals, and they are difficult to count.

THE LINEAR EMBRYO

For a time, namely, until about the 36th hour of incubation, the chick embryo is linear, and except for the heart it is symmetrical. After about the 36th hour it begins to twist and curl in a manner presently to be described. As a linear embryo it is comparable to the young amphibian embryo described in Chapter 6. However, the presence of the huge mass of yolk accounts for significant differences.

The Neural Tube

The right and left neural folds come into contact with each other at about the 5-somite stage in the region which is to become the mid-

brain. The contact gradually progresses forward and back, but for a time (until about the 13-somite stage) an opening remains at the anterior end of the neural tube known as the *anterior neuropore* (Fig. 8–1). At the posterior end the neural folds remain open for a considerably longer time. Later, because of its shape, this opening is called the *sinus rhomboidalis*. When it does close (at about 20 somites) the neural folds enclose the primitive node and anterior part of the primitive streak.

The progressive closure of the neural tube is readily observed in living chick embryos which have been removed from the yolk, floated into a watchglass of warm salt solution, and observed by reflected light (Fig. 8–1). Embryos which have been fixed, stained, cleared, mounted whole on glass slides, and then viewed by transmitted light show the sidewalls of the neural tube as dark longitudinal bands (Fig. 8–2). By studying mounted cross sections of a 24-hour chick embryo, beginning at the posterior end and working forward, one gets an idea of the progressive closure of the neural folds (Fig. 8–3).

The anterior half of the neural tube becomes the brain. It shows the same three divisions—*forebrain, midbrain,* and *hindbrain*—which were described in connection with the amphibian embryo (Fig. 8–4). The hindbrain tapers gradually into the *spinal cord.* The early bulging of the *optic vesicles* from the sides of the forebrain is a notable feature of chick development. By the 10-somite stage they reach the full width of the head and appear to be adherent to the epidermis. A broad downpocketing of the forebrain below the vesicles is the *embryonic infundibulum.*

Almost from the first the neural tube presents a series of lateral swellings of the brain known as *neuromeres.* Their exact number and their relation to the subdivisions of the brain is a matter of controversy. Do they represent a primitive segmentation of the vertebrate brain? Soon they reach their maximum development and disappear.

The Notochord

The notochord as seen in whole mounts of linear embryos is a narrow median strand of cells underlying the neural tube (Figs. 8-2 and 8-4). It is entirely too weak to function as a skeletal structure. Nevertheless, it has an important morphogenic function, for around it forms the axial skeleton. If it is removed by experimental surgery, the somites of the two sides may unite across the midline.

The notochord begins at the forward end of the hindbrain and extends back to the *notochordal bulb* on the underside of the primitive node (Fig. 8-2). It elongates at its anterior end until it comes close to the infundibulum on the undersurface of the forebrain. However, it is difficult at this early stage of development to distinguish the most anterior notochordal cells from the prechordal mesoderm anterior to it.

The Foregut

In Chapter 7 we noted that at first the foregut extends to the forward tip of the head fold; but the neural tube bulges forward and soon overshadows it. Its growth in length is at its posterior end and is accomplished by the closing in of the walls of the anterior intestinal portal beneath it (Fig. 8-3). As a consequence of this concrescence the anterior intestinal portal recedes posteriorly. The process is readily seen when a living blastoderm is removed to a watchglass of warm salt solution, turned ventral side up, and observed by re-flected light. In lightly stained whole mounts the margins of the foregut and portal appear as delicate double lines (Fig. 8-2). One line is the endoderm, the other is the accompanying splanchnic mesoderm.

Near its anterior end, the floor of the foregut (endoderm) is in contact with ectoderm of the roof of the subcephalic pocket. This is the location of the oral plate (Fig. 8-5). The blind extension of the foregut anterior to the oral plate is the *preoral gut* or Seessel's pocket. It is temporary.

The Mesoderm

Mesoderm is present between ectoderm and endoderm throughout the embryo, but at first it is not abundant within the head fold. Here it consists of mesenchyme, that is, loose cells derived from prechordal mesoderm and from the forward extension of axial mesoderm, together with some contribution from the neural crest.

The axial mesoderm is seen in whole mounts as dark bands on each side of the neural tube. At about the 21st hour of incubation a cleft appears across each band a short distance anterior to the primitive node (Fig. 7-8, *C*). It marks the division between the 1st and 2d somites. Approximately one hour later a second cleft appears posterior to the first and divides the 2d somite from the future 3d somite. The process repeats itself approximately once an hour for 20 hours. After that the somites continue to appear at a somewhat slower rate. The most posterior somites are, of course, the youngest.

The intermediate mesoderm undergoes little development during the period of the linear embryo. It may be recognized in stained whole mounts as the less dense region lateral to the

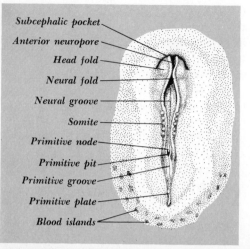

Subcephalic pocket
Anterior neuropore
Head fold
Neural fold
Neural groove
Somite
Primitive node
Primitive pit
Primitive groove
Primitive plate
Blood islands

FIGURE 8–1 Dorsal view of a 25-hour chick (5 somites, stage 8 +) as seen by reflected light. × 11.

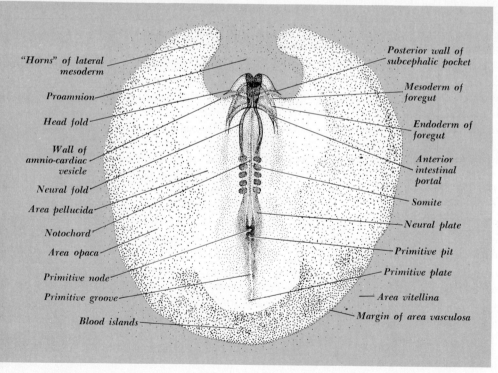

"Horns" of lateral mesoderm
Proamnion
Head fold
Wall of amnio-cardiac vesicle
Neural fold
Area pellucida
Notochord
Area opaca
Primitive node
Primitive groove
Blood islands

Posterior wall of subcephalic pocket
Mesoderm of foregut
Endoderm of foregut
Anterior intestinal portal
Somite
Neural plate
Primitive pit
Primitive plate
Area vitellina
Margin of area vasculosa

FIGURE 8–2 Whole mount of a 25-hour chick viewed by transmitted light. × 17. The blastoderm of the proamnion and area vitellina is shown in gray.

somites and medial to the more darkly stained lateral mesoderm (Fig. 8–4). In cross sections it appears as "necks" of cells connecting the somites with the lateral mesoderm (Fig. 8–3, *E*).

The lateral mesoderm is present at the sides of the embryo and posterior to it. Its future is partly embryonic and partly extra-embryonic, but at this early stage the two parts merge with each other without any indication of a demarcation between them.

Beneath the head fold and anterior to it, there is an area into which mesoderm has not yet entered and which is known as the *proamnion* (Fig. 8–4). The extra-embryonic lateral mesoderm pushes forward on each side of the proamnion as "horns." These can be seen in lightly stained whole mounts. Ultimately they join in front of the embryo and completely surround the proamnion.

Everywhere the lateral mesoderm splits into the two layers: *somatic mesoderm* next to the ectoderm, and *splanchnic mesoderm* next to the endoderm. The body cavity, or coelom, lies between them. Part of it will become the body cavity of the embryo; but the larger part is *extra-embryonic coelom,* or exocoel, and will remain outside the embryo.

The behavior of the splanchnic mesoderm in the region of the foregut is of special interest and importance. Here, as everywhere else, it hugs the endoderm closely. It moves in from the sides as the anterior intestinal portal shifts posteriorly. As a result, two coelomic pockets or bays are formed, one on each side of the foregut, which in time unite to become the pericardial cavity. Since at this early stage they are still widely open laterally to the extra-embryonic coelom, they are best called *amnio-cardiac vesicles* (Fig. 8–6). Each is bounded anteriorly by the subcephalic pocket, dorsally by

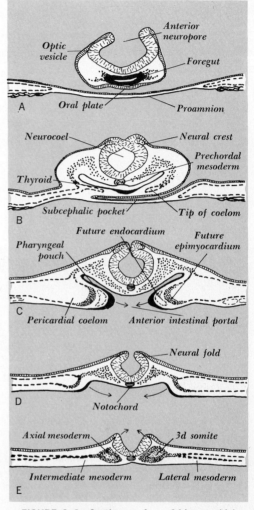

FIGURE 8–3 Sections of a 24-hour chick illustrating the closure of the neural tube and foregut. × 57.

the floor of the foregut, posteriorly by the anterior intestinal portal, and ventrally by the splanchnopleure which covers the yolk, namely, the yolk sac.

According to DeHaan, this movement of the

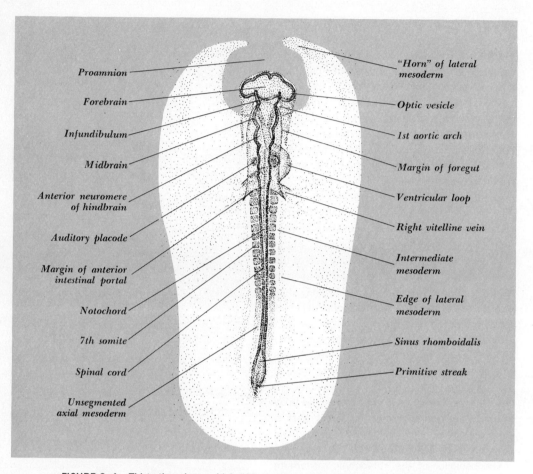

FIGURE 8–4 Thirty-three-hour chick (13 somites, stage 11) seen as a transparent object. × 20. At this stage the embryo proper is not distinguished from the outer blastoderm. (See Fig. 8–2 for the shape of the blastoderm.)

splanchnic mesoderm in the region of the fore-gut is not just a passive one (Fig. 8–7). The mesoderm is not simply carried along with the endoderm as the latter folds in and forms the floor of the foregut. An actual migration of mesodermal cells takes place. Clusters of cells which at first are at the sides of the head process (stage 5) are seen, by means of slow-motion photography, to move about in an apparently random manner (Fig. 8–7, A). Then, at about the time the head fold appears (stage 6), they begin to migrate in an oriented fashion forward and toward the midline until they form a crescent around the anterior intestinal portal (Fig. 8–7, B). What orients the mesodermal cells in their migration? DeHaan suggests an explanation. He notes that a change of shape of the endodermal cells takes place at about the time that the migration of mesodermal clusters becomes orderly. At first the endoderm forms an irregular squamous epithelium. But at about stage 6 its cells become spindle-shaped and oriented toward the intestinal portal. He suggests that this ordered surface of the endoderm provides "contact guidance" to the migrating mesoderm. However this may be, as the anterior intestinal portal shifts posteriorly, more and more heart-forming mesoderm moves into place beneath the floor of the foregut and there organizes a heart (Fig. 8–7, C).

Extra-embryonic Blood Vessels

The lateral mesoderm spreads outwardly from the primitive streak until it reaches the area opaca. It then pushes some distance into the area opaca, and as it does so it appropriates to itself some of the yolk granules of the

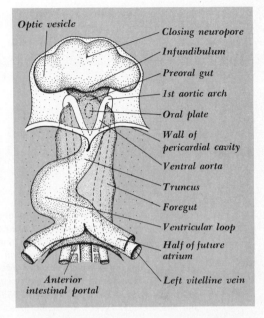

FIGURE 8–5 Ventral view of the anterior end of a 33-hour chick (diagrammatic). Approximately × 34.

FIGURE 8–6 Diagrams from Patten illustrating the development of the chick heart from paired primordia. A. 27 hours. B. 29 hours. C. 36 hours. (Redrawn from B. M. Patten, 1964, "Foundations of Embryology," McGraw-Hill Book Company, New York, 2d ed., Figs. 9–7 and 9–8.)

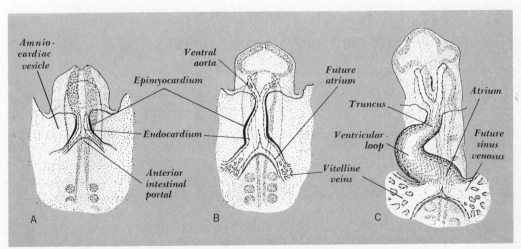

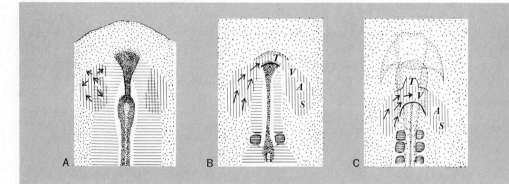

FIGURE 8–7 Diagram of cell movements during the production of the chick heart. A. Head-process stage (stage 5). B. 1-somite stage (stage 7). C. 7-somite stage (stage 9). Vertical lines indicate heart-forming material. Horizontal lines show the axial meso- derm. Small arrows indicate the direction of movement of clusters of cells which are random at first, but which become directed toward the anterior intestinal portal. T, V, A, and S refer to the future truncus, ventricle, atrium, and sinus venosus, re- spectively. *(Based on R. L. DeHaan, 1963, in "Biological Organization at the Cellular and Supercellular Level," Academic Press Inc., New York, p. 155.)*

area, possibly even some of the cells. Where it overlaps the area opaca it forms cords of cells containing hemoglobin which are known as *blood islands* (Fig. 8–1). The source of the cells is in dispute, but they may be derived from mesenchyme which migrated outwardly in ad- vance of the lateral mesoderm.

As the blood islands extend, they unite to form a network of cords lying between the en- doderm and mesoderm of the extra-embryonic splanchnopleure. Their surface cells become the endothelia of the vitelline (yolk sac) blood vessels. Their central cells become free, and most of them are red blood corpuscles known as embryonic erythroblasts. That part of the area opaca which is thus invaded and vascu- larized is known as the *area vasculosa* (Fig. 8–2). Farther out, the area opaca remains free of mesoderm and blood vessels and is called the *area vitellina*.

A network of endothelial vessels develops also in the splanchnopleure of the area pel- lucida. Possibly it is the result of ingrowth of vessels toward the embryo from the area vas- culosa. At any rate, the vessels soon join endo- thelial cords which grow outward from the embryo. Thus there is established a continuity of vessels between the embryo and the vascu- lar region of the blastoderm.

Blood Vessels within the Embryo

Most of the information given in Chapter 6 concerning the development of the blood ves- sels of the young amphibian embryo applies, with modifications, to the chick. It will be re- called that in the amphibian there are two pairs of primitive blood vessels, one pair dorsal to the gut and one pair ventral to the gut (Fig. 6–20, *A*). Since the chick embryo is spread out flat upon the surface of a huge yolk, the picture of the blood vessels is of necessity different. The dorsal vessels are seen first in the splanchnopleure immediately lateral to the axial mesoderm (Fig. 8–8, *D*, and *E*). They become the *dorsal aortas* and their forward extensions to the forebrain, the *internal carotid arteries* (Fig. 8–13). Posteriorly they connect with a network of vessels of the area vasculosa, the forerunners of the *vitelline arteries* (Fig. 8–9).

The vessels of the chick which correspond to the primitive ventral blood vessels of the amphibian are less easily recognized. At first they are represented by the blood islands and vascular net of the area vasculosa. Blood flows forward in this net until it reaches the horns of the area vasculosa. Then it returns as the *vitelline veins* to the folds which border the anterior intestinal portal (Fig. 8–9). As the splanchnic mesoderm migrates medially beneath the foregut, it seemingly pushes strands of endothelia before it (Fig. 8–6). These strands become the *endocardium* of the heart and, farther forward, the *ventral aortas*. The latter pass forward on each side of the oral plate and then curve dorsally around the forward border of the foregut as the first pair of *aortic arches*.

The Heart

The layers of splanchnic mesoderm which enfold the endocardium become the muscular layer of the heart, namely, the *myocardium,* and its outer serous covering, the *epicardium.* At this early stage of development, however, the myocardium and epicardium together constitute a single layer, the *epimyocardium* (Fig. 8–8, *C*). Dorsal to the heart, that is, between it and the underside of the foregut, the layers of splanchnic mesoderm meet each other and

form a mesentery, the *dorsal mesocardium.* This is present in the linear embryo but soon breaks through. Beneath the heart the same two layers of splanchnic mesoderm also come together, but here they break down immedi-

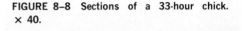

FIGURE 8–8 Sections of a 33-hour chick. × 40.

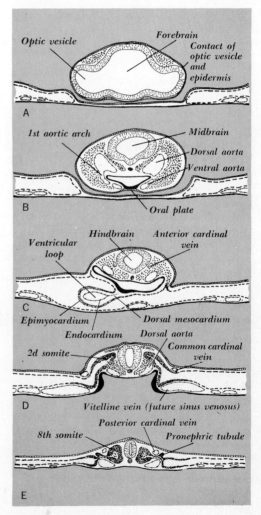

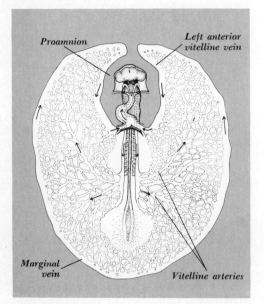

FIGURE 8–9 Ventral view of a chick embryo showing the primary blood vessels and the entire circulation. The drawing is schematized in that it pictures a 33-hour chick embryo while showing the external circulation of a somewhat later stage. The blastoderm outside the area vasculosa is gray.

venosus are still represented by paired primordia in the folds of the portal. A few hours later the atrial rudiments join in the midplane and give rise to a single *atrium;* but it is nearly two days before the primordia of the *sinus venosus* unite and form one organ.

The heart begins to twitch as soon as a part of the ventricle has taken form. It beats faster and at a more regular rate when the atrium comes into being. When the sinus venosus is added, it speeds up again and the sinus venosus takes over from the atrium the function of setting the pace. This was shown in 1920 by Florence Sabin, who also pointed out that the beat of the heart originates before nerve fibers have entered it. The beat is therefore of myogenic (muscular) origin, and the impulse is conducted from the sinus venosus forward by muscle tissue.

The heart grows in length faster than the walls of the cavity within which it lies. As an adaptation to this, it bows to the right and forms a loop, the *ventricular loop.*

ately, so that a *ventral mesocardium* is at best a transitory structure. As a result, the right and left amnio-cardiac vesicles join each other both above and below the heart and form a single *pericardial cavity.* The heart is now free to twist and spiral without involving the body wall.

The first chambers of the heart which take form beneath the foregut are the *truncus* and the anterior part of the *ventricle* (Fig. 8–7, *C*). As the foregut grows in length the anterior intestinal portal recedes, and the heart is added to at the rear. At the 33d hour (13 somites) the ventricle is complete, but the atrium and sinus

THE FETAL MEMBRANES

A principal innovation in the case of animals which lay their eggs on land is the development of fetal membranes, that is, membranes which are derived from the egg but which do not become a part of the embryo proper. There are four of these in the higher vertebrates (amniotes) (Figs. 8–10 and 8–11):

1 The *chorion* or *serosa,* an outer covering which encloses all (the word chorion has a somewhat different meaning in mammal embryology)
2 The *amnion,* a thin membranous sac which surrounds the embryo and provides

a sort of private aquarium to protect it from pressure, abrasion, irritation, and loss of water

3 The *yolk sac*, a bag which encloses and digests the yolk and which, like the liver at later stages, serves as a place of origin of blood cells

4 The *allantois*, by origin a precocious bladder, in which waste accumulates and

which becomes the respiratory organ of the embryo.

The Chorion and Amnion

These two fetal membranes are derived from the extra-embryonic somatopleure. About the 30th hour of incubation, a transverse crescent-shaped thickening of the ectoderm rises up

FIGURE 8–10 Diagrammatic lateral views of a chick embryo illustrating the development of the fetal membranes. *A.* At 2 days. *B.* At 3 days. The embryo proper is shaded with diagonal lines. The numerals show the continuity in the posterior region.

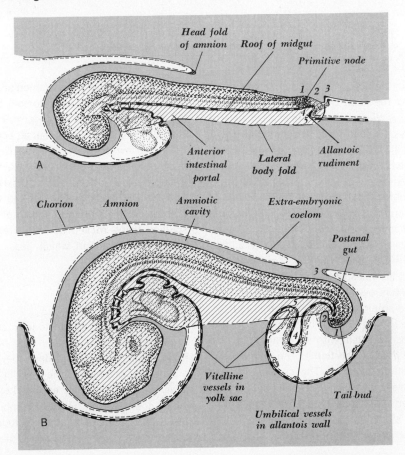

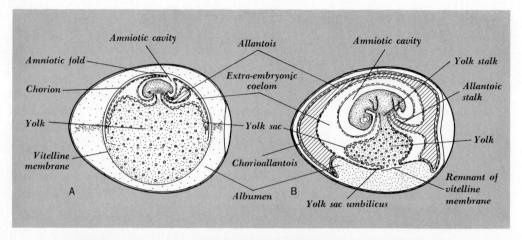

FIGURE 8–11 The later development of the fetal membranes. *A.* At 4 days. (However, at 4 days the amniotic folds are normally closed.) *B.* At 12 days.

anterior to the head and begins to move backward over the head of the embryo. It is the *head fold of the amnion.* At first it consists of ectoderm only, but soon mesoderm enters it from the sides. The outer limb of the amnion fold is the *chorion.* It has no direct connection with the embryo. The inner limb which lies next to the embryo is the *amnion.* The head fold of the amnion moves posteriorly over the embryo as lateral amniotic body folds move in from the sides (Figs. 8–12 and 8–19). Tardily a tail fold of the amnion makes its appearance until ultimately all the folds come together and form a complete tent over the embryo. The space between the amnion and the embryo is the *amniotic cavity* (Fig. 8–11). It is lined with ectoderm and is filled with a clear fluid, the *amniotic fluid,* which bathes the delicate epidermis of the embryo. The cavity gives the embryo freedom to move its members. Later, preliminary to hatching, the embryo drinks from the amniotic fluid and passes alimentary waste, meconium, into the surrounding cavity.

The space between the amnion and the chorion is the extra-embryonic coelom or exocoel. It is lined with somatic mesoderm. The outside layer of the chorion is, of course, ectoderm.

After a few days, smooth-muscle fibers appear in the amnion and begin rhythmic contractions. It is indeed an impressive sight to watch a live chick embryo of the 6th to the 10th day of incubation in a dish of warm salt solution being rocked as in a cradle by the waves of contraction which pass across the amnion.

The Yolk Sac

The margins of the blastoderm continue to grow outward over the surface of the yolk mass. By the end of two days they cover approximately one-half of the yolk. Several days later, the yolk is entirely enclosed except for a small opening, the yolk sac umbilicus (Fig. 8–11, *B*). In this process the ectoderm grows outward somewhat in advance of the endoderm, while the mesoderm lags behind. Wherever the mesoderm pushes between the ectoderm and the endoderm, it splits into its two layers: somatic mesoderm and splanchnic mesoderm. The vitelline blood vessels are located between the splanchnic mesoderm and endoderm.

The extra-embryonic endoderm and the splanchnic mesoderm which accompanies it constitute the *yolk sac*. The function of the endoderm is to secrete enzymes which digest the yolk and then to absorb the products of digestion into the vitelline blood vessels of the yolk sac. But the yolk sac is also the first source of blood cells and blood vessels of the embryo. Even at this early stage the vitelline blood transports oxygen and carbon dioxide. The yolk sac also performs biochemical functions which are later taken over by the liver.

The Allantois

About the third day of incubation, the region of the future floor of the hindgut begins to bulge as a precocious bladder, namely, the *allantois* (Fig. 8–10). The outpocketing slowly enlarges and invades the extra-embryonic coelom. Wherever the mesoderm of the outpocket comes into contact with the mesoderm of the chorion, it adheres. The result is a single membrane, the *chorioallantois*. In time the expanding chorioallantois bursts through the vitelline membrane and pushes outward toward the shell membrane. As it does so, it progressively envelops the albumen and so becomes a sac filled with albumen (Fig. 8–11, *B*). It aids in the absorption of water and albumen.

A network of abundant blood vessels, the *umbilical circulation,* develops in the wall of the allantois; and before the end of the first week of incubation, it has begun to serve as the primary respiratory organ of the embryo. Wastes from the embryo enter the cavity of the allantois as soon as the embryonic kidneys become functional. By the 15th day of incubation, uric acid, a relatively insoluble nitrogenous waste, begins to precipitate out. This is an important adaptation for an embryo which develops in air, for a more soluble waste, such as urea, would diffuse back into the embryo and poison it.

TWO- AND THREE-DAY CHICK EMBRYOS

Flexion and Torsion

While the folds of the amnion are descending like a curtain over the head of the embryo, the embryo itself is bending and twisting so that it soon becomes C-shaped with its left side turned toward the surface of the yolk (Fig. 8–12). The process starts at the head end and gradually progresses posteriorly. It may be simulated by a person standing and facing a wall (the yolk); then turning (torsion) by twisting his head to the right while at the same time bowing lower and lower (flexion); then increasingly turning first his shoulders and then his trunk until finally he is curled up with the entire left side of his body next to the wall.

The first flexure is in the region of the midbrain and is known as the *cranial flexure*. It is well advanced by the end of the second day. By this time also the head, but not the trunk, has revolved to the right. One day later strong forward bending takes place in the cervical region posterior to the hindbrain, the *cervical flexure,* and a flexure has begun at the far posterior end, the *caudal flexure* (Fig. 8–19).

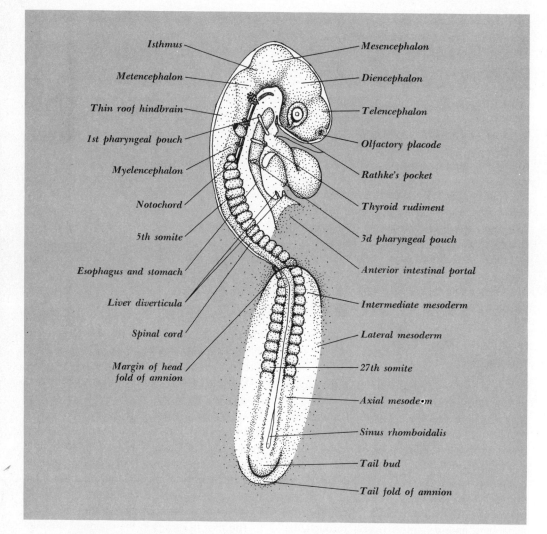

Figure 8–12 Forty-eight-hour chick (25 somites, stage 15) seen by reflected light. × 17.

During the fourth day of incubation, flexion and torsion become complete.

These bends and twists make mounted cross sections of chick embryos difficult for the beginner to interpret. The student will do well to have a whole mount of the same age at hand (or a figure of the same) and to constantly refer to it for the location of the sections he is studying.

Why does the bird embryo twist and turn?

Reptile and mammal embryos do the same. Is it not so that the embryo will fit compactly into the amnion or egg membrane or uterus? Not until the time of hatching (or birth) does the embryo straighten out.

The Nervous System and Sense Organs

The statements which were made concerning the nervous system of the young amphibian embryo apply also to the 2- and 3-day chick. The primary brain vesicles—forebrain, midbrain, and hindbrain—are even more clearly defined in the chick than in an amphibian of comparable age. By the 48th hour the forebrain is divided into telencephalon and diencephalon (Fig. 8–12).

Olfactory placodes are present in the epidermis of the 48-hour chick at the sides of the

FIGURE 8–13 Forty-eight-hour chick viewed as a transparent object. × 16.

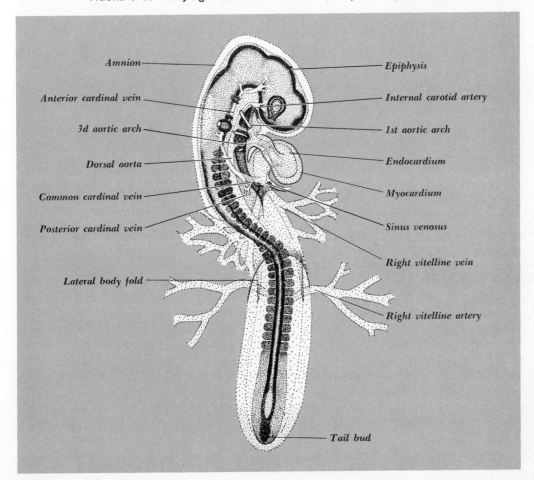

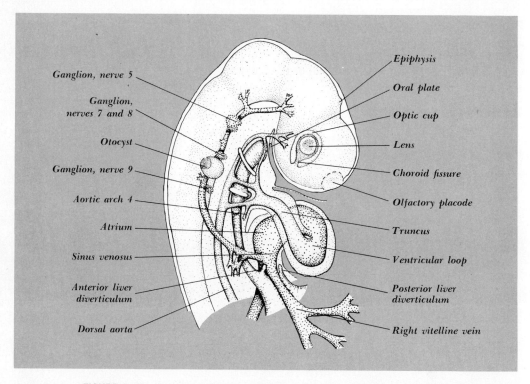

Ganglion, nerve 5

Ganglion,
nerves 7 and 8

Otocyst

Ganglion, nerve 9

Aortic arch 4

Atrium

Sinus venosus

Anterior liver
diverticulum

Dorsal aorta

Epiphysis

Oral plate

Optic cup

Lens

Choroid fissure

Olfactory placode

Truncus

Ventricular loop

Posterior liver
diverticulum

Right vitelline vein

FIGURE 8–14   Anterior region of a 48-hour chick from the right side. × 27.

telencephalon (Fig. 8–15, *C*). By the 72d hour these placodes have shifted ventrally and have caved in to form the *olfactory pits* or vesicles (Fig. 8–25, *A*). By the 72d hour, also, the walls of the telencephalon adjacent to the placodes have begun to bulge laterally and to form the *cerebral hemispheres.*

The *optic vesicles* which pushed laterally from the diencephalon have caved in and formed the *optic cups* (cupping began at about the 40th hour). They are connected with the diencephalon by narrowed *optic stalks* (Fig. 8–24, *B*). It will be noted that each optic cup is incomplete on its ventral side. The cleft which is present there is known as the *choroid fissure*

(Fig. 8–14). It permits blood vessels to enter and leave the cavity of the cup.

Lenses also make their appearance at about the 40th hour as thickenings of the epidermis where the epidermis is in contact with the expanded optic vesicles. By the 48th hour each lens placode has sunk inward and has become a pit open to the exterior (Fig. 8–15, *B*). By the 72d hour it has separated from the epidermis and has formed a vesicle which occupies the center of the optic cup (Fig. 8–24, *B*).

The *epiphysis* appears at about the 48th hour, or soon thereafter, as a small upward pocket from the roof of the diencephalon (Fig. 8–14). The *embryonic infundibulum* similarly

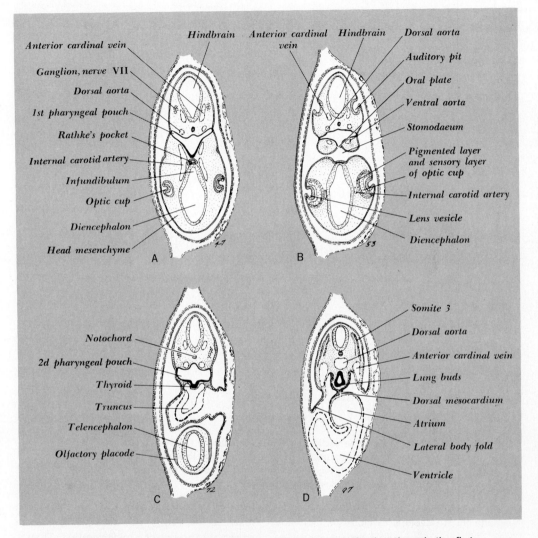

FIGURE 8–15 Transverse sections of a 48-hour chick. *A.* Section through the first aortic arch and Rathke's pocket. *B.* Section through the eyes and the auditory pits. *C.* Section through the truncus and olfactory placodes. *D.* Section through the ventricular loop and the 3d somites. × 25.

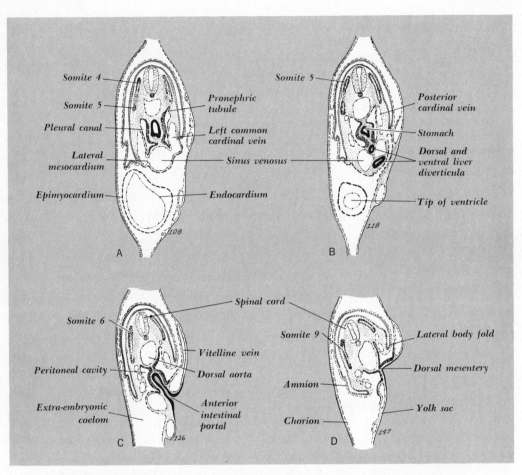

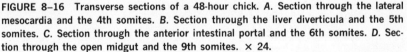

FIGURE 8–16 Transverse sections of a 48-hour chick. *A.* Section through the lateral mesocardia and the 4th somites. *B.* Section through the liver diverticula and the 5th somites. *C.* Section through the anterior intestinal portal and the 6th somites. *D.* Section through the open midgut and the 9th somites. × 24.

is a downward pocket from the floor of the diencephalon. It develops in close contact with Rathke's pocket (embryonic hypophysis) (Fig. 8–15, *A*).

The midbrain (mesencephalon) of the young chick embryo is large and rounded in anticipation of its importance later as the center for reflexes based on sight. The cranial flexure takes place in the region of the midbrain. The result is that transverse sections of 48-hour chicks cut through the midbrain first. Then in sequence the serial sections cut simultaneously anteriorly toward the tip of the forebrain and posteriorly toward the midbrain and spinal cord. The same principle applies to the 72-hour chick except that in this case, because of the cervical bending, the transverse sections first encounter the thin roof of the hindbrain.

FIGURE 8–17 Transverse sections of a 48-hour chick. A. Section through pronephric tubules and the 14th somites. B. Section through the anterior limb buds, mesonephric tubules, and the 18th somitès. C. Section through the vitelline arteries and the 22d somites. D. Section through the newly formed 27th somites. × 44.

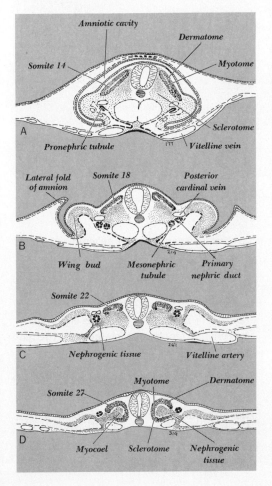

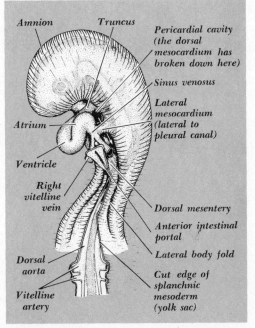

FIGURE 8–18 The 48-hour chick within the amnion as seen from the ventral side. The yolk sac has been cut away at the borders of the anterior intestinal portal. The heart is seen protruding into the extra-embryonic coelom. A small white arrow is shown passing through the left pleural canal (medial to the left lateral mesocardium).

At the 72d hour it is possible to recognize the roots of the third cranial nerves, the *oculomotor nerves,* emerging from the floor of the midbrain (Figs. 8–19 and 8–23, *B*).

The *isthmus* is a narrowed region of the brain which marks the division between the midbrain and the hindbrain.

The roof of the hindbrain is thin and shaped like a kite. Its place of greatest width marks the division of the hindbrain into metencephalon and myelencephalon. The sidewalls of the

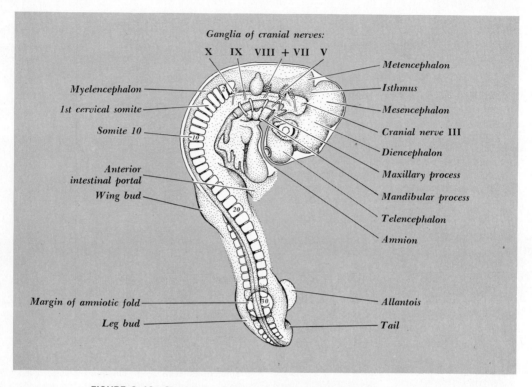

FIGURE 8–19   Seventy-two-hour chick embryo from the right side. × 12.

hindbrain are thick and show a series of vertical folds known as neuromeres, to which reference has already been made. These seem to have some relation to the series of cranial ganglia which flank the hindbrain, but this relation is far from clear. The most anterior of these ganglia are those of the great fifth or *trigeminal nerves* (Fig. 8–19). At the 72d hour each of these is seen to be composed of three divisions: the *ophthalmic nerve* to the region of the optic cup, the *maxillary nerve* to what will be the upper jaw region, and the *mandibular nerve* to the region of the lower jaw.

Posterior to the ganglia of the fifth nerve are the combined ganglia of the seventh and eighth nerves, the *facial* and *auditory nerves,* respectively (Fig. 8–23, *A*). Later in development these ganglia separate; the ganglion of the seventh nerve supplies fibers to the second branchial or hyoid arch, while the eighth ganglion remains closely associated with the *otocyst.*

At the 48th hour the future otocyst is a pit of epidermis at the side of the hindbrain, open widely to the exterior (Fig. 8–15, *B*). But at the 72d hour its opening has closed or almost closed (Fig. 8–23, *A*).

A short distance posterior to the otocyst is

the ganglion of the ninth or *glossopharyngeal nerve* (Figs. 8–19 and 8–23, *A*). It supplies sensory nerve fibers for the third branchial arch. Next comes the ganglion of the tenth or *vagus nerve* whose fibers go to the remaining branchial arches.

The hindbrain narrows gradually to the spinal cord, which is oval in cross section and has thick lateral walls. Its roof plate is thin, and its floor plate is narrow. At the sides of the cord are rows of *spinal ganglia,* arranged at regular intervals like a series of saddlebags. These are not easy to demonstrate at 48 hours, for they are still in the process of formation from the neural crest; but by the 72d hour they are clearly seen lying between each somite and the neural tube (Fig. 8–25, *B*).

The posterior neuropore (sinus rhomboidalis) closes (or ceases to exist) about the 40th hour, and from that time on the entire neural tube ends posteriorly at the tail bud.

The Stomodaeum

The stomodaeum of the linear embryo, if one can be said to be present at all, is a shallow recess of ectoderm on the underside of the head fold immediately beneath the oral plate. As the forebrain bends downward the stomodaeum deepens. The proliferation of the mesoderm of the first branchial arches produces swellings on each side of the oral plate and thus further increases the depth of the stomodaeum. By the second and third day a transverse *oral cleft* cuts across the underside of the head and partially splits the lateral mesodermal swellings into *maxillary processes* (anteriorly) and *mandibular processes* (posteriorly). These become the upper and lower jaws, respectively.

The oral plate is still present in 48-hour chicks, but by the 72d hour it has broken down. Soon it disappears without trace (Fig. 8–23, *B*).

The Pharynx

At first the pharynx underlies the entire brain; but as a result of the bulging forward of the forebrain and the subsequent formation of the cranial flexure, it comes to underlie the hindbrain only (Fig. 8–19). For most of its length it is broad from side to side, but it tapers to an apex posteriorly. Along each side of the pharynx are the *pharyngeal pouches,* five in number; however, the fifth is rudimentary, little more than an appendage of the fourth. As each pouch pushes laterally through the head mesenchyme it comes into contact with the epidermis of the side of the head. The epidermis responds by forming a *branchial groove* or furrow (Fig. 8–23, *B*). Some of the closing plates (gill plates) between the pouches and grooves break through and become open *branchial clefts* (gill slits). At the 72d hour the first pouch has a dorsal opening, the second pouch has both a dorsal and a ventral opening, while the third pouches have not yet broken through. The fourth pouches never open in the chick.

The *thyroid rudiment* is a downpocketing from the floor of the pharynx, situated approximately midway between the second pair of pharyngeal pouches close to where the ventral aorta from the heart divides into the second and third aortic arches (Figs. 8–15, *C,* and 8–24, *A*).

The floor of the pharynx posterior to the fourth pouches develops a ventral longitudinal groove, the *laryngotracheal groove.* From the

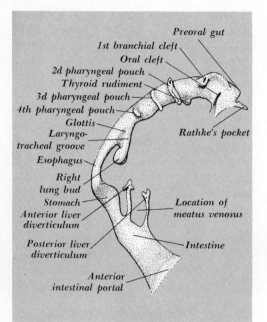

Preoral gut
1st branchial cleft
Oral cleft
2d pharyngeal pouch
Thyroid rudiment
3d pharyngeal pouch
4th pharyngeal pouch
Glottis
Laryngo-
tracheal groove
Esophagus
Right
lung bud
Stomach
Anterior liver
diverticulum
Posterior liver
diverticulum
Anterior
intestinal portal
Rathke's pocket
Location of
meatus venosus
Intestine

FIGURE 8–20 The foregut of a 72-hour chick. × 25.

posterior end of this groove the right and left lung buds push out laterally (Fig. 8–24, *B*). Gradually, beginning at its posterior end and progressing forward, the groove pinches away from the dorsal part of the pharynx and becomes the *trachea;* but an opening into the pharynx remains at the anterior end, namely, the *glottis* (Fig. 8–20).

Esophagus, Stomach, and Liver Diverticula

The foregut continues to grow in length by the retreat of the anterior intestinal portal posteriorly. By the 33d hour most of what will be the pharynx has been formed, and the portal is a broad opening from side to side (Fig. 8–5).

By the 48th hour the future *esophagus* and *stomach* have been added, and the portal has narrowed to a vertical slit on the undersurface of the blastoderm (Figs. 8–16, *C,* and 8–18).

At this latter stage of development (48th hour), two outpocketings are to be found on the forward wall of the intestinal portal: a dorsal one and a ventral one. These become the *anterior* and *posterior liver diverticula* (Figs. 8–12, 8–16, *B,* and 8–20). The vitelline veins, which run in the lateral folds of the anterior intestinal portal, come together and unite between the diverticula (Fig. 8–14). By the end of the fourth day, the duodenum has been added to the posterior end of the foregut. The liver diverticula open into it. The diverticula, moreover, have given rise to a basket of anastomosing cords which encircle the venous channel (Fig. 8–25, *B*).

The Hindgut

The hindgut forms within the tail fold in much the same way that the foregut forms within the head fold. At the 48th hour a pocket of endoderm is just beginning to take form (Fig. 8–10, *A*). Actually it is the beginning of the future allantois. But soon the true hindgut appears and grows in length very much as the foregut grows. However, in this case the migration of the *posterior intestinal portal* is forward (Fig. 8–10, *B*).

The Mesoderm

The importance of the mesoderm is great. Starting as a relatively small part of the embryo, it increases tremendously in volume until it exceeds the other germ layers. All the systems of the body, with the exception of the nervous system, are predominantly meso-dermal in their origin. Yet the principal claim of the mesoderm to preeminence is not its vol-ume nor its numerous important functions, but the fact that it plays the primary role in the establishment of the embryo. It tells the other germ layers what to do.

This is as true of the chick embryo as it is of the amphibian. As the chorda-mesoderm of the primitive streak migrates forward past the primitive node, its regions become specified for their respective fates. When it comes to rest in its final location, it determines the fu-ture of the cells of the other germ layers with which it is associated. It is true that the rela-tions between the germ layers are reciprocal, for ectoderm and endoderm are necessary for the normal differentiation of mesoderm. But their roles are different. The mesoderm takes the initiative. The ectoderm and endoderm feed back their influences on the mesoderm.

Head Mesenchyme

The mesenchyme of the head is at first meager, but it increases rapidly, and by the end of the second day of incubation it exceeds the other tissues of the head in volume (Fig. 8-15). It appears to be homogeneous, yet its cells are derived from several embryonic sources: prechordal mesoderm, axial meso-derm anterior to the somites, neural crest, and possibly even the anterior tips of the foregut and notochord. The term mesenchyme, there-fore, should be thought of as designating a histological structure rather than a particular embryonic origin. Mesenchyme cells are loose, star-shaped cells capable of migration—al-though they do not necessarily migrate. In fact they are commonly united into a spongelike

mesh in the interstices of which there is a tis-sue fluid, a sort of thin jelly containing physio-logical nutrients which serves as a fit medium for cell life.

The Axial Mesoderm (Epimeres)

The axial mesoderm gives rise to somites. At first each somite is a block of columnar cells which radiate out from a central cavity, the *myocoel* (Fig. 8-17, *D*). This cavity, however, is more or less filled with loose cells.

The derivatives of each somite are three "-tomes": dermatome, myotome, and sclero-tome.

1 The *dermatome* is derived from the lateral wall of each somite, that is, from the wall adja-cent to the epidermis. It retains its laminar character for a considerable time, but ulti-mately (about the fourth day in the case of the anterior somites) it breaks down into mesenchyme and gives rise to the dermis and subcutaneous tissues of the back of the em-bryo.

2 The dorsal portion of the median wall of each somite constitutes the *myotome*. At first the myotomes are small, but as development proceeds they swing outwardly to a position under the dermatomes. Here they grow, and give rise to the axial muscles of the dorsal side of the body (Fig. 8-17, *D* to *A*). (The earliest stages are the most posterior.) Their cells, known as myoblasts, elongate longitudinally until they extend the full length of each somite.

3 The ventral portion of the inner wall of each somite, namely, the wall which is adja-cent to the notochord and the ventral half of the neural tube, forms the *sclerotome*. As early as the second day of incubation its cells, beginning with the anterior somites, lose their epithelial character and become mesenchyme which migrates around the neural tube, notochord, and even around the aorta. They multiply rapidly and soon make up a large part of the cross section of the embryo. Their fate is to give rise to the axial skeleton.

The Intermediate Mesoderm (Mesomeres)

The anterior part of the intermediate mesoderm becomes segmented and gives rise to a series of *nephrotomes,* that is, to necks of cells which connect the somites with the lateral mesoderm (Fig. 8–8, *E*). In amphibians some of these produce functional *pronephric tubules,* but in the chick the pronephric tubules are vestigial and seemingly never perform as excretory organs (Fig. 8–17). They are to be seen as tiny rudiments, as early as the 33d hour, lateral to somites 8 to 15. Each tubule begins as a bud which grows outward from the somatic layer of the nephrotome and then hollows out (Fig. 8–8, *E*). Its central end retains an opening into the body cavity known as a *nephrostome,* while its lateral end turns backward and unites with the tubule which is next posterior to it. Thus a continuous duct, the *primary nephric duct,* is formed (known also as the pronephric, mesonephric, or Wolffian duct). It grows backward in the space between the somites and the somatic layer of lateral mesoderm. At the 48th hour it has reached the level of about the 25th somite (Fig. 8–17). By the 72d hour it has made contact with the hindgut (cloaca). Shortly afterward it has entered it (Fig. 8–22). The pronephric tubules degenerate, but the primary nephric ducts, to which they give rise, endure.

The intermediate mesoderm posterior to the pronephric tubules does not become seg-mented into distinct nephrotomes; nevertheless, it is nephrogenic. During the third day of incubation it gives rise to *mesonephric tubules* (Fig. 8–17, *B*). These become the functioning excretory tubules of the young embryo. The process can be seen in sections of a 72-hour chick (Fig. 8–26). The most posterior tubules, as usual, are the youngest and consist of spherical condensations of nephrogenous material (cells derived from the mesomere) on the medial side of each primary nephric duct. Farther forward the condensations have become hollow balls of cells. Still farther forward each ball of cells has elongated and become S-shaped, with its lateral end in contact with the primary nephric duct and its medial end encapsuling a tuft of capillaries, the *glomerulus.*

Three features thus distinguish mesonephric tubules from the preceding pronephric tubules: (1) Mesonephric tubules are not segmentally arranged; or rather, several tubules arise from each somite so that segmentation is not apparent. (2) Mesonephric tubules make use of the primary nephric duct which the pronephros produced. (3) Mesonephric tubules begin in capsules which enfold glomeruli, whereas pronephric tubules begin in nephrostomes which open from the body cavity. If there are knots of blood vessels—and they do occasionally occur in the chick—they are in the body cavity external to the tubule. (These matters and the later development of the metanephros are considered in Chapter 18.)

The Lateral Mesoderm (Hypomeres)

The lateral mesoderm gives rise to the linings of the body cavities—the pericardium, pleura, and peritoneum—but it does far more than this: The somatic layer produces the dermis, muscles, and connective tissues of the lateral and ventral body walls, and also the limb buds. The splanchnic layer gives rise to the epimyocardium of the heart, the walls of the lungs and gut, the cortex of the adrenal gland, and the sex glands. (These matters are discussed further in the chapters which deal with the development of these several organs.)

The Coelom

The development of the coelom is of considerable interest. Figure 8–18 is a ventral view of a 48-hour chick in which the entire splanchnopleure (yolk sac) has been cut away external to the borders of the anterior intestinal portal. This exposes the inner surface of the somatic mesoderm. The lateral body folds are seen pushing ventrally; but as yet they are widely separated, and the heart is shown protruding between them into the extra-embryonic coelom. By the 72d hour the right body fold has pulled all the way across the ventral side of the heart and has fused with the left body fold. As a result, the pericardial cavity is almost completely cut off from the extra-embryonic coelom (Fig. 8–10, *B*). The dorsal mesocardium has broken through above the

FIGURE 8–21 Diagram of the circulatory system of a 72-hour chick. × 14.

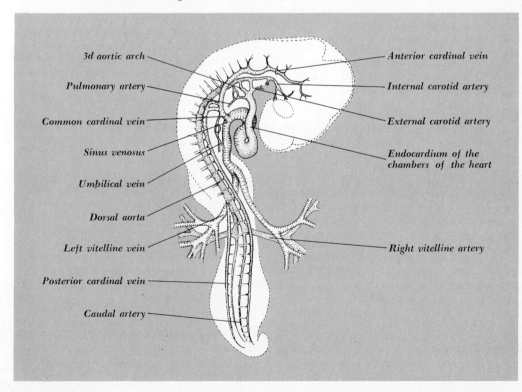

3d aortic arch

Pulmonary artery

Common cardinal vein

Sinus venosus

Umbilical vein

Dorsal aorta

Left vitelline vein

Posterior cardinal vein

Caudal artery

Anterior cardinal vein

Internal carotid artery

External carotid artery

Endocardium of the chambers of the heart

Right vitelline artery

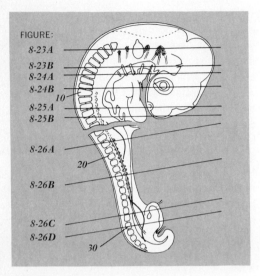

FIGURE:
8-23A
8-23B
8-24A
8-24B
   10
8-25A
8-25B

8-26A
      20
8-26B

8-26C
8-26D
   30

FIGURE 8–22 Index figure to the sections illustrated in Figures 8–23 to 8–26.

heart, so that the right and left pericardial coeloms communicate with each other.

Lateral to the sinus venosus, the folds of the body wall fuse with the wall of the heart and form bridges, known as the *lateral meso-cardia,* between the somatic and splanchnic mesoderms. Across these bridges, the common cardinal veins transport blood from the body wall to the sinus venosus of the heart (Figs. 8–16, *A,* and 8–25, *A*).

Dorsal and median to each lateral mesocardium, the pericardial and peritoneal coeloms remain in communication with each other by what we shall call the *pleural canals.* Later, the lung buds grow into them, and they become the pleural cavities. Communications between the pericardial and peritoneal coeloms remain open for a brief time, ventral to

the lateral mesocardia. But ultimately they close.

Farther posterior, the vitelline veins course within the side folds that bound the anterior intestinal portal. They then enter the sinus venosus (Fig. 8–18). The dorsal and ventral hepatic diverticula, which have outpocketed from the forward wall of the anterior intestinal portal, push between the veins, and together the diverticula and endothelia of the veins give rise to the liver. When the body wall finally becomes complete in this region, it fuses with the underside of the liver and thus forms what is essentially a ventral mesentery (Fig. 8–10, *B*). In other words there arises secondarily in the chick a structure (the ventral mesentery) which is primary and original in the lower vertebrates. The lateral mesocardia plus the ventral mesentery are equivalent to the *septum transversum* of mammal embryos.

The Heart and Circulation

During the second and third days of incubation, the area vasculosa continues to expand outwardly across the surface of the yolk. It is bounded at the periphery by the *marginal vein* (Fig. 8–9). Its network of capillaries develops channels which become the vitelline arteries and veins. The arteries diverge to the right and left from the dorsal aorta in the midtrunk region. They transport blood laterally toward the marginal vein and also forward. The blood, having gathered in the forward regions of the area vasculosa, returns posteriorly by way of the *anterior vitelline veins* until it reaches the sidewalls of the anterior intestinal portal. It then turns medially and enters the heart. Later, *lateral vitelline veins* develop, which more or less parallel the vitelline arteries. They lie on the mesodermal side of the arteries. The anterior vitelline veins and the marginal veins then gradually disappear.

By the 48th hour of incubation, the anterior intestinal portal has retreated sufficiently for the vitelline veins to unite in the midline and form the sinus venosus (Figs. 8–14 and 8–18). A day later, the veins have joined behind the sinus venosus to form a single channel between the dorsal and ventral hepatic diverticula. It is known as the *ductus venosus* (Fig. 8–25, *B*). Now, the ductus venosus of the chick is not the same vessel as the ductus venosus of the mammal embryo. It is derived

from the vitelline veins and carries yolk sac blood. The mammal vessel of the same name is a new formation within the liver which carries umbilical blood. After several days, the bird's ductus venosus closes, and the vitelline blood on its way to the heart is forced to detour through sinusoids (capillaries) between the hepatic cords of the liver. The proximal portion of the vitelline veins thus becomes a part of the hepatic portal system of the adult. (A portal system is one which begins in capillaries and breaks again into capillaries before the blood proceeds on to the heart.)

The heart of 2- and 3-day chicks continues

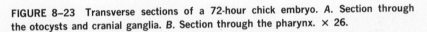

FIGURE 8–23 Transverse sections of a 72-hour chick embryo. *A.* Section through the otocysts and cranial ganglia. *B.* Section through the pharynx. × 26.

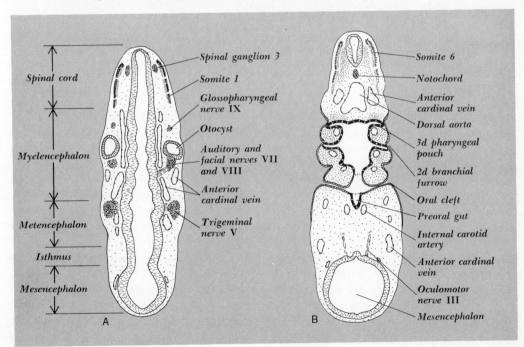

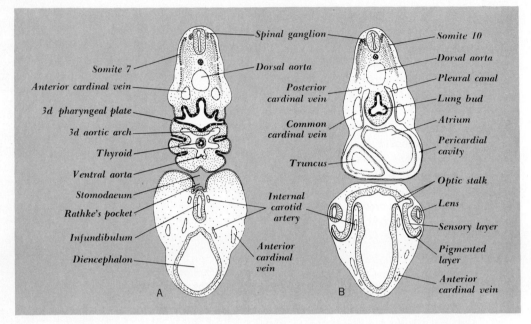

FIGURE 8–24 Transverse sections through the 72-hour chick. A. Section through
Rathke's pocket and the thyroid rudiment. B. Section through the eyes and lung buds.
× 25.

to elongate, and as it does so, it becomes S-shaped. The ventricular loop, which at first is oriented transversely, swings posteriorly to a position ventral to the sinus venosus. At the same time the truncus comes to lie ventral to the atrium. At this stage, the heart may be described as a counterclockwise spiral (Fig. 8–14).

The truncus leads forward to the roots of the several aortic arches. At the 48th hour, the first aortic arches are still functioning, although the second arches are now carrying most of the blood. The third arches have just begun to operate. By the 72d hour, the first arches are practically closed, the second and third arches are in full operation, and the fourth

arches are beginning to function (Fig. 8–21). Only sprouts tell us where the sixth arches will form. (The story of the arches is stated more fully in Chapter 17.)

The blood which flows dorsally through the aortic arches divides into anterior and posterior streams. The anterior streams follow the *internal carotid arteries* forward to the region of the eyes. Here the blood spreads out in a net of capillaries next to the surface of the forebrain (Fig. 8–21). The rest of the blood flows posteriorly in the two *dorsal aortas*. At the level of the posterior end of the pharynx, the right and left aortas unite in the midline to form a single median dorsal aorta. The process of union progresses posteriorly. At 48 hours, it is complete as far back as the 10th somite. By the 72d hour, it has reached the 20th somite (Fig. 8–21). Ultimately, the entire aorta posterior to the pharynx is one.

The branches of the dorsal aorta correspond to the three divisions of the mesoderm (Fig. 8–27): (1) Dorsal branches grow into the axial mesoderm between adjacent somites and supply the body wall, including the limbs. These are known as *segmental* (intersegmental) *arteries* (Fig. 8–21). (2) Short lateral branches, which we shall call *nephric arteries,* grow from the aorta into the intermediate mesoderm and supply the organs, notably the glomeruli, which are derived from it. (3) Ventral branches go to the splanchnopleure and connect with the capillary net of the area vasculosa. At the

level of the 22d somites, they organize into definite channels, the vitelline arteries (Fig. 8–21).

Blood returns to the heart by four circulatory "arcs" (Fig. 8–28):

1 The first of these is the vitelline or splanchnic arc already described. The blood begins to flow in this arc at about the 40th hour of incubation. While out on the yolk sac, it gains cells from the blood islands, absorbs nutrients from the digested yolk, and probably also exchanges gases—a function later taken over by the allantois.

2 The second arc, the arc of the anterior cardinal vein, is somatic. It consists of a capillary bed of blood vessels, lateral to the brain. It develops from angioblasts in the head

FIGURE 8–25 Transverse sections through the 72-hour chick. *A.* Section through the olfactory pits and lateral mesocardia. *B.* Section through the liver diverticula and cerebral hemispheres. × 25.

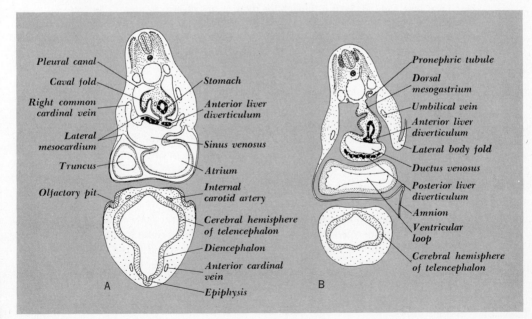

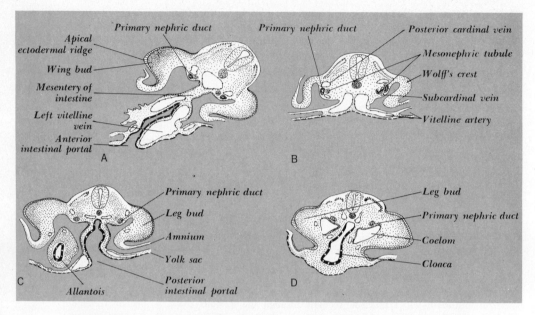

FIGURE 8–26 Transverse sections of a 72-hour chick. (These are drawn from a different embryo than the previous sections.) *A*. Section through the anterior intestinal portal and wing buds. *B*. Section through the roots of the vitelline arteries. *C*. Section through the posterior intestinal portal and the leg buds. *D*. Section through the cloaca. Approximately × 25.

mesenchyme. The anterior cardinal veins carry blood posteriorly to about the level of the heart. Here, joined by the posterior cardinal veins, they become the common cardinal veins, or ducts of Cuvier. The common cardinals then cross the lateral mesocardium and enter the sinus venosus of the heart.

3 The third arc, also a somatic arc, is that of the posterior cardinal veins. At first they bring blood forward from the body wall and tail region (Fig. 8–21). In the trunk, each vein runs just at one side of the intermediate mesoderm and receives blood from the somites and nephric tissue. It carries the blood forward to the level of the sinus venosus, where, as we have noted, it joins the anterior cardinal veins and, as the common cardinal, enters the heart.

4 The elements of the fourth or umbilical arc consist at first of a net of blood vessels in the lateral folds of the body wall, ventral to the posterior cardinal veins (Figs. 8–21 and 8–28). Hence, in origin they are somatic vessels. When the allantois (which is splanchnic) develops with its broad attachment to the ventral body wall, its capillaries tap the vessels of the ventral body wall. Thus the blood discovers a direct route to the heart. The allantoic veins correspond to the umbilical veins of mammals. Although they are splanchnic in origin, they go forward to the heart by way of the somatopleure.

EXPERIMENTS ON "CYTODIFFERENTIATION"

The young chick embryo is a favored subject for experimentation. Some results of the work which has been done will be referred to in the chapters on the organ systems (Chapters 11 to 18). For the present we shall give attention to certain studies concerned with "cytodifferentiation," that is, with the processes by which cells become visibly and functionally different from one another. Actually differentiation has already taken place when cells commence cytodifferentiation, for although the cells of an organ rudiment may appear to be undifferentiated, in fact they are already committed to their fates. Cytodifferentiation, therefore, is only the last step in the series of steps by which cells become specialized.

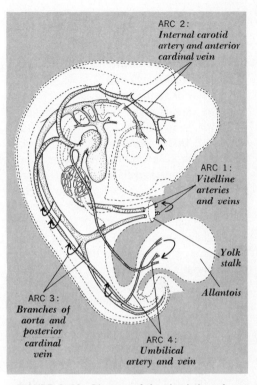

ARC 2:
Internal carotid artery and anterior cardinal vein

ARC 1:
Vitelline arteries and veins

Yolk stalk

Allantois

ARC 3:
Branches of aorta and posterior cardinal vein

ARC 4:
Umbilical artery and vein

FIGURE 8–28 Diagram of the circulation of a 96-hour chick.

FIGURE 8–27 Diagrammatic section through the trunk to illustrate the location of the principal veins and the branches of the aorta.

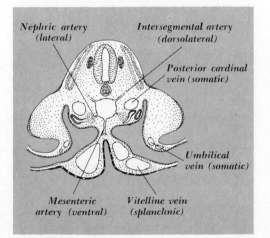

Nephric artery (lateral)

Intersegmental artery (dorsolateral)

Posterior cardinal vein (somatic)

Umbilical vein (somatic)

Mesenteric artery (ventral)

Vitelline vein (splanchnic)

In 1952 Moscona treated cells of chick and mouse embryos with calcium-free solutions. He then subjected them briefly to weak solutions containing trypsin or other protein-digesting enzyme. The treatment loosened the bonds between the cells, with the result that they rounded up and became free (page 53). After rinsing them in physiological salt solution, he dispersed them in a fluid culture medium containing equal parts of salt solution, chick embryo juice, and chicken serum. The cells settled to the bottom of the culture dish and moved about by means of pseudopod-like processes. When by chance they came into contact one with another, they adhered and drew together. The result was that the bottom of the dish became covered with numerous clusters of squirming cells.

For a day or so the cells which were derived from the various organ rudiments behaved in much the same manner. But then differences appeared. Under suitable culture conditions some of the cells continued to differentiate. Thus future muscle cells produced myofibrils, and prospective nerve cells sprouted axons. It must be noted, however, that clusters of cells of a single kind do not produce organized multicellular structures. Even if they already show some specialization, they usually lose their special character. This is not a true dedifferentiation, for they remain committed to their fates and cannot ordinarily be made to redifferentiate in a new direction.

What happens when cells of two or more different kinds are intermingled? Moscona mixed kidney-forming and cartilage-forming cells in the same cell suspension. For 24 hours they remained mingled in the same clusters. Then they began to sort out and differentiate according to kind. It looked as though each cell knew its own kind and knew what to do. Steinberg interprets the sorting out in terms of quantitative differences in the strengths of intercullar adhesive forces (page 54).

The positions which different kinds of cells assume when they sort out reflect the relative strengths of their adhesiveness. When cartilage cells are mixed with liver cells, the cartilage cells migrate from the surface and become an inner mass, and the liver cells remain as an outer coat. When liver cells and cells of the intestinal epithelium are mixed, the liver cells become the inner mass, and the epithelial cells remain on the outside. Thus there is a sort of hierarchy of the strength of adhesion—cartilage > liver > epithelium.

The sorting out not only restores spatial arrangements which often resemble the arrangement of tissues in an embryo, but it results in inductive interactions. As a result, cytodifferentiations take place which do not occur when cells are isolated. We shall refer to several of these interactions in the pages which follow: the interdependence of ectoderm and mesoderm in the production of a limb (page 248); the interaction between endoderm and the investing mesenchyme of the gut (page 319); the mutual response of epidermis and dermal mesenchyme, by reason of which the skin becomes organized (page 336); and the interaction of nephric epithelium and nephric mesenchyme in the production of the finer structure of kidney tissue (page 391).

What happens when cells of different species of animals are mixed together? One might expect that they would separate and each go its own way. This, indeed, is what happens when cells of different species of sponges are intermingled. But Moscona found otherwise when he mixed chick and mouse kidney cells or chick and mouse cartilage cells. Separation according to tissue took place, but segregation according to species did not occur. On the contrary, cells of like type cooperated in the production of "chimeric" structures in which some of the cells were chick cells and some were mouse cells. It is clear that the signals (epigenetic factors) which lead to differentiation are the same in birds and mammals. Wilde obtained similar results when he mixed myoblasts (future muscle cells) of a mouse and chick. The myoblasts fused together to form multinucleate striated muscle fibers. Some of the nuclei in each fiber were chick nuclei, and some were mouse nuclei (page 346).

Experiments on the reaggregation of disso-ciated cells make us aware of activity at two levels of organization: the cellular level and the supracellular, or organismic, level. At the cellular level individual cells are the actors. Yet we must not forget that individual cells ac-

TABLE 8-1. "NORMAL STAGES" OF CHICK DEVELOPMENT
As defined by V. Hamburger and H. L. Hamilton, 1951 (J. Morphol. 88:49–92).

	Approximate Age in Hours	
	At 100–101°F	At 103°F
Primitive Streak Stages:		
Stage 1. Embryonic shield (Fig. 7–6, *A*)	4–5 hours	4 hours
Stage 2. Initial streak (Fig. 7–6, *B*)	6–7	6
Stage 3. Intermediate streak (Fig. 7–6, *C*)	12–13	12
Stage 4. Definitive streak (Fig. 7–8, *A*)	18–19	18
Head Formation:		
Stage 5. Head process (Fig. 7–8, *B*)	19–22	19
Stage 6. Head fold (Fig. 7–8, *C*)	23–25	20
Early Somite Stages:		
Stage 7. First somite	23–26	21
Stage 8. 4 somites (Figs. 8–1 and 8–2)	26–29	24
Stage 9. 7 somites	29–33	27
Stage 10. 10 somites	33–38	30
Stage 11. 13 somites	40–45	33
Stage 12. 16 somites	45–49	36
Flexion and Torsion:		
Stage 13. Beginning of head torsion; 19 somites	48–52	40
Stage 14. Head fully turned; 22 somites	50–53	44
Stage 15. Midbrain farthest forward; 25 somites		
(Figs. 8–11 and 8–12)	50–55	48
Stage 16. Rear of midbrain farthest forward; 28 somites	51–56	54
Stage 17. Front of hindbrain forward; 31 somites	52–64	60
Stage 18. Entire roof of hindbrain forward; 35 somites		
(Fig. 8–18)	72	72
Limb-bud Stages:		
Stage 19. Length of limb bud $\frac{1}{4}$ of its breadth	3 days	
Stage 20. Length of limb bud $\frac{1}{2}$ of its breadth	3$\frac{1}{2}$ days	
Stage 21. Length of limb bud equal to its breadth;		
torsion complete (Fig. 11–4, *A*)	4 days	

Note: In reckoning the age of an embryo allowance must be made for the time it takes the egg to warm to incubation temperature. This may amount to 6 to 8 hours or more depending on the characteristics of the incubator. The ages given for 100–101°F (37$\frac{1}{2}$–38$\frac{1}{2}$°C) are from Ham-burger and Hamilton and indicate the great variability in the rate of development. The ages indicated for 103°F (39–39$\frac{1}{2}$°C) are roughly those which have been used in the past in most textbooks of embryology. They are quite arbitrary.

quired their distinctive characters when they were unspecialized parts of a larger whole. Moreover, when cells are isolated they behave as though they are still parts of a whole which no longer is present. Indeed, they fulfill their destinies only when they act and interact within the framework of the greater community. The community is as real as the cell, and the language which describes it is as valid. The fact that the organism disappears when it is analyzed does not negate this; so does the cell disappear when it is reduced to its parts.

SELECTED READINGS

For general accounts of chick development see readings referred to in connection with Chapter 7. Experiments on young chick embryos are listed in the chapters on the organ systems (Chaps. 10 to 18).

Experiments on Fetal Membranes of the Chick

Pierce, M. E., 1933. The amnion of the chick as an independent effector. J. Exptl. Zool., **65**:443–473.

Zwilling, E., 1946. Regulation in the chick allantois. J. Exptl. Zool., **101**:445–453.

Adamstone, F. B., 1948. Experiments on the development of the amnion of the chick. J. Morphol., **83**:359–371.

Randles, C. A., Jr., and A. L. Romanoff, 1950. Some physical aspects of the amnion and allantois of the developing chick embryo. J. Exptl. Zool., **114**:87–101.

Bellairs, Ruth, 1966. Biological aspects of the yolk of the hen's egg. Advan. Morphogenesis, **5**:217–272.

Experiments on Cell Differentiation

Moscona, A., 1952. Cell suspensions from organ rudiments in chick embryos. Exptl. Cell Res., **3**:535–539.

—— and H. Moscona, 1952. The dissociation and aggregation of cells from organ rudiments of the early chick embryo. J. Anat., **86**:287–301.

——, 1957. The development in vitro of chimeric aggregates of dissociated embryonic chick and mouse cells. Proc. Natl. Acad. Sci., **43**:184–194. Reprinted in Bell (ed.), 1965. Molecular and Cellular Aspects of Development. Harper & Row, Publishers, Incorporated, New York. Pp. 55–64.

——, 1961. How cells associate. Sci. Am., **205** (September):142–162.

Steinberg, M. S., 1963. Reconstruction of tissues by dissociated cells. Science, **141**:401–408. Reprinted in Bell (ed.), 1965. Molecular and Cellular Aspects of Development. Harper & Row, Publishers, Incorporated, New York. Pp. 64–76.

Trinkaus, J. P., 1965. Mechanism of morphogenetic movements. In DeHaan and Ursprung (eds.), Organogenesis. Holt, Rinehart and Winston, Inc., New York. Pp. 55–104.

chapter 9

THE
EARLY
DEVELOPMENT
OF THE
MAMMAL

Viviparity, in which an embryo or fetus is nurtured within its mother's body, is characteristic of mammals. Exceptions are the primitive monotremes of Australia and Tasmania, namely, the platypus, *Ornithorhynchus,* and the spiny anteater, *Echidna.* Viviparity is an adaptation to adverse environmental conditions such as cold, dryness, and seasonal change. It permits the mother freedom to move about, to feed, and to migrate, freedom which is not enjoyed by a bird parent during its task of incubation. Viviparity also is found in some elasmobranchs, teleosts, and reptiles. In most of these, however, the egg is supplied with abundant yolk, but develops within organs of its mother's body until hatching. Birds apparently have not developed vivi-

parity because of the burden of carrying the unborn young when in flight. However, bats, which are mammals, carry their offspring in flight.

The Ovarian Follicle

At the time of birth, the ovary of a typical female mammal contains a hundred thousand or more oogonia (potential eggs) (Fig. 9–1). Some are still in the process of growth and mitosis. But soon mitoses cease, and during the remaining life of the female, no further multiplication of germ cells takes place.

At first the oogonia are associated with the mesodermal epithelium (germinal epithelium) which covers the ovary. Whether they are ac-

FIGURE 9–1 Diagram illustrating the sequence of events in the mammalian ovary. (From B. M. Patten, 1964, "Foundations of Embryology," McGraw-Hill Book Company, New York, 2d ed., Fig. 2–11.)

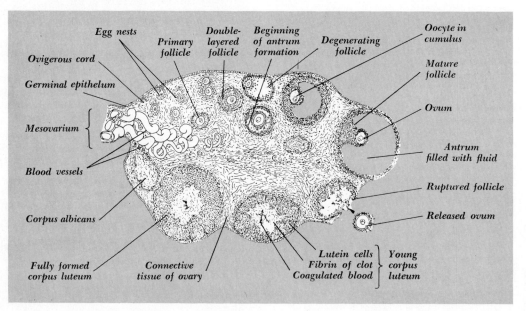

tually derived from the germinal epithelium has been debated (see Chapter 18). At any rate, cords of cells from the epithelium, including germ cells, push inward into the connective tissue of the ovary (ovigerous cords). Here they break up into "nests" of cells, each consisting of a central germ cell—now termed an oocyte—and, surrounding it, a single layer of follicle cells. The entire cluster is a *primary ovarian follicle.*

Most of the oogonia do not reach this stage, and of those that do, only a very few ever complete the process and become ripe ova. In the entire life span of a human female, from the time of puberty (about 11 years of age) to the menopause 30 or 40 years later, normally only one ovum matures each month. This makes a total of 400 to 500 ova. Some mammals, on the other hand, release several ova at each ovulation.

The primary ovarian follicle increases greatly in size as a result of the multiplication of the follicle cells and the growth of the oocyte at its center. When its wall becomes several cells thick, fluid-filled vesicles appear in the midst of the follicle cells and begin to coalesce. Finally the mature follicle resembles a blister on the surface of the ovary. It is then known as a *Graafian follicle,* after the seventeenth-century physiologist who first described it. Such a follicle consists of an outer layer of follicle cells surrounding a central cavity filled with follicular fluid. On one side of the cavity, attached to its wall, is a mass of inner follicle cells (the cumulus) with a maturing ovum at its center.

Ovulation and Fertilization

At the proper time in the ovarian cycle, the mature follicle bursts. Follicular fluid flows out,

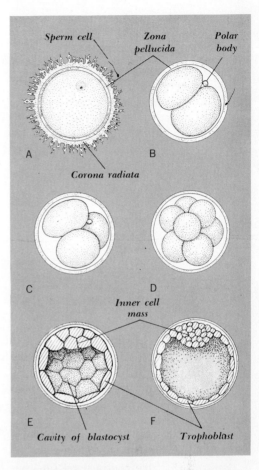

FIGURE 9–2 Early stages in the development of the mammal egg. *A.* Unfertilized egg immediately after ovulation. *B.* Two-cell stage. *C.* Three-cell stage, the result of one blastomere dividing before the other. *D.* "Morula" stage. *E* and *F.* Blastocysts in median section. *(Adapted mostly from L. B. Arey, 1954, "Developmental Anatomy," W. B. Saunders Company, Philadelphia, 6th ed., pp. 37, 68.)*

sweeping the ovum along with it. The follicle cells which accompany the ovum form what is known as the *corona radiata* (Fig. 9–2, *A*).

They soon disappear. Normally, the ovum is swallowed directly into the mouth of the oviduct (uterine or fallopian tube) and is carried by ciliary and muscular action toward the uterus. If sperm are present, they will have been propelled rapidly to the upper end of the oviduct by the contraction of the muscles of the uterus and tube. The capacity of the sperm to swim apparently plays but little part, for dead sperm have been found to arrive at their destination as rapidly as living sperm. Fertilization takes place as the egg enters the tube and begins its journey to the uterus.

Occasionally an egg may fail to enter the oviduct but instead remains in the body cavity. It may even be fertilized within the body cavity. In this case the placenta, or organ of exchange between the embryo and the mother, forms against the peritoneum of the body cavity.

The ability of eggs to find and enter the mouth of an oviduct is remarkable. They may even cross the body cavity from one side to the other. In an experiment on a sow, an ovary was removed on one side, and the oviduct was tied (ligated) on the other side. Not only did several eggs cross the body cavity and enter the oviduct of the contralateral side, but when they reached the sow's two-horned uterus, they spread apart and spaced themselves in both horns.

The Mammal Egg

The mammalian ovum which bursts from the ovary is typically a cell from 70 to 200μ in diameter. (One micron, or μ, is one-thousandth of a millimeter.) The mouse egg is 70μ; the human egg is twice that size, 140μ. It is a speck which is barely big enough to be seen with the unaided eye. Yet as cells go, it is large.

A red blood corpuscle, for example, is 8μ in diameter.

The egg is surrounded by a tough outer membrane, the *zona pellucida* (not to be confused with the area pellucida of the hen's egg). Probably the zona pellucida is a product of both the cytoplasm of the oocyte and the surrounding follicle cells.

A typical mammal egg possesses little, if any, yolk. Hence it must be classed as microlecithal, or isolecithal, since what little yolk it has is evenly distributed. Its nucleus is central. Its cleavage is complete (holoblastic). Yet, in its later development, a mammal egg resembles the development of the yolk-rich cleidoic eggs of reptiles and birds. This can be explained by the proposition that the mammal of today is descended from ancestors which had yolk-rich eggs.

The most primitive mammals which are still living, the monotremes, lay yolk-rich eggs which are surrounded by albumen and a shell secreted in the oviduct. Their cleavage is incomplete (meroblastic) as in the case of the eggs of birds and reptiles. The eggs of some marsupial mammals, the opossum, for example, also receive a secretion of albumen as they descend the oviduct. They even possess a delicate shell membrane. Albumen is also deposited around the eggs of some placental mammals, such as the rabbit. Even in man the oviduct secretes a nutrient fluid which nourishes the egg until it has opportunity to implant itself in the uterus.

Cleavage

Cleavage takes place as the egg moves slowly down the oviduct, driven along by cilia and muscular contraction of the wall of the duct. The journey takes about four days in the case of the human species.

The first cleavage is slightly unequal, and the plane which separates the first two blastomeres does not necessarily pass through the pole where the polar bodies were given off (Fig. 9-2, *B*). In the case of the pig, according to Streeter, the smaller of the first two blastomeres gives rise to cells which become the embryo proper. The larger blastomere becomes the extra-embryonic ectoderm (trophoblast).

The significance of these observations has not been experimentally confirmed. Whatever the normal fate of the first two blastomeres, there is good reason to believe that they have like capacities. Nicholas and Hall found that in the rat, if the first two blastomeres are separated, each can give rise to a whole young embryo. It is even possible to combine two fertilized eggs and obtain a single embryo. Others have noted that the early blastomeres are loose and roll around within the confines of the zona pellucida. Furthermore, a single mammal egg may give rise to two or more embryos. Facts such as these make it uncertain as to whether the first two blastomeres differ morphogenetically from each other. Certainly their fates have not been irreversibly determined.

As a result of early cleavages, the egg becomes a solid ball of cells known as a *morula* (Greek for "little mulberry") (Fig. 9-2, *D*). It is not a blastula, for it lacks a blastocoel. Rather it should be compared to the blastodisc of a bird's egg. It is as though the yolk of the bird's egg (during cleavage) had been removed and the cells compacted into a ball.

The Blastocyst

Soon the cells of the morula rearrange themselves and develop a cavity in their midst. The structure is known as a *blastocyst,* or blastodermic vesicle (Fig. 9-2, *E* and *F*). What is the

cavity? In the past it has generally been considered to be a blastocoel. But it is not comparable to the blastocoel of an amphibian blastula, for the latter separates prospective ectoderm of the animal hemisphere from prospective endoderm of the vegetal hemisphere. The cavity of the mammal blastocyst, on the other hand, separates cells which will become the embryo proper—the so-called *embryonic disc* or inner cell mass—from cells which will form no part of the embryo whatever. Instead, the latter cells become a thin outer layer, the *trophoblast,* which soon comes into contact with the wall of the maternal uterus.

The cavity of the mammalian blastocyst is comparable to the yolk region of a chick egg; or more precisely, it corresponds to the subgerminal cavity that separates the blastoderm from the mass of yolk beneath it (Fig. 9-3). The embryonic disc of the mammal egg represents the central region of the chick blastoderm (the area pellucida). The trophoblast corresponds to the ectoderm of the outlying regions of the blastoderm, namely, to the ectoderm of the area opaca.

The size of the embryonic disc varies from species to species. In some mammals, notably, in some marsupials, it is at first not recognizable at all. In others, for example, in man, a knob protrudes into the cavity of the blastocyst —hence the name, "inner cell mass." In this case, only the deeper cells of the inner cell mass are actually the embryonic disc (Fig. 9-4). The surface cells which cover the mass are extra-embryonic like the trophoblast.

The Development of the Fetal Membranes

Great variety exists among the mammals regarding the manner in which the blastocyst gives rise to the embryo and its fetal membranes. Typically, the embryonic disc (inner

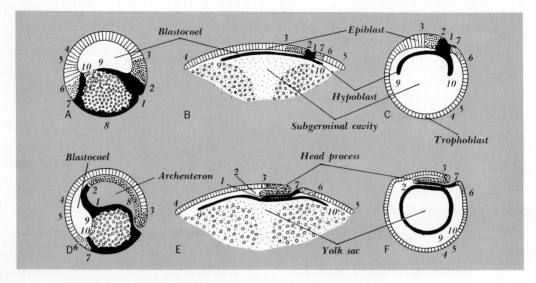

FIGURE 9–3 An interpretation of the homologies of blastulas and gastrulas. *A* and *D.*
Blastula and gastrula of an amphibian. *B* and *E.* The primitive streak and head-process
stages of a duck or reptile. *C* and *F.* Comparable stages in mammal development.

cell mass) splits (delaminates) into two layers in the same way in which the blastoderm of the chick splits. The underlayer is the *hypoblast.* The overlayer is the *epiblast.* The cells of the hypoblast spread laterally, as in the chick, and finally close to form a complete yolk sac.

In discussing the chick, we noted that two interpretations of delamination are possible (page 164): (1) According to the usual view, delamination is a mode of gastrulation. The hypoblast, which splits away from the epiblast, represents the roof and wall of the gut. (2) According to the other view, delamination is a method of blastulation (Fig. 9–3, *A* to *C*). The cleft which forms between the epiblast and hypoblast corresponds to the blastocoel of an amphibian egg. If this latter interpretation is correct, the hypoblast represents the floor, rather than the roof, of an amphibian archenteron.

Cell movements of the embryonic disc take place in the same way that they do in the chick. Cells of the epiblast stream toward the posterior quadrant of the disc. Here they heap up as the *primitive streak.* From the undersurface of the streak, mesoderm pushes out laterally between the epiblast and hypoblast. Soon—in some cases at its very beginning—the mesoderm splits into two layers. The outer layer is somatic mesoderm. The inner layer is splanchnic mesoderm. The extra-embryonic coelom lies between the two. The trophoblast and somatic mesoderm together give rise to the chorion and amnion. The splanchnic mesoderm and hypoblast produce the yolk sac and allantois.

The *chorion,* since it is the outer layer, is the place where exchange of substances occurs between embryonic tissue and the maternal environment. To facilitate the ex-

change, various adaptations have been acquired. In the pig, the chorion elongates tremendously (Fig. 9–5). Its great area affords
abundant contact with the uterus wall.

In most mammals, the chorion develops
more or less complex outgrowths known as
villi. These penetrate into the tissue of the
uterus wall and greatly increase the area and
intimacy of contact between fetal and maternal
tissues. Usually there is a specialized region of
the chorion where exchange between the
mother and embryo is most efficient. Such an
area is known as a *placenta*. (We discuss this
topic in more detail in Chapter 10.)

The *amnion* may be thought of as a membranous tent around the embryo. It contains a
fluid, the *amniotic fluid,* which bathes the outside of the embryo. Its function is to free the
embryo from pressures, permit freedom of
motion, and protect the outer surface of the
embryo from abrasion.

In more primitive blastocysts, such as those
of the pig, the amnion arises in the same way it
arises in birds; that is, it is formed by the rising
up of folds from the extra-embryonic somatopleure. These close in above the embryo (Fig.
9–5). In some of the more advanced mammals,
on the other hand, the amnion seems to form
in a different and more direct manner. The
amniotic cavity makes its appearance as a
cleft in the inner cell mass. This method of
forming the amnion is known as *cavitation*
(Fig. 9–4, *B* and *C*). This is a sort of shortcut
in development which accomplishes the same
result as folding. The cells which are above the
cavity correspond to the amniotic folds
(chorion and amnion). The cells which are below the cleft, that is, between the cleft and the
yolk sac, are the embryonic disc proper. They
alone give rise to the embryo. In the guinea pig,
the amnion first forms by cavitation; then it
opens out and exposes the embryonic disc;
finally, it closes in again by folding.

In all cases the amniotic cavity expands until
it completely surrounds the embryo, except, of
course, where the umbilical cord is attached.

The *yolk sac* also varies from species to
species. But why should a mammal have a yolk

FIGURE 9–4   The development of the fetal membranes in mammals. A. In the pig,
the amnion is formed by folding and the yolk sac by outgrowth, much as in the chick.
B. In most groups of mammals, the amniotic cavity and the yolk sac are formed by
cavitation within the inner cell mass. C. In man, even the extra-embryonic coelom is
formed by cavitation. In each figure, the embryo proper (at open neural plate stage)
lies just beneath the asterisk. (*Adapted from several sources.*)

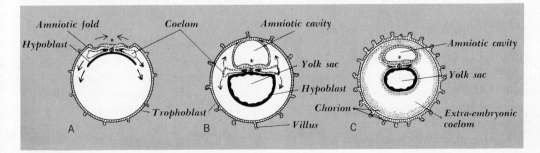

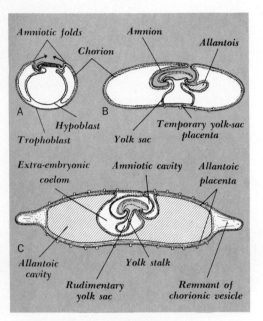

FIGURE 9–5   The development of fetal membranes in the pig. Highly diagrammatic. Ectoderm is shown by a laddered line, mesoderm
by stippling, and endoderm by a solid line.
The cavity of the allantois is lined diagonally.
A. Formation of amnion by folding. B. Early
stage with a temporary yolk-sac placenta. C.
Fully developed allantoic placenta.

sac at all? Little or no yolk is present in the egg
to be digested and absorbed. One answer, no
doubt, is an evolutionary one. The ancestors of
present-day mammals had yolk sacs full of
yolk, and the mammals today recapitulate the
developmental history of their forebears.

This, however, is an inadequate answer. The
yolk sacs of modern mammals do have a function. To a varying degree they are organs of
exchange of substances between the embryos
and the maternal fluids which surround them.
Even in the case of the highest mammals, the
yolk sac probably carries on some respiratory

exchange of gases before the allantois becomes functional. The yolk sac at first is close
to maternal tissue. It is separated from the
maternal blood by the thin ectoderm and mesoderm of the chorion. In most marsupials, a
yolk-sac placenta is formed. This is a specialized region where the vitelline blood vessels
of the yolk sac carry on exchange of substances with the mother. Even in some placental mammals, the pig for example, the yolk
sac, for a time, is large and probably functional
(Fig. 9–5, *B*). It enfolds the embryo and its amniotic covering. In most other placental mammals—and this includes man—the yolk sac is
small and is separated from the chorion by
the extra-embryonic coelom (Fig. 9–6).

Even when the yolk sac is not an organ of
exchange with the mother, it has important
functions. In mammals, as in birds and reptiles, it is the organ in which the first blood
cells are formed. For a time, it probably, also,
has certain synthetic and storage functions
similar to those which are later performed by
the embryonic liver.

Finally, the *allantois* varies greatly in different mammals. When describing the development of the allantois of the chick, we compared
it to a precocious urinary bladder which early
pushes out from the hindgut of the bird embryo and spreads around the embryo next to
the shell membranes. Its mesoderm fuses with
that of the chorion. Its rich plexus of umbilical
blood vessels acquires the role of transporting
respiratory gases.

In placental mammals and in some marsupial mammals, the allantois similarly fuses
with the chorion. Together they produce an
allantoic placenta for the effective exchange of
substance with the maternal environment. In
the pig, the allantois expands tremendously
(Fig. 9–5, *C*). Its cavity serves as a sac in which

wastes from the embryo accumulate. In man, on the contrary, the cavity of the allantois is rudimentary (Fig. 9–6). In fact, the human placenta is so efficient that the wastes of the embryo pass into the maternal bloodstream and are excreted by the mother's kidneys. Yet, even though the endodermal part of the human allantois is small, its splanchnic mesoderm is large and develops precociously. In fact, it is present and already fused with the mesoderm of the chorion when the cavities of the amnion and extra-embryonic coelom first appear.

In embryology there are many examples of advanced animals which have adopted short-cuts in their development. The development of the fetal membranes is an instance in point. Primitive vertebrates produce the membranes by the more laborious and circuitous routes of folding. Higher forms have adopted the more direct method of cavitation. Of course, we do not know what morphogenetic movements may have taken place before the different cavities appear. Certainly, if actual morphogenetic movements do not take place now in the development of the individual, they did occur in the past phylogenetic history of the race.

The Development of the Embryo Proper

The great variety which is shown in the development of the fetal membranes is not duplicated in the development of the embryo proper. The processes of neurulation and embryo formation of all mammal embryos are remarkably alike. In fact, embryo formation in the mammal is so similar to that in the chick, as described in Chapters 7 and 8, that it will not be necesary to repeat the story in detail.

The embryonic disc of the mammal is comparable to the area pellucida of the unincubated chick egg. Both give rise to the embryo proper. In both, the mass movement of the epiblast is toward the posterior quadrant. In both, also, this movement leads to the formation of the primitive streak. Again, in both, the cells of the primitive streak sink inward and spread as mesoderm between the epiblast and hypoblast (Fig. 7–6).

Of special importance is the *head process* which pushes forward from the anterior end of the primitive streak. In the mammal, as in the duck and many reptiles (but not in the chick), the head process develops a central canal (notochordal canal) which ends posteriorly at

FIGURE 9–6 A human embryo surrounded by its membranes (schematic). Note that the allantois (endodermal) is rudimentary although its mesoderm is greatly developed. *(Adapted from several sources.)*

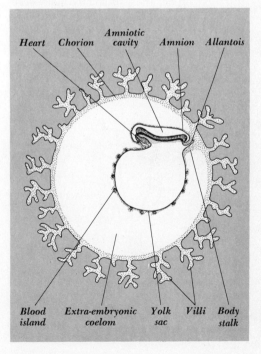

the primitive pit (Fig. 9–3, *E,* and *F*). It is adherent to the hypoblast beneath it. Shortly its floor disappears, and its sidewalls open out laterally. Thus it gives rise to the central region of the roof of the archenteron. Its longitudinal axis is the notochord.

The neural plate forms, and the neural folds rise up and become the neural tube in the mammal, just as they do in the chick (Fig. 9–7). The neural tube pushes forward in the head fold and grows backward in the tail bud, in like fashion. The endoderm develops a foregut, midgut, and hindgut as in the chick. The mesoderm differentiates in the same manner.

The most obvious differences between the mammal embryo and the chick embryo have to do with the relative proportions of some of the organs. The anterior part of the chick embryo is precocious as compared with its posterior portion. The eyes and the brain, especially the midbrain, are exaggerated. In mam-

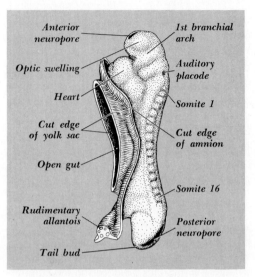

FIGURE 9–8 A 16-somite human embryo (2.5 mm, approximately 26 days) seen from the left side. The amnion and the yolk sac have been cut away. At this age the heart has begun to beat. Approximately × 24. *(Adapted from C. H. Heuser's 14-somite embryo, 1930, Carnegie Contrib. Embryol., 22:135–154, Plate 1; and W. J. Atwell's 19-somite embryo, 1930, ibid., 21:1–24, Plate 1.)*

FIGURE 9–7 A 7-somite human embryo (2.2 mm, approximately 22 days) seen in dorsal view. × 20 *(Redrawn from F. Payne, 1924, Carnegie Contrib. Embryol., 16:115–124, Plate 1.)*

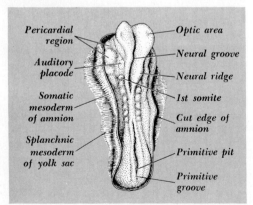

mals, the development of the heart and liver are surprisingly accelerated. The heart of a 2.5-mm human embryo is bigger than the brain and has already begun to beat (Fig. 9–8). In the chapters which follow, other contrasts between the chick and the mammal will be noted.

Twinning in the Mammal

Twinning in the mammalian embryo is interesting and important. Human mothers not infrequently give birth to more than one offspring at the same time. Often these are "fraternal twins" (heterozygous twins). They are actually not true twins at all but "litter mates." They are related to each other as brothers and sisters born on separate occasions. Each develops from a separate egg fertilized by a separate sperm. Although they may be situated close together in the uterus, their membranes are separate (Fig. 9–9, B).

True, or "identical" twins, also termed homozygous twins, are the products of a single egg fertilized by a single sperm. They have the same chorion and placenta, but their amnions and umbilical cords are usually separate (Fig. 9–9, C). Since they come from the same fertilized egg, they have an identical heredity.

How do twins of this sort arise? They are probably not the result of a separation of the first two blastomeres, or of the subdivision of the entire blastocyst into two parts. Instead, two embryos arise from the same blastocyst. Either (1) two distinct embryonic discs develop independently on the same blastocyst; or (2) one embryonic disc subdivides and becomes two; or (3) one embryonic disc gives rise to two embryos. In cases (1) and (2) the embryos will have separate amnions and umbilical cords. In case (3) it is likely that one amnion will enclose both embryos (Fig. 9–9, D). Separate identical twins are thought to be usually the result of (1) or (2). Conjoined twins, such as Siamese twins, and the various double-headed

FIGURE 9–9   Twinning in man. A. Human uterus with a single embryo. B. Fraternal "twins" (from two eggs) with separate chorions and placentas. C. Identical twins (from one egg) with one chorion and one placenta but with separate amnions and yolk sacs (the latter are not shown). This results when two inner cell masses arise from one blastocyst. D. Identical twins with one set of fetal membranes. This presumably is the result of two embryonic axes forming from one inner cell mass. (Based in part on diagrams in L. B. Arey, "Developmental Anatomy," W. B. Saunders Company, Philadelphia, 6th ed., pp. 188, 189.)

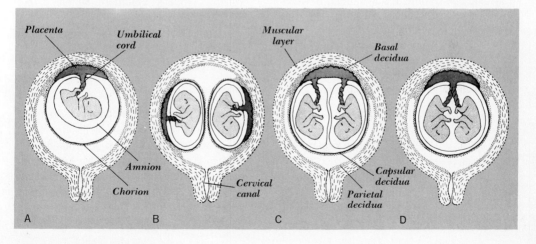

monsters, are probably the consequence of (3), that is, the incomplete subdivision of one embryonic disc.

The armadillo is peculiar among mammals in that it regularly gives birth to multiple embryos from one blastocyst. It is instructive to think of twinning in vertebrates as a form of asexual reproduction.

SELECTED READINGS

See the Appendix for general references on mammalian development.

Stages of Normal Development (Man)

Arey, L. B., 1965. Developmental Anatomy. W. B. Saunders Company, Philadelphia. 7th ed. Chap. 6, Embryonic and fetal stages. The table facing p. 106 is especially informative.

Hamilton, W. J., J. D. Boyd, and H. W. Mossman, 1962. Human Embryology. The Williams & Wilkins Company, Baltimore. 3d ed. Chap. 7, Growth of the embryo: development of external form; estimation of embryonic and foetal age.

Streeter, G. L., 1942, 1945, 1948, and 1951. Developmental horizons in human development. Carnegie Contrib. Embryol., **30**:211–245; **31**:26–63; **32**:133–203; and **34**:165–196.

Ovarian Follicle; Ovulation; Fertilization

Arey, L. B., 1965. Developmental Anatomy. W. B. Saunders Company, Philadelphia. 7th ed. Chap. 3, The reproductive organs and sex cells (especially pp. 31–40). Chap. 4, The discharge and union of the sex cells.

Nelsen, O. E., 1953. Comparative Embryology of Vertebrates. McGraw-Hill Book Company, New York. Chap. 2, The vertebrate ovary and its relation to reproduction. Chap. 5, Fertilization.

Pincus, G., 1936. The Eggs of Mammals. The Macmillan Company, New York.

———, 1951. Fertilization in mammals. Sci. Am., **184** (March):44–47.

Chang, M. C., 1955. The maturation of rabbit oocytes in culture, and their maturation, activation, fertilization, and subsequent development in the Fallopian tubes. J. Exptl. Zool., **128**:379–405.

Austin, C. B., 1961. The Mammalian Egg. Blackwell Scientific Publications, Ltd., Oxford.

Blandau, R. J., 1961. Biology of eggs and implantation. In Young and Corner (eds.), Sex and Internal Secretions. The Williams & Wilkins Company, Baltimore. 3d. ed., 2 vols. See especially pp. 797–837.

Edwards, R. G., 1966. Mammalian eggs in the laboratory. Sci. Am., **215**(August):72–81.

Cleavage and the Blastocyst

Arey, L. B., 1965. Developmental Anatomy. W. B. Saunders Company, Philadelphia. 7th ed. Chap. 5, Cleavage and gastrulation.

Hamilton, W. J., J. D. Boyd, and H. W. Mossman, 1962. Human Embryology. The Williams & Wilkins Company, Baltimore. 3d ed. Chap. 4, Fertilization, cleavage, and formation of the germ layers.

Balinsky, B. I., 1965. An Introduction to Embryology. W. B. Saunders Company, Philadelphia. 2d ed. Chap. 11, Embryonic adaptations (especially pp. 272–295, dealing with the early development of mammals).

Blandau, R. J., 1961. Biology of eggs and implantation. In Young and Corner (eds.), Sex and Internal Secretions. The Williams & Wilkins Company, Baltimore. 3d ed., 2 vols. See especially pp. 838–865.

The Fetal Membranes of Mammals

Arey, L. B., 1965. Developmental Anatomy. W. B. Saunders Company, Philadelphia. 7th ed. Chap. 7, The fetal membranes.

Hamilton, W. J., J. D. Boyd, and H. W. Mossman, 1962. Human Embryology. The Williams & Wilkins Company, Baltimore. 3d ed. Pp. 447–458.

Patten, B. M., 1964. Foundations of Embryology. McGraw-Hill Book Company, New York. 2d ed. Chap. 15, The extra-embryonic membranes of mammals, and the relations of the embryo to the uterus.

Twinning

Newman, H. H., 1917. The Biology of Twins. The University of Chicago Press, Chicago.

Newman, H. H., 1923. The Physiology of Twinning. The University of Chicago Press, Chicago.

Morgan, T. H., 1934. Embryology and Genetics. Columbia University Press, New York. Chap. 9. Twins and twinning.

See also brief accounts in Arey, pp. 191–198; Hamilton, Boyd, and Mossman, pp. 151–156; and Nelsen, pp. 380–386. (See full references above.)

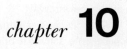

chapter **10**

THE MAMMAL EMBRYO AND ITS MOTHER

This chapter concerns mammalian parents and their role in the production of offspring. It deals mainly with the adjustments of the mother's body to the embryo and the embryo to the mother.

The Reproductive Hormones

The story begins when, for some reason, the anterior lobes of the pituitary body of the parents begin to secrete and release increased amounts of gonad-stimulating hormones into the bloodstreams. The time is at puberty, when the young animals commence to become sexually mature. In many mammals the process repeats itself annually, usually in the spring. In any case, it is the prelude to sex behavior. In the case of the female, if reproduction does not take place, the cycle usually begins again at regular intervals.

What causes the anterior lobe of the pituitary to become active? A full answer cannot as yet be given. In some mammals and in birds the increased daylight of spring stimulates the retinas and through them the hypothalamus. This is the part of the forebrain which is close to the pituitary body. Cells of the hypothalamus in turn synthesize a neurohumor which is absorbed into the bloodstream and carried directly to the anterior lobe. The journey is a short one. It involves a set of hypophyseal portal veins which are only about 12 mm long (in man) (Fig. 10–1).

When the neurohumor reaches the anterior lobe, it stimulates the lobe to secrete hormones of its own into the blood. There are several of these, each with its own "target organ." The growth hormone, for example, acts on supportive tissues throughout the body, activating them to more extensive growth. The thyrotropic hormone acts on the thyroid gland, caus-

ing it to synthesize and release an increased amount of thyroxin into circulation. The thyroxin then stimulates the metabolism of the body. The adrenocorticotropic hormone, ACTH, acts on the cortices of the adrenal bodies, with the result that they secrete hormones which resist stress.

Our present concern is with the pituitary hormones which affect the reproductive functions, namely, the gonadotropic hormones. If the pituitary body of an immature animal is removed, the gonads fail to mature. If the operation is performed after maturity has been attained, the gonads retrogress, and ripe germ cells and sex hormones are not produced. The entire body reverts to a less sexually differentiated state. If, on the contrary, adult pituitary tissue is implanted into a juvenile animal, or extracts of the anterior lobe are injected beneath the skin, or the pituitary develops a tumor, the gonads promptly enlarge and produce germ cells and sex hormones.

The hormones of the anterior lobe of the pituitary which influence reproduction are three in number. Their names and their effects in the female are as follows:

1 The *follicle-stimulating hormone, FSH,* brings about the growth of one or more ovarian follicles and the maturing within each follicle of a ripe ovum. It also stimulates the production by ovarian tissue (interstitial tissue) of the female sex hormone *estradiol.* (The general term for a female sex hormone is *estrogen.*)

2 The *luteinizing hormone, LH,* controls ovulation and the subsequent development of a corpus luteum (see later). Usually it acts in conjunction with FSH. It also stimulates the production and release by the corpus luteum of the hormone of pregnancy, *progesterone.*

3 *Prolactin* is required for the maintenance of the above secretions during pregnancy. It

also stimulates the growth of the mammary glands and activates parental behavior.

The same three hormones are active in the male, and with comparable effects. In other words, the gonadotropic hormones are not specific with regard to sex.

1 FSH increases the growth of the testes and stimulates the production of sperm cells within the seminiferous tubules.

2 LH causes the interstitial cells of the testes (the cells which lie between the semi-niferous tubules) to increase in number and to produce the male sex hormone, *testosterone*.

3 Prolactin affects the male, as it does the female, by calling forth parental behavior. A rooster under its influence will accept and nurture baby chicks.

Biochemists have determined that these several pituitary hormones are, in all cases, proteins or mucoproteins. The hormones of the gonads (and of the adrenal cortex), on the other hand, are steroids. Hormones of a sim-

FIGURE 10–1 Diagram of an adult pituitary body (median section) illustrating its nerve and blood supply. The adenohypophysis (derived from Rathke's pocket) is shown striated vertically in the case of pars tuberalis, obliquely in pars intermedia, horizontally in pars distalis or the anterior lobe. The neurohypophysis (parts derived from the embryonic infundibulum) is shown stippled. The pituitary is controlled from nuclei (nerve centers) in the hypothalamus. The control of the anterior lobe is indirect: Nerve tracts first carry neurohumors (or their precursors) to the median eminence and pars tuberalis; then hypophyseal portal vessels transport the neurohumors to the pars distalis (anterior lobe) which responds by secreting hormones. The posterior lobe stores and releases neurohumors (hormones) brought to it by nerve tracts from the hypothalamus. *(Based on G. W. Harris, 1955, "Neural Control of the Pituitary Gland," Edward Arnold (Publishers) Ltd., London, Figs. 2, 7, and 8.)*

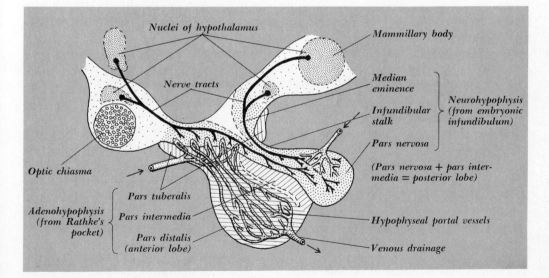

ilar kind are present in all vertebrates, although their effects vary somewhat.

We come now to an important principle concerned with the regulation of body processes— the principle of negative feedback. The production of a tropic hormone is antagonized by the hormone whose production it stimulates. Thus thyrotropin is repressed by thyroxin. ACTH is opposed by the adrenocortical hormones. In the same way, the gonadotropic hormones are reduced when the sex hormones of the gonads increase in the circulation. The maximum occurrence of FSH, LH, and prolactin in the blood is found in castrated animals, for in them there are no sex hormones to inhibit the production by the pituitary.

Negative feedback mechanisms result in rhythmic activity, especially if there is a lag between the stimulus and the feedback.

The Ovarian Cycle

The ovarian cycle is an example of a rhythm produced by negative feedback. We begin the story at the time when the reproductive organs of the female are at a low level of activity. The ovary, uterus, vagina, and mammary glands are in as nearly a nonreproductive state as they ever are. No follicle in the ovary has begun to grow. The endometrial lining of the uterus is thin, its blood vessels are reduced, and its secretory activity is at a low ebb. It is then that the sex hormones in the blood are minimal in amount. In primates, including the human female, this state is approximated following the menses. In these, therefore, the cycle is called the *menstrual cycle.* For mammals in general, it might be called the female cycle. (This is not the same as the estrus cycle. See page 230.)

The events of the cycle, as they take place in the ovary, are illustrated in Fig. 9–1, taken from Patten. The figure is schematic, for it shows the state of the ovary stage by stage in a clockwise fashion, beginning at the upper left. Actually, there is no such orderly arrangement of stages in an ovary.

The ovarian cycle divides naturally into two phases (Fig. 10–2): a preovulation phase and a postovulation phase. Each phase is further divisible into stages.

Phase I. The preovulation or follicular phase

STAGE A Cords of cells bud inward from the germinal epithelium of the ovary into the subjacent connective tissue stroma and there break into "egg nests." Each nest consists of a central oogonium surrounded by a single layer of follicle cells, the *stratum granulosum.* The entire structure is a *primary ovarian follicle.* This process takes place about the time of birth. It is as far as the story goes before puberty.

STAGE B As puberty approaches, and again at the beginning of each female reproductive cycle, the pituitary body, free of negative feedback, increases its production of gonadotropic hormones, notably, FSH. The FSH stimulates a few—only a few—oogonia to continue their growth as oocytes. Their nuclei continue the process of meiosis as described in Chapter 2. At the same time, the surrounding follicle cells multiply until the stratum granulosum has become several cells thick. The majority of the follicles do not complete the process. Instead, they atrophy. In the human female usually only one follicle matures each month.

STAGE C Under the continuing action of FSH, a cavity appears in the midst of the follicle cells, partially separating an outer layer of cells from an inner mass, the *cumulus,* which lies attached at one side. In the center of the mass is the oocyte. At the same time a sheath of connective tissue cells, the *theca,* forms around the growing follicle and serves to protect the follicle. Its cells also are the principal source of the female sex hormone, estradiol. The latter accumulates in the follicular fluid of the cavity and feeds back to the pituitary gland, decreasing its production of FSH. At the same time, it increases the discharge of LH by the pituitary.

FIGURE 10–2 Schema which illustrates the regulation of the ovarian and uterine cycles (human) by hormones of the pituitary body and ovary. 1. The cycle begins with the secretion of FSH (broken black arrow) by the pituitary. This stimulates the growth of an ovarian follicle. 2. The growing follicle secretes estradiol (solid black arrow), which, feeding back on the pituitary body, stimulates the production of LH (broken gray arrow). 2'. The estradiol also stimulates the growth of blood vessels and glands in the uterus. 3. Under the influence of FSH and LH the ovarian follicle matures and produces increasing amounts of estradiol. 4. About the middle of the cycle ovulation occurs, and the follicle cells which remain in the ovary begin to transform into a corpus luteum. The latter secretes the hormone progesterone (solid white arrow). This causes the pituitary to release a third hormone, prolactin (broken white arrow), which prepares the uterus for a possible implantation. 5 and 6. This continues for a time, but if implantation does not take place, the corpus luteum regresses and the uterine mucosa is cast off (menstruation), thus ending the cycle. The breadth of the arrows suggests the relative amounts of hormones released.

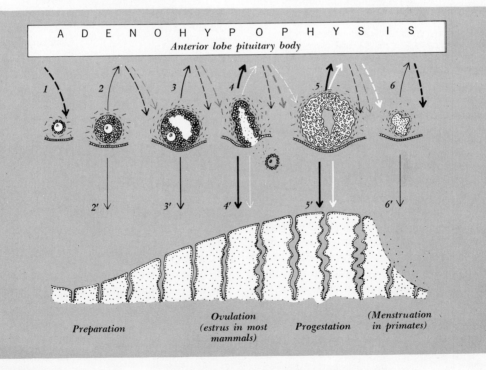

STAGE D Under the combined influences of FSH and LH, *ovulation* takes place. This is accomplished as follows: The follicle expands until it bulges at the surface of the ovary. Its outer wall becomes thin and finally bursts. The oocyte, surrounded by a *corona radiata* of follicle cells, is swept into the body cavity. In the human female, this ovulation takes place at about the 14th day of the 28-day (menstrual) cycle which starts with the beginning of the menses. (Further history of the ovum is given in Chapter 9.)

In most mammals, the ripening of an egg is accompanied by a period of heat or *estrus*. Hence the entire cycle from one estrus to the next is termed an *estrus cycle*. Note that this cycle is not the same as the primate menstrual cycle. In some mammals, notably, the cat and rabbit, ovulation takes place only during copulation. Apparently, in these animals an increase of nervous tension is necessary to cause the pituitary body to discharge sufficient LH to bring about ovulation.

Phase II. The postovulation or luteal phase

STAGE E The increased production of LH and associated FSH by the pituitary body induces the follicle cells which have remained behind in the ovary to multiply, enlarge, and become a temporary gland of internal secretion, known as a *corpus luteum* (a name derived from the Latin, *lutum,* a reddish-yellow dye.) The corpus luteum secretes the hormone *progesterone.*

STAGE F What follows next depends upon whether or not the egg becomes fertilized and the embryo becomes implanted in the wall of the uterus. If implantation *does* occur, the corpus luteum continues to grow and produce progesterone. As long as the latter hormone is produced, the embryo is retained, no new follicles mature, and ovulation does not again

take place. We shall return to this matter under the heading of pregnancy. But if, on the other hand, fertilization and implantation *do not* take place, the growth and activity of the corpus luteum soon reaches a peak, and the secretion of progesterone falls off. The corpus luteum then shrinks to a small mass of scar tissue known as a *corpus albicans.*

These events repeat themselves throughout the reproductive life of the female mammal. In some species they take place without intermission, in others only during the breeding season. The ovary of a young human female at first contains hundreds of thousands of primary follicles. But with each cycle, the number diminishes until at the end of the reproductive period of life very few follicles remain. The rest have atrophied.

Uterine Cycle

The reproductive hormones of the ovary control the uterine cycle and the cycles of the other reproductive organs. There are two phases corresponding to the preovulation and postovulation phases of the ovary (Fig. 10–2).

Phase I During the preovulation phase the uterus responds to the estradiol produced by the ovary by preparing for a possible pregnancy. This phase can be artificially induced by treatment with estrogenic hormones even when the ovaries have been removed or have ceased to function. The endometrium, i.e., the lining of the uterine cavity, thickens as a result of cell proliferation. The blood vessels of the uterine wall multiply, and the glands elongate and become columnar.

This *preparatory phase* continues as long as estradiol is supplied, whether naturally or artificially. If continued long, the uterine tissues hypertrophy (become overdeveloped). It ceases, and the uterus regresses within 6 to 10 days if the supply of estradiol or other estrogen is withdrawn.

Phase II The postovulation phase of the uterus takes place while progesterone is present in the bloodstream. The two hormones, progesterone and estradiol, acting together, bring about a state which may be thought of as the beginning of a pregnancy, even if, in the absence of an implanted embryo, the pregnancy does not continue. The endometrium grows thicker; the blood vessels enlarge and become distended with blood; and the uterine glands fold back on themselves, become contorted, and secrete a "uterine milk" which is able to nourish an early embryo for several days while it is free within the uterus. For this reason, the postovulation phase is sometimes called the *secretory phase* of the uterus. In the opossum, implantation of embryos does not occur; the embryos depend upon uterine secretions for their growth.

It is of interest that in some mammals the hormones progesterone and estradiol may so sensitize the endometrium of the uterus that it will respond to a small foreign object by engulfing it.

These changes in the uterus are progressive rather than sudden. In primates they reach a climax at about 10 days following ovulation. If implantation does not take place, the corpus luteum regresses, the level of progesterone in the blood falls, and within two or three days the uterus suddenly returns to its resting condition. This regression is accompanied by menstruation, namely, the sloughing away of the endometrial lining. There is some loss of blood and glandular secretions. The menses may be held in abeyance by the continued use of progesterone. Estradiol or other estrogen alone will not prevent it. These facts have been demonstrated experimentally in animals from which the ovary has been removed.

Pregnancy

Pregnancy begins when an egg is fertilized and the embryo implants itself in the wall of the uterus. It ends at childbirth. Its duration is spoken of as the "period of gestation." A uterus which is carrying young is termed a "gravid uterus."

In nature, the cycle which includes pregnancy is the usual one. The estral and menstrual cycles are adaptations to a failure of fertilization and implantation.

The timing of the events of the ovarian and uterine cycles is well coordinated (Fig. 10–3). It takes a few days following ovulation for the ovum to descend the oviduct and reach the uterus. During this time the ovum undergoes cleavage. It takes a few more days for the early embryo to reach the blastocyst stage and for its outer protective covering, the zona radiata, to dissolve away. By the time this is accomplished, the endometrium is ready to receive the embryo.

There are a few mammals, such as the armadillo and marten, in which implantation normally is delayed for weeks or even months. In this case, the blastocyst remains inactive in the uterus, waiting for the endometrium to become ready to receive it. One is reminded of the delay, often of several days, which takes place in the bird's egg between the time of laying and the beginning of incubation.

The implantation of the blastocyst in the wall of the uterus has a profound effect on the mother's body. The mechanism of this reaction seems to be that the trophoblast (outer ecto-

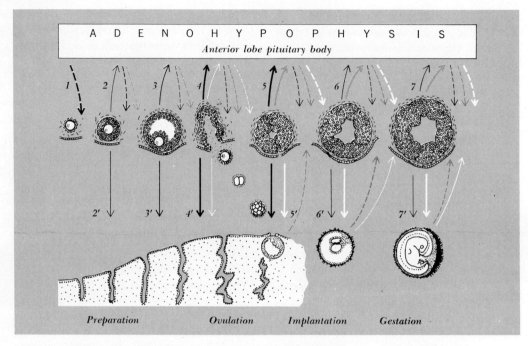

FIGURE 10–3 Diagram illustrating the events which take place in the human ovary and uterus during early pregnancy. 1 to 5. See legend under Fig. 10–2. 4'. If fertilized, the ovum undergoes cleavage as it descends the oviduct. By the 6th day after ovulation, it is a blastocyst ready to implant into the lining (endometrium) of the uterus. 5'. Implantation occurs, and the trophoblast of the embryo secretes a pituitary-like hormone (comparable to LH) (broken gray arrow). This augments the growth of the corpus luteum. 6'. As the placenta grows it produces, not only a pituitary-like hormone, but hormones comparable to estradiol and progesterone (solid white arrow). 7'. Toward the end of pregnancy—later than diagrammed here—the secretion of estradiol and progesterone by the ovary declines, but the production of like hormones by the placenta continues until the time of birth.

derm of the blastocyst) secretes a hormone which acts on the mother's corpus luteum or on her pituitary body. At any rate, the corpus luteum grows until it is several times the size of the nonpregnant corpus luteum. It secretes increasing amounts of progesterone. The production of estradiol also increases. If the corpus luteum is surgically removed before implantation occurs, implantation does not take place. If the operation is delayed until after implantation has already occurred, the blood vessels of the wall of the uterus regress, the uterine muscles become irritable, and the embryos are rejected.

As the embryo grows, the uterus also grows, as much as twentyfold. Its smooth-muscle fibers increase in number and in size. It is remarkable that this stretching of the muscles of the wall of the uterus does not cause the fibers to contract, for smooth muscle is notoriously subject to stimulation by stretching. The progesterone keeps them quiescent. If progesterone is withdrawn, as by removing the corpus luteum, the muscle fibers become irritable, and the embryo is aborted. This is especially the case during the early stages of pregnancy.

In the late stages of pregnancy in man and monkeys, the removal of the corpus luteum does not result in abortion. Why is this so? Lacking the progesterone of the corpus luteum, what keeps the uterine muscle quiet? The answer seems to be that the placenta itself becomes an endocrine gland which secretes progesterone, and not only progesterone, but an estrogen and a pituitary-like hormone as well. The latter presumably aids in maintaining the production of hormones by the mother's ovary.

The Placenta

A placenta is an organ of attachment between embryonic tissue and maternal tissue. Across it, nutrients and gases are exchanged between the mother and the embryo. In well-developed placentas, wastes are transferred also.

Placentas of a sort are present in viviparous fish and in certain reptiles. As a rule, in these cases the yolk sac comes into close relation with the maternal bloodstream, and the vitelline circulation carries the materials to and from the embryo. Yolk-sac placentas are developed in most marsupial mammals and are present also at an early stage in some placental mammals. In occasional reptiles, in a few marsupials, and—most importantly—in all placental mammals, the allantois provides the

definitive placenta. It is therefore the umbilical circulation of the allantois which transports substances between the mother and the fetus.

In most instances the area of contact between the mother's uterus and the chorion which encloses the embryo and its membranes is made extensive and intimate by the development of *villi*. These are rootlike processes which grow out from the chorion into the adjacent maternal tissue. Indeed, placentas are commonly classified according to the distribution of these villi (Fig. 10-4).

1 In pigs, horses, and various other mammals, the villi cover most of the surface of the chorion. These animals are said to possess *diffuse placentas*.

2 In cattle, sheep, and other ruminants, the villi are gathered into clusters or rosettes known as cotyledons. Such placentas are termed *cotyledonary placentas*.

3 In cats, dogs, and carnivores generally, the placentas have the form of girdles which surround the embryo and come into contact

FIGURE 10-4 Types of placentas. **A.** Diffuse placenta (pig). **B.** Cotyledonary placenta (cow). **C.** Zonary placenta (cat). **D.** Discoid placenta (human). At first the entire surface of the human blastocyst is covered with villi.

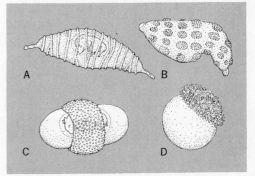

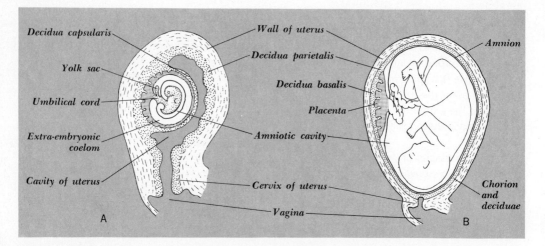

FIGURE 10–5   The gravid human uterus. A. At five weeks (8-mm embryo). B. Advanced fetus. *(Adapted from L. B. Arey, 1954, "Developmental Anatomy," W. B. Saunders Company, Philadelphia, 6th ed., Figs. 103 and 113.)*

with the inner wall of the uterus. These are *zonary placentas.*

4 In rodents and primates, including man, the villous area is disc-shaped; hence the term, *discoidal placentas.*

Placentas differ in the intimacy of the contact between the embryonic and maternal tissues. At one extreme, in the pig for instance, the blastocyst elongates and enlarges tremendously. Its entire outer surface comes into contact with the uterine wall and develops villi; but the villi are simple and minute. The relation between the chorionic sac and the endometrium (lining of the uterus) is one of apposition only. In this case, the nutrients and gases which reach the embryo must diffuse across six tissues: (1) the endothelium of the maternal blood vessels, (2) mesenchyme, (3) the epithelial lining of the uterus, (4) the ectoderm of the chorion, (5) fetal mesenchyme, and finally (6) the endothelium of the umbilical

capillaries. When at birth a *nondeciduate placenta* of this sort is cast off, there is no loss of maternal tissue or blood.

At the other extreme are the placentas of rodents and primates in which the blastocyst erodes deeply into the endometrium and becomes buried in the uterine wall. The chorion forms complex villi which invade the uterine tissue and rupture maternal blood vessels. When such a placenta is cast off at the time of birth, there is loss, not only of embryonic membranes, but of encapsuling maternal tissue as well. Such a placenta is called a *deciduate placenta.*

The maternal tissues which are expelled at birth in the case of deciduate placentas are called deciduae (Fig. 10–5, *A*). There are three regions: (1) The part which lies between the chorionic vesicle and the muscles of the uterus wall is the *decidua basalis.* It alone shares in the development of the placenta. (2) The part which surrounds the chorionic sac and separates it from the cavity of the uterus is the *decidua capsularis*. It is the least developed. (3) The part which forms the inner lining of the rest of the uterus is the *decidua parietalis*.

As the human fetus grows, the cavity of the uterus is squeezed out of existence. The parietal and capsular deciduae come together and fuse, and the latter becomes thin and practically ceases to exist (Fig. 10–5, *B*). The amnion becomes closely applied to the chorion. Thus, there is finally only one cavity in the gravid uterus, namely, the amniotic cavity which surrounds the fetus.

Figure 10–6 illustrates the development of the human placenta. Note that the outer ectodermal layer of the blastocyst, namely, the trophoblast, actively destroys maternal tissue as it eats its way into the wall of the uterus. Its outer cells lose their cellular character and become syncytial. Its inner cells, however, remain as a cellular epithelium. The syncytial layer gives rise to complex and branching villi. Within the villi irregular spaces (lacunae) develop and then unite together to form a labyrinth of channels. Through these channels, blood of the ruptured maternal blood vessels slowly flows.

FIGURE 10–6 Development of the human placenta (highly diagrammatic). *A.* The trophoblast (ectoderm of the chorion) has an inner cellular layer (cytotrophoblast) and an outer syncytial layer (syntrophoblast). The latter produces villi which actively erode the endometrial lining of the uterus. *B.* The villi acquire mesodermal cores which are supplied with capillaries carrying fetal blood. Lacunae in the syntrophoblast unite and become intervillous spaces filled with maternal blood. The spaces divide the chorion into a "chorionic plate" and an outer "basal plate" next to the uterine tissue. *C.* The villi become very complex (bushy). The trophoblast covering the villi finally degenerates until very little remains separating the maternal and fetal bloodstreams except the endothelium or the villous capillaries.

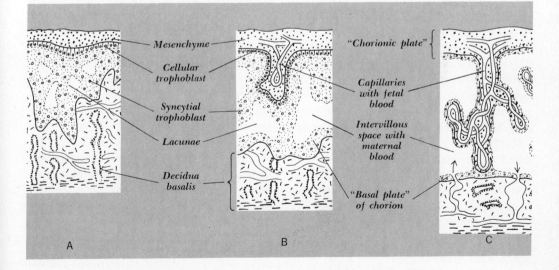

The lacunae split the human placenta into two plates: an outer or *basal plate,* and an inner or *chorionic plate.* Cores of epithelium and mesoderm of the chorionic plate invade the syncytial villi and bring umbilical capillaries into close relation with the maternal blood which flows among the villi between the plates. Late in pregnancy even the syncytial tissue largely disappears, so that little more than the endothelial walls of the umbilical vessels remain to separate the maternal and fetal bloodstreams. This, in brief, is the structure of the human placenta. Details must be sought in more extensive works. The physiology of the placenta is also a matter of great and present interest. It cannot be considered here.

Parturition

Parturition is the separation of the fetus and its membranes from the mother's body at the time of birth. Those aspects of birth which have to do with the newborn, such as the first breath and the reorganization of the circulation, will be described in connection with the respective organ systems (pp. 321 and 377). We are now concerned with the mother.

A hormone-like substance termed *relaxin* appears in the bloodstream as the period of gestation draws to a close. This substance seems to be a product of the uterus or of the placenta, and it serves to prepare the mother's body for the work of childbirth. Hisaw discovered it first in guinea pigs as being a substance which relaxes the pubic symphysis. It has since been reported in our domesticated animals and in man. In man it renders the cervix of the uterus more readily dilatable.

As gestation draws to a close, rhythmic contractions of the smooth muscle of the uterus begin. Soon they are accompanied by reflex and voluntary activity of skeletal muscle. The result is that internal pressure forces the fetus, normally head first, into the cervix of the uterus. The fetal membranes (combined deciduae, chorion, and amnion) then bulge through the cervix and burst. Amniotic fluid is discharged. The child then enters the vagina and is born. Some 15 minutes later, the uterus again goes into rhythmic contraction, and an "afterbirth" consisting of the fetal membranes and deciduae is expelled. Surprisingly enough, there is usually only a minor amount of bleeding. Within the next few days, the endometrium is restored and the uterus returns to nearly its original dimensions.

The mechanisms of birth are complicated and not well understood. Nervous reflexes of the lower cord are involved. Nevertheless, parturition will take place even when the spinal cord is transected. Higher centers are also involved. *Oxytocin,* a hormone of the posterior lobe of the pituitary gland, when given hypodermically, augments the contraction of uterine muscle at childbirth; yet delivery will take place even in animals from which the pituitary body has been removed. It is not entirely certain, therefore, that oxytocin is normally involved in the birth of the child.

What activates the nerve centers? What decrees that the time for birth has come? The answer is hormonal. Supposedly there is a decrease in the amount of progesterone in the bloodstream. We have seen that during late pregnancy progesterone is produced by the placenta. It maintains the uterus in a quiescent state. If the fetuses and placentas of a pregnant rat are removed and replaced by paraffin balls, the balls are expelled within two days. But if some of the placentas are left in place, parturition does not take place until the normal time. If the fetuses of a mouse or monkey are destroyed within the uterus without damaging the placentas, the placentas are retained and continue to produce placental hormones until full term. These experiments indicate that the placenta has a definite span of life.

Genetic studies on cattle indicate that the hormones of the fetus (including its placenta) rather than those of the mother set the time for delivery. A cow which regularly gives birth to normal calves at the proper time—280 to 290 days—will carry a certain type of abnormal fetus for months beyond the expected period of gestation. The abnormality in question results when the fetus inherits a double dose of a recessive gene. It shows itself in defects of the pituitary body and deficiencies in the pituitary-like hormones of the fetus. There is still much to be learned, however, about the factors which bring about childbirth.

Lactation

The mammary glands of juveniles and males are mere rudiments. During adolescence and under the influence of estrogens, the glands of young females grow by proliferation of the ducts. There is some waxing and waning with each recurring female cycle.

During pregnancy, the mammary glands increase greatly and become functional. The first half of each pregnancy is marked by the multiplication of the ducts and the development of acini, namely, the terminal secretory sacs. This takes place in response to various pituitary and ovarian hormones but especially the progesterone secreted by the corpus luteum. Yet the progesterone of the placenta is even more important, for if the placentas of a pregnant rat are removed, recession of the mammary glands soon follows. Recession does not occur if the corpora lutea or even the fetuses are taken away and the placentas are left in place.

During the second half of pregnancy, the acini become distended with secretion. The agent operating here is prolactin, produced by the pituitary. It is of interest that under the influence of prolactin the crop glands of a pigeon produce a "milk." Prolactin also stimulates parental behavior and nest building in birds and mammals.

By treatment with a proper sequence of hormones, male mammary glands can be made to secrete milk and the male animal caused to accept and suckle young.

What inhibits the flow of milk from the mammary glands before birth, and what brings forth the flow of milk at the time of birth? Probably gonadotropic hormones are responsible, especially prolactin. But the nervous reflexes associated with suckling are also important.

SELECTED READINGS

General References

Arey, L. B., 1965. Developmental Anatomy. W. B. Saunders Company, Philadelphia. 7th ed. Chaps. 8 and 9.

Hamilton, W. J., J. D. Boyd, and H. W. Mossman, 1962. Human Embryology. The Williams & Wilkins Company, Baltimore. 3d ed. Chaps. 3 and 5.

Parkes, A. S. (ed.), 1956, 1960. Marshall's Physiology of Reproduction. Longmans, Green & Co., Ltd., London. 3d ed., 2 vols. (Treats reproduction from the comparative viewpoint.)

Young, W. C., and G. W. Corner (eds.), 1961. Sex and Internal Secretions. The Williams & Wilkins Company, Baltimore. 3d ed., 2 vols.

Hormones and the Female Cycle

Arey, L. B., 1965. Developmental Anatomy. W. B. Saunders Company, Philadelphia. 7th ed. Chap. 9, Reproductive cycles and their hormonal control.

Hamilton, W. J., J. D. Boyd, and H. W. Mossman, 1962. Human Embryology. The Williams & Wilkins Company, Baltimore. 3d ed. Chap. 3, Cyclic changes in the female genital tract.

Everett, J. W., 1961. The mammalian female reproductive cycle and its controlling mechanisms. In Young and Corner (eds.), Sex and Internal Secretions. The Williams & Wilkins Company, Baltimore. 3d ed., 2 vols. Pp. 497–555.

Implantation and the Placenta

Arey, L. B., 1965. Developmental Anatomy. W. B. Saunders Company, Philadelphia. 7th ed. Chap. 8, Placentation.

Hamilton, W. J., J. D. Boyd, and H. W. Mossman, 1962. Human Embryology. The Williams & Wilkins Company, Baltimore. 3d ed. Chap. 5, The implantation of the blastocyst, and the development of the foetal membranes, placenta, and decidua. Also Chap.

16, Comparative vertebrate development (especially pp. 458–467).

Wislocki, G. B., 1929. On the placentation of primates, with a consideration of the phylogeny of the placenta. Carnegie Contrib. Embryol., **20:**51–80.

———— and Helen Padykule, 1961. Histochemistry and electron microscopy of the placenta. In Young and Corner (eds.), Sex and Internal Secretions. The Williams & Wilkins Company, Baltimore. 3d ed., 2 vols. Pp. 883–957.

Wilkins, P. G., 1965. Organogenesis of the human placenta. In DeHaan and Ursprung (eds.), Organogenesis. Holt, Rinehart and Winston, Inc., New York. Pp. 743–769.

Mossman, H. W., 1965. The principal interchange vessels of the chorioallantoic placenta of mammals. In DeHaan and Ursprung (eds.), Organogenesis. Holt, Rinehart and Winston, Inc., New York. Pp 771–786.

Gestation, Parturition, Lactation

Zarrow, M. X., 1961. Gestation. In Young and Corner (eds.), Sex and Internal Secretions. The Williams & Wilkins Company, Baltimore. 3d ed., 2 vols. Pp. 958–1031.

Cowie, A. T., and S. J. Folley, 1961. The mammary gland and lactation. In Young and Corner (eds.), Sex and Internal Secretions. The Williams & Wilkins Company, Baltimore. 3d ed., 2 vols. Pp. 590–642.

chapter **11**

THE
FORM
OF
THE
BODY

The early stages in the development of the form of the body are similar in all vertebrates. The neural plate folds to form the neural tube. The *head fold* rolls forward and the *tail bud* grows backward, apparently owing to the elongation of the neural tube.

In the chick, the lengthening of the embryo is accompanied by what appears to be a pinching of the embryo away from the extra-embryonic regions of the blastoderm. Actually, there is a zone around the embryo proper which does not enlarge as the embryo and extra-embryonic membranes grow. First the *subcephalic pocket* forms beneath the head fold at the anterior end, then *lateral body grooves* appear. Finally a *subcaudal pocket* develops beneath the tail bud. While this is going on, the chick embryo becomes C-shaped because the dorsal side elongates more rapidly than the ventral side.

The story of the mammal embryo is similar, except that in some mammals—and this in-cludes man—there exists for a brief time a reversed curvature in the midbody region (Fig. 9–8). Ultimately all mammal embryos are C-shaped.

The surface features of a young embryo reveal the locations of the developing organs within: the vesicles of the brain; the olfactory organs, eyes, and otocysts; the gills and somites; the heart, liver, and mesonephroi (Fig. 11–1). But there are other external features such as the face and limb buds which result from the proliferation of mesenchyme in the head and body wall. These are now our main concern.

THE HEAD AND NECK

The Development of the Face

The face has its origin in mesenchymal swellings along the borders of the nasal plac-

FIGURE 11–1   Six-mm pig embryo seen from the left side. × 10.

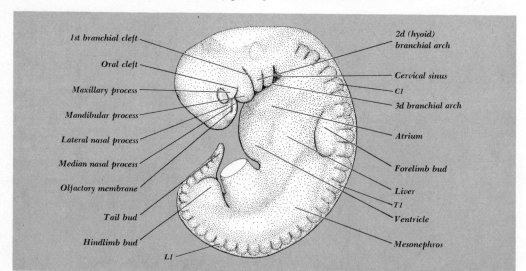

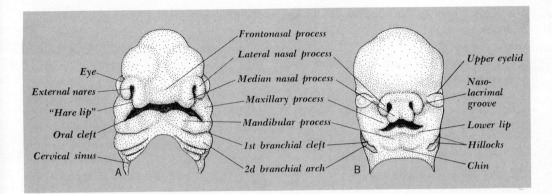

FIGURE 11–2 Development of the human face. A. 10-mm embryo (5½ weeks). B. 15-mm embryo (6½ weeks). Approximately × 7½. (Adapted from several sources.)

odes, oral plate, and first branchial cleft (Fig. 11–2). These are, respectively, the *median nasal* and *lateral nasal processes,* the *maxillary* and *mandibular processes,* and several small hillocks which become the external ear.

The median and lateral nasal processes taken together are horseshoe-shaped and surround the olfactory placodes except on their ventrolateral sides. As they grow forward they give rise to the nose, or snout. At the same time, they deepen the nasal cavities and recess the olfactory epithelia. The bridge of the nose is derived from the mesenchymal tissue which lies between the two median nasal processes.

The maxillary and mandibular processes are derived from the first branchial arches. They become separated from each other when the oral cleft invades the first branchial arches (Fig. 11–2). They surround the stomodaeum and grow forward as the upper jaws (in part) and the lower jaws, respectively.

In fishes except lungfish, the nasal placodes are anterior and separated from the stomodaeum, and the nasal pits to which they give rise are blind pockets with no other function than a sensory one. In them the upper jaws are formed by the maxillary processes alone. But in those vertebrates which breathe air—and this includes the lungfish—the nasal openings are farther back. They become pinched when the median nasal and maxillary processes join together to form the upper jaw (Fig. 11–3). As a result, each nasal pit now has two openings: an *external nares* open to the outside, and a primary *internal nares,* or *choana,* which opens well forward on the roof of the oral cavity. This is true, at least in principle, even though in mammals the internal openings are at first closed by membranes. Later they break through and form choanae. Primary internal nares are present in adult amphibians and reptiles (except alligators).

Occasionally a median nasal process and maxillary process fail to completely unite. The cleft which remains between them is known as a "harelip."

In mammals a groove is present between each maxillary process and lateral nasal process. It deepens and in time closes to become the *nasolacrimal duct* (Fig. 11–2), which drains tears from the inner (anterior) corner of the eye cavity to the lateral wall of the nasal cavity.

The mandibular processes grow forward and become the two halves of the lower jaw. Their union at the forward end forms the chin. A "cleft chin" results when the union is incomplete.

The first branchial furrow, namely, the furrow which lies between the first and second branchial arches, gives rise to a pit which becomes the *external auditory meatus* (Fig. 13–8). The closing plate between the furrow and the first pharyngeal pouch (endodermal) becomes the eardrum. Actually, in mammals, the original furrow fills in and is later replaced by the meatus. In mammals, mesenchymal hillocks form along the borders of the first branchial furrow (Fig. 11–2). They expand outwardly in such a manner that they give rise to the pinna (auricle) or lobe of the visible ear.

The Nasal and Oral Cavities

The several processes which form the face grow forward, and the stomodaeum deepens. Then, in birds and mammals, shelves—we shall call them the *palatine shelves*—develop on the median side of each maxillary and median nasal process (Fig. 11–3). At first, the shelves of the maxillary processes are directed downward like the fallen leaves of a folding table. At this time they are separated from each other by the tongue which has pushed upward from below (Fig. 14–2). Later, the tongue is withdrawn (or squeezed out), and the shelves mold themselves upward until each unites with its fellow of the opposite side. The secondary palate which is thus formed divides the stomodaeum into a nasal region above and an oral region below. As a result of the formation of the palate, the primary internal nares are moved posteriorly and become secondary internal nares. Thus a newborn child can suck,

FIGURE 11–3 Development of the internal nares and palate of the pig. Ventral views after the lower jaw has been cut away. *A.* 15 mm. *B.* 20 mm.

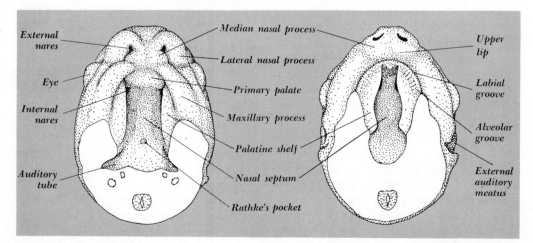

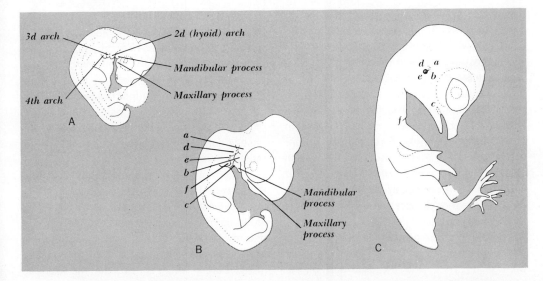

FIGURE 11-4   The fate of the branchial arches in the chick. A. 4-day embryo (stage 22). B. 6-day embryo (stage 27). C. 10–12-day embryo (stage 36). The letters *a* to *f* refer to swellings ("hillocks") bordering the first branchial cleft. *(Adapted from V. Hamburger and H. L. Hamilton, 1951, J. Morphol., 88:49–92, Plates 7, 8, and 12.)*

or an adult can feed with the mouth, while at the same time breathing through the nose.

While the palatine shelves are closing together to form the palate, a median partition, the *nasal septum,* grows downward from the roof of the stomodaeum. It divides the nasal region into a right and left nasal cavity (Fig. 14-2). Later, the lateral walls of the nasal cavities become complicated by the formation of scroll-like folds, the *nasal conchae.* Recesses from the nasal cavities grow into the lateral head mesenchyme and form the *nasal sinuses.* This process, however, is not completed until after birth. When, as sometimes happens, the palatine shelves fail to come together in the midline, the result is an embryonic defect known as a "cleft palate." In this case the tongue remains in contact with the underside of the nasal septum, as in the young embryo.

The Branchial Arches and Neck

The lower vertebrates which respire in water have no neck. Instead, their heads are joined broadly to their trunks. The region in them which corresponds to a neck is occupied by branchial arches and clefts. Their hearts, moreover, are located well forward beneath the pharynx, as in the embryos of the higher vertebrates.

The first branchial arches, commonly called the *mandibular* arches, give rise to the skeleton and muscles of the jaws. The second, or hyoid, arches are important as the source of the hyoid skeleton and musculature (Fig. 16–5). In bony fish, the second arches form an operculum which grows over the more posterior arches and protects the gills. In a somewhat similar fashion, the second arches of higher vertebrates enlarge and encroach upon the posterior arches. The higher vertebrates breathe air and hence have no need of gills. In mammals the posterior arches regress; but for a time they leave a depressed region on each side of the future neck known as a *cervical sinus* (Fig. 11–1). Normally the cervical sinuses disappear, although occasionally one remains as an embryological defect.

As the bird or mammal embryo grows older, the body wall posterior to the second branchial arches constricts, and the rear part of the pharynx, together with the structures associated with it, become stretched out. The heart recedes into the trunk (thorax). In this manner a neck is formed which permits freedom of motion to the head without an accompanying movement of the trunk. The formation of the neck is especially well shown in the bird embryo (Fig. 11–4).

THE TRUNK

The Body Wall

The dorsal side of the embryo, from the head to the tail bud, is marked by the ridge made by the spinal cord. On each side of the ridge is a series of mounds made by the somites which flank the cord. The ventral side of the embryo is at first incomplete, and the body cavity is widely open to the extra-embryonic coelom. Lateral body folds mark the line where embryo and amnion join. In the chick, the heart protrudes for a time between the body folds into the extra-embryonic coelom

(Fig. 8–18). Gradually the body folds push ventrally. First they envelop the heart so that the pericardial cavity becomes separate from the extra-embryonic coelom. Then they fuse with the liver to form the septum transversum. Finally they close in around the *umbilical stalk,* or *cord.* Even so, for a considerable time the ventral body wall continues to be weak, consisting only of an epidermis and a thin layer of somatic mesoderm.

The strengthening of the lateral and ventral body walls is, in part, the result of a downward extension of the somites. The dermatome and sclerotome supply connective tissues, and the myotome contributes muscle. But there is also a recruitment of cells from the somatic mesoderm. In fact, it appears that the strengthening of the ventral body wall is the result of a descending wave of proliferation and differentiation, rather than a migration of tissue.

The Limb Buds

Before the limb buds appear, a thickened strand of somatic mesoderm is to be seen on each side of the embryo, ventral to the somites. It extends all the way from the posterior margin of the head to the tail bud (Fig. 8–26, *B*) and is known as "Wolff's crest." Its anterior and posterior regions become limb buds. The intervening regions regress until they are indistinguishable from the mesoderm of the rest of the body wall. At this early stage of development there is nothing to distinguish the ectoderm of the future limb from the general epidermis of the flank.

The limb buds are at first moundlike swellings of mesenchyme, somewhat elongated in an anteroposterior direction (Fig. 11–1). Soon, in birds and mammals, an *apical ectodermal ridge* appears which extends forward and backward along the border of each bud (Figs. 11-8, *A,* and 11-14). In amphibians there are no ridges, but instead each limb bud is covered by a thickened ectodermal cap.

The limb buds of different vertebrates vary in their location. The forelimbs of salamanders are ventral to the 3d to 5th somites. In mammals, they lie below the 8th to 13th somites. In birds they are still farther back, namely beneath the 16th to 20th somites. This difference in location is related, at least in part, to the varying number of somites which contribute to the posterior part of the skull and neck.

The position of the hindlimb buds also varies, being dependent on the number of trunk somites. The hindlimbs are approximately ventral to somites 12 to 14 in salamanders, 28 to 34 in pigs, and 26 to 32 in chicks.

The limb buds of birds and mammals grow rather directly outward from the sides of the trunk and then turn downward and posteriorly. Soon they become paddle-shaped as a result of the flattening of their distal portions. The original ventral surfaces of the paddles face inward toward the sides of the body and become the palms or soles. Flexion (bending) soon takes place at what will be the elbow or knee joint. Then a little later it occurs also at the wrist or ankle joint. At about the same time, digits make their appearance around the borders of the paddles. They appear first as thickenings along the apical ectodermal ridge.

Let us call the position of the limbs which we have just described, namely, the position in which the limbs project laterally and bend ventrally, the "primitive position" (Fig. 11-5, *A*). This is approximately the position of the limbs of a salamander or turtle when it is sprawled upon the ground. In this position

homologies between the limbs are easy to point out: The dorsal surface of each limb is the extensor surface, and the ventral surface is the flexor surface. The preaxial (anterior) borders are the radial and tibial borders, as the case may be, with the thumbs and big toes at the distal ends. The postaxial (posterior) borders are the ulnar and fibular borders. In this primitive position the elbow and knee joints point laterally, and flexion is toward the midplane of the body.

During the evolution of four-footed runners —this is notably true of mammals—the limbs revolved beneath the body so as to raise the body high above the ground and provide for speed and a longer stride. Torsion is already apparent in a 25-mm pig embryo (compare Fig. 11-5, *B*). The forelimbs become appendages which pull. In adaptation to this function, the forelimbs revolve backward so that the elbows are directed posteriorly. This would bring the back of the hand toward the rear (an impossible position for running) were it not that an accompanying rotation (pronation) of the forearm also takes place. The preaxial (radial) border crosses over from the lateral side to the medial side, anterior to the postaxial (ulnar) border. As a result, the palms of the hands face backward. The hindlimbs, on the contrary, revolve forward so that the knees point forward. The soles of the foot then face backward, and no further rotation is necessary. The hindlimbs become appendages which push.

If the student were to place his own limbs in the "primitive position" and then rotate them in the typical mammalian manner, he would discover that what is usually called "flexion" of the ankle joint is actually hyperextension. If now he were to go farther and assume the standing position, the so-called

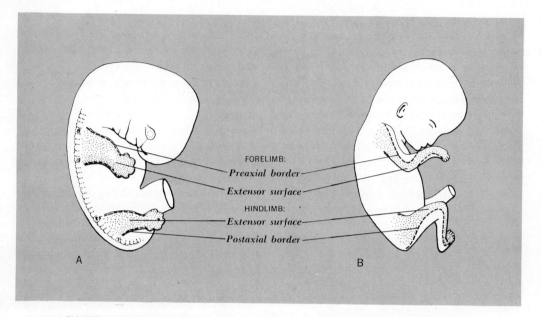

FORELIMB:
Preaxial border
Extensor surface
HINDLIMB:
Extensor surface
Postaxial border

A   B

FIGURE 11–5 Diagram illustrating the rotation of the limbs. *A.* 12-mm human embryo showing the limbs in their "primitive position." The dorsal or extensor surface is stippled. *B.* 60-mm human embryo after rotation of the limbs has taken place.

anatomical position of human anatomy, he would find that the extensor side of the arm is dorsal, while that of the thigh is ventral; the elbow joint points backward, and the knee joint, its homologue, points forward; the radius (at the elbow) is lateral to the ulna, while the tibia, its homologue, is medial to the fibula. Clearly, homologies can not be drawn when the limbs are in the anatomical position.

Experiments on Limb Buds

The determination of a vertebrate limb is a process of three phases:

1 *Initial determination,* by which a region of the embryo is set apart as a self-differentiating limb bud. Probably this includes also the *specification* as to whether the bud will become a forelimb or hindlimb.

2 *Axiation,* by which the axes of the future limbs are decided.

3 *Regionalization,* in which the regions of the limb bud are determined.

1 Initial determination or specification The prospective limb buds, that is, the cells which are destined to form a limb given normal development, have been identified as early as the beginning gastrula stages of amphibian development; but at this early stage they are not yet capable of self-differentiation into a limb. If explanted or transplanted to new surroundings, they develop according to their new surroundings. Hence they are undetermined.

Chaube has identified the prospective limb material of the chick before the limb buds appear. Her method was to prick exposed blastoderms with fine glass needles which were dipped into powdered colored chalk. She then followed the movements of the spots which she had produced. She found that at the 2-somite stage the forelimb material is located in the prospective lateral mesoderm immediately posterior to the primitive node (Fig. 11–6). The material of the hindlimb follows close behind it. By the 13-somite stage, the node has regressed, and the limb-bud material has stretched out in a longitudinal direction until all the limb-forming cells are anterior to the node. It is likely that the limb material undergoes its determination about the time that it moves forward past the node.

What accounts for the initial determination of a limb? Very little of the vast amount of experimental work which has been done on the development of limbs bears on this question. We may, however, make a surmise. Wolff's crest, which was referred to previously, seems to arise as a response of somatic mesoderm to some sort of influence which emanates downward from the ventral borders of adjacent somites. Now, any part of the mesoderm of Wolff's crest is potentially a limb bud. In Chapter 6, we noted that Balinsky found that a part of the flank of an amphibian embryo, if sufficiently stimulated, may form a limb, even though it be a supernumerary limb. Normally, however, only the anterior and posterior regions of Wolff's crest produce limb buds: the anterior bud becomes a forelimb; the posterior, a hindlimb. The intermediate parts regress. Why is this so? We suspect that the answer lies in differences in the inductive action of the ventral borders of the somites.

The specification must take place early, for

Saunders and associates found, in experiments on the 72-hour chick embryo, that if the mesoderm of a wing bud is replaced by the meso-

FIGURE 11–6 The limb-forming regions of the chick embryo according to the work of Chaube. An X marks the primitive node. A. 22 hours (stage 7). B. 33 hours (stage 11). C. 40 hours (stage 13). D. 72 hours (stage 18). (Based on S. Chaube, 1959, J. Exptl. Zool., 140:29–78, Fig. 3.)

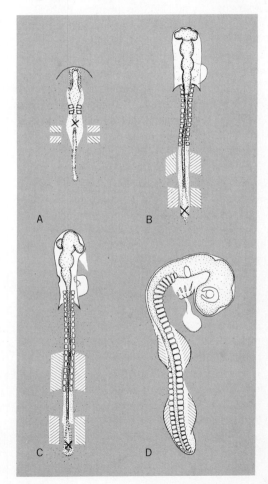

derm of a leg bud, the limb which develops has a skeleton, scales and digits of the hind-limb type. Zwilling found that if the mesoderm comes from the leg bud of a duck and is grafted to a chick, the foot which develops tends to have the webs of a duck's foot.

2 Axiation The classic experiments on the determination of limb buds are those which Harrison and his students performed on amphibian embryos (page 148). You will recall that they found, by transplanting limb discs at the tail-bud stage, that the initial determination had already been accomplished. The mesoderm of the disc of the forelimb already has the capacity to self-differentiate into a fore-limb. But it is still a "harmonious equipotential system" in that any part of the disc has the capacity to form a whole appendage. These investigators found, furthermore, that the fixation of the axis of the future limb takes place by steps.

Similar observations have been made by Chaube on the chick. When a prospective wing bud from the same side (homopleural) is grafted to the flank of a chick embryo close behind the normal bud, it develops into a wing. If it is placed in normal orientation, the second wing is *normal* in posture (Fig. 11–7, *A*). It "points" posteriorly and ventrally (pv), and produces a wing with the preaxial (radial) border forward. What happens when the graft is rotated 180° and reimplanted? The answer depends on its state of development. If the bud is from a 13-somite embryo or younger, it gives rise to a *reversed* wing which points forward and has its preaxial border posteriorly (Fig. 11–7, *B*). This indicates that the anteroposterior axis (AP axis) of the bud is already determined. But typically the wing from the rotated bud

points downward (av) as is normal, showing that the dorsoventral axis (DV axis) at this stage is yet undetermined and still subject to the controlling influences of adjacent tissues. After the 13-somite stage, however, both axes have been decided, and the same operation results in a *revolved* wing which points both anteriorly and dorsally (ad) (Fig. 11–7, *D*).

If the bud is from the opposite side (hetero-pleural) and is grafted with its AP axis reversed (DV axis normal), the resulting wing is *reversed* (Fig. 11–7, *B*). But if it is grafted so that its AP axis is not reversed but instead its DV axis is inverted, the result depends on the age of the bud. Buds of embryos with less than 13 somites produce wings which point backward and downward in the normal manner (pv). Those which have been taken from a 13-somite embryo or older produce wings which are *inverted*, that is, which point backward and upward (pd) (Fig. 11–7, *C*). These experiments indicate that from the stage of 13 somites onward the DV axis as well as the AP axis is decided.

3 Regionalization Even after the axes of the limb buds have been determined, the differentiation of the regions of the buds continues. At the present time extensive experiments are being conducted on the limb buds of 3- and 4-day chick embryos to discover how this regionalization is accomplished. At three days of incubation, a chick limb bud has already passed beyond the stages of specification and axiation, but the assigment of regions is incomplete. The material for the proximal regions of the limb has been laid down; but the mesoderm for the distal parts, namely, the hand and the digits, has not yet been proliferated (Fig. 11–8, *A* and *B*).

Saunders, by operations on 3-day chicks, showed that the apical ectodermal ridge plays a far more vital role in the determination of a limb than had previously been suspected. He found that if the ectodermal ridge is removed at this stage, only the proximal parts of the limb are formed, namely, the shoulder girdle and a stump of the humerus (Fig. 11–8, *C*).

If the operation is performed a day later, the forearm also develops, but carpals and digits are still lacking (Fig. 11–8, *D*). If the operation is delayed beyond four days of incubation, a whole limb including digits may form. Saunders found further that when a part of the apical ectodermal ridge is removed, a reduction in the number of digits takes place. If

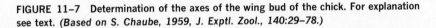

FIGURE 11–7 Determination of the axes of the wing bud of the chick. For explanation see text. *(Based on S. Chaube, 1959, J. Exptl. Zool., 140:29–78.)*

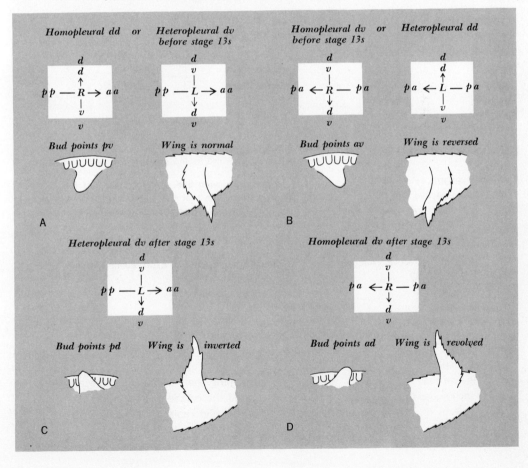

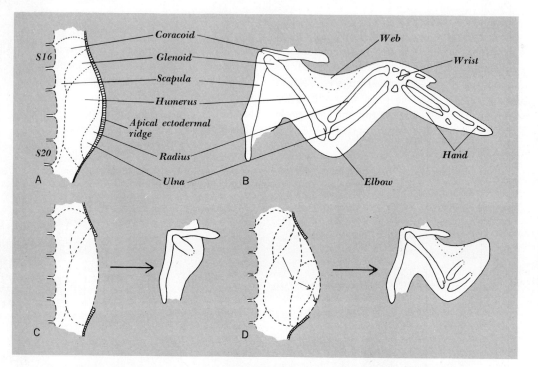

FIGURE 11–8 Saunder's experiment on the chick wing bud. *A.* "Fate map" of the prospective regions of a 72-hour chick wing bud. Note that the material which is to form the wrist and hand has not yet been proliferated. *B.* Wing of an older chick embryo showing skeletal features. *C.* If the apical ectodermal ridge of a 72-hour chick wing bud is removed, the distal parts of the limb fail to develop. *D.* If the apical ectodermal ridge is removed at the 96th hour, more of the limb develops, but the wrist and hand are still defective. *(Adapted from J. W. Saunders, Jr., 1948, J. Exptl. Zool., 108:363–403, Fig. 6.)*

two apical ridges are grafted onto the same limb bud, one along its preaxial border and the other along its postaxial border, supernumerary digits result.

Zwilling carried the analysis further. By suitable techniques he separated the ectoderm and mesoderm of chick limb buds and then recombined them in various ways. The ectoderm could be peeled off after a brief (5-minute) interval in a trypsin solution. The mesoderm was freed of ectoderm in a solution containing the chelating agent, versene, which removes calcium ions from the solution by trapping them within its molecule. When he lifted the ectoderm of a limb bud, revolved it 180°, and replaced it, he found that the AP polarity of the limb remained as it was. It was rooted in the mesoderm. When, however, he revolved the mesoderm and replaced it, the polarity of the resulting limb was reversed. Yet the completeness of the limb depended on the adequacy of the apical ectodermal ridge. In other experiments, he showed that a part of the mesoderm of the limb bud, when overlaid by a whole ectodermal ridge, becomes a whole limb, but that a whole mesoderm overlaid by a part of the ectodermal ridge becomes a defective limb. The wholeness seems to depend on the ectoderm.

From these experiments we conclude: (1) The initial determination of the future limb is in the mesoderm. This includes specification and, later, axiation. (2) The mesoderm induces the ectoderm which overlies it to form an apical ridge. (3) The apical ectodermal ridge in turn controls, or at least conditions, the development of the distal regions of the limb.

For a time the ectodermal ridge continues to depend on the underlying mesoderm for its existence. If older mesoderm, or mesoderm taken from elsewhere in the embryo, is substituted for specific limb mesoderm, the ridge flattens down and becomes indistinguishable from other epidermis. In Zwilling's words, the mesoderm of the limb bud supplies an "apical ectoderm maintenance factor."

So far the evidence seems clear, but there have been conflicting findings. Bell and his collaborators have found that, under special circumstances, a limb will develop even though the apical ridge has been removed. A full explanation of the discrepancy has not yet been found; but the results remind us that it is primarily the mesoderm which differentiates into successively more distal levels of the limb. The ectodermal ridge, which is itself a product of the inductive capacity of the mesoderm, acts solely as a stimulating or releasing mechanism.

The Mammary Ridge

In mammal embryos, a shallow ridge of epidermis is seen on each side of the trunk soon after the limb buds appear. This is the *mammary ridge* or "milk line." Soon it gives rise to a series of *mammae* (nipples) (Fig. 11–9). In human and elephant embryos, only a single pair in the region of the forelimbs normally develops; but in hogs, dogs, and many other animals several pairs form along the abdomen. In horses and cows, two or three pairs mature in the inguinal region.

In each case, epidermal cells invade the underlying mesoderm in the manner of glands. There they branch and ramify in the loose connective tissue. This occurs in both sexes but is most extensive in the female at the time of puberty and again when pregnancy takes place. The enlargement of the mammary glands is controlled by hormones (page 237).

FIGURE 11–9 Twenty-mm pig embryo seen from the left side.

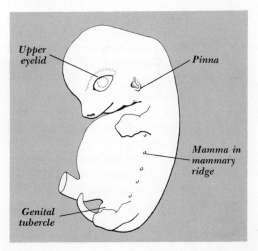

THE TAIL

The Tail Bud

The phylum Chordata is characterized by gill clefts, a notochord, a dorsal tubular nerve cord, and a postanal body or *tail*. The tail serves water vertebrates as a powerful propelling organ, and with its help they are able to drive through their medium with a speed not attained by any of the invertebrates. In land vertebrates, the tail has tended to lose importance although it serves various functions.

A tail, unlike other organs of the embryo, is a structure of indeterminate growth like the shoot of a plant. Its tip is a region of proliferating embryonic cells, a blastema comparable to the blastema of a regenerating appendage. It buds off somites much as the apical meristem of a plant shoot buds off leaves and lateral branches. The youngest segments are always those nearest the apex.

The tail bud continues the same processes of proliferation and differentiation which were begun by the lips of the blastopore. In amphibians, as long as the blastopore remains open, the cells of the blastoporal lips involute and give rise to the notochord, axial mesoderm, and lateral mesoderm. After the blastopore has closed to a slit and involution on the ventral side has largely ceased, further lateral mesoderm is not produced; but the production of notochord and axial mesoderm continues. The dorsal and dorsolateral lips transform into the rudiment of the tail bud and continue to proliferate axial structures (Fig. 6-23). Correspondingly, in the chick, the primitive node becomes the tail bud (Fig. 8-12).

The dorsal and ventral lines of the tail bud are sutures (seams) where the neural folds came together. In lower vertebrates, the neural-crest cells of the tail bud produce skeletal structures and induce the development of the dorsal and ventral tail fins.

ADDENDUM: THE ANATOMY OF A 6-MM PIG EMBRYO

Frequent references will be made in the chapters which follow to the anatomy of a 6-mm pig embryo. It is therefore appropriate at this point to introduce a series of drawings of sections which will aid the student in his laboratory work.

In most respects a 6-mm pig is farther along in its development than a 72-hour chick. Only the eyes and perhaps the forebrain appear to be less developed. The heart of the pig has begun to divide into right and left chambers. The liver is large and functional. The mesonephroi are well developed. Mesenchymal cells have multiplied and fill the spaces between the germ layers. In fact, most of the mass of the embryo is now mesoderm.

The following figures are not intended to take the place of a thorough laboratory study of a complete series of cross sections. Nothing in embryology can substitute for patiently going over serial sections while trying to visualize the embryo as a three-dimensional object.

Figure 11-10 is an index figure which has been drawn to indicate the location of the sections shown in Figs. 11-11 to 11-15. For details and labeling the reader is referred to other figures in this book:

Figure 11-1 for external features of a 6-mm pig

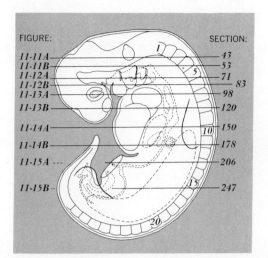

FIGURE:		SECTION:
11-11A		43
11-11B		53
11-12A		71
11-12B		83
11-13A		98
11-13B		120
11-14A		150
11-14B		178
11-15A ---		206
11-15B		247

FIGURE 11–10 Six-mm pig embryo from the left side. Index diagram for Figs. 11–11 to 11–15. The series from which this is reconstructed numbers 322 sections, approximately 19 μ per section. × 9.3.

FIGURE 11–11 Sections of a 6-mm pig embryo. (For location of sections see Fig. 11–10.) × about 20.

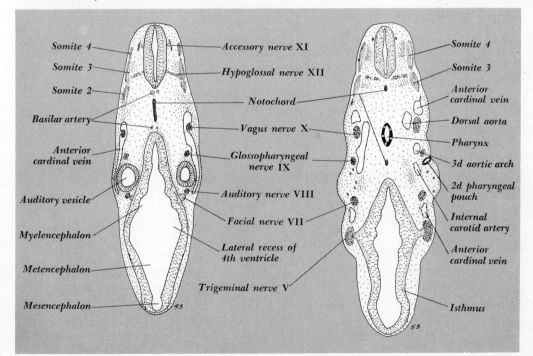

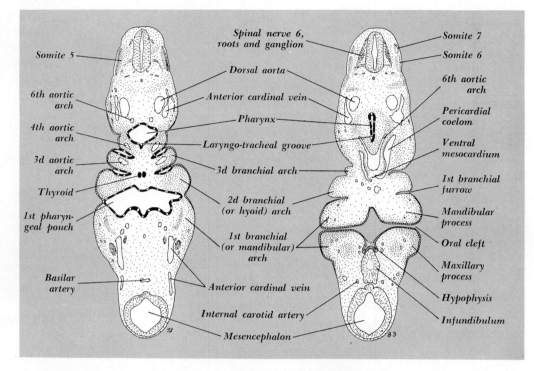

FIGURE 11–12   Sections of a 6-mm pig embryo.

Figure 12–5 for the central nervous system

Figure 13–2 for the peripheral nervous system (redrawn from Streeter's figure of a 10-mm human embryo)

Figure 14–1 for the digestive and respiratory systems

Figure 15–4 for the coelom and mesenteries

Figure 17–6 for the heart and arteries

Figure 17–11 for the veins

Figure 18–1 for the excretory organs

Unfortunately, the posterior parts of a 6-mm pig embryo are greatly twisted, making the in-terpretation of sections in this region difficult. Therefore, for the study of the cloacal area and tail, sections of a 10-mm pig are to be preferred.

Students who have had a course in comparative anatomy will be impressed by the similarities between the anatomy of a 6- to 10-mm pig and that of an adult shark: the parts of the brain and cranial nerves, the structure of the pharynx, the pathway of blood through the heart and the arrangement of the blood vessels, the excretory organs—these and many other features are remarkably similar.

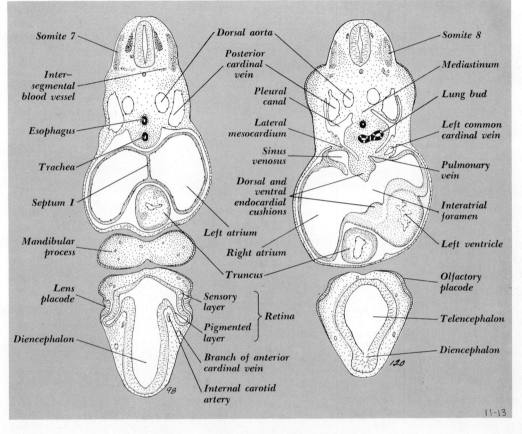

FIGURE 11–13   Sections of a 6-mm pig embryo.

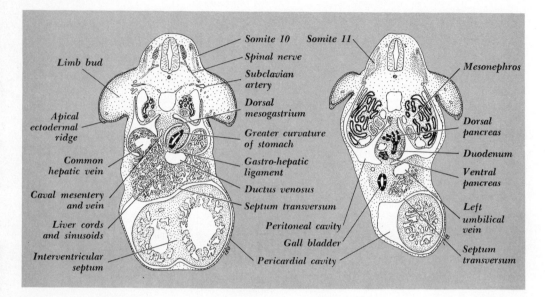

FIGURE 11–14   Sections of a 6-mm pig embryo.

FIGURE 11–15   Sections of a 6-mm pig embryo.

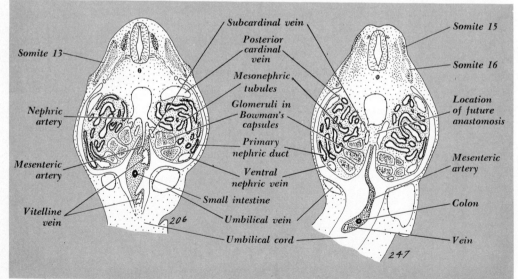

SUGGESTED READINGS

The Face and Oral Region

Patten, B. M., 1964. Foundations of Embryology. McGraw-Hill Book Company, New York. 2d ed. Chap. 19, The face and oral region.

Arey, L. B., 1965. Developmental Anatomy. W. B. Saunders Company, Philadelphia. 7th ed. Pp. 199–207, 226–232, 548–550.

Streeter, G. L., 1922. Development of the auricle of the human embryo. Carnegie Contrib. Embryol., 14:111–137. 6 plates.

Fraser, F. C., 1960. Some experimental and clinical studies on the causes of congenital clefts of the palate and of the lip. Arch. Pediat. 77:151–156.

The Limbs

References to experiments on amphibian limbs will be found in Chapter 6.

Saunders, J. W., Jr., 1948. The proximo-distal sequence of origin of the parts of the chick wing, and the role of the ectoderm. J. Exptl. Zool., 108:363–403.

Cairns, J. M., and J. W. Saunders, Jr., 1954. The influence of embryonic mesoderm on the regional specification of epidermal derivatives in the chick. J. Exptl. Zool., 127:221–248.

Saunders, J. W., Jr., J. M. Cairns, and M. T. Gasseling, 1957. The role of the apical ridge in the differentiation of the morphological structure and inductive specificity of limb parts in the chick. J. Morphol., 101:57–87.

Chaube, S., 1959. On axiation and symmetry in transplanted wing of chick. J. Exptl. Zool., 140:29–78.

Zwilling, E., 1961. Limb morphogenesis. Advan. Morphogenesis, 1:301–330.

Milaire, J., 1962. Histochemical aspects of limb morphogenesis in vertebrates. Advan. Morphogenesis, 2:183–209.

Amprino, R., 1965. Aspects of limb morphogenesis in the chicken. In DeHaan and Ursprung (eds.), Organogenesis. Holt, Rinehart and Winston, Inc., New York. Pp. 255–281.

Milaire, J., 1965. Aspects of limb morphogenesis in mammals. In DeHaan and Ursprung (eds.), Organogenesis. Holt, Rinehart and Winston, Inc., New York. Pp. 283–300.

Anatomy of Mammalian Embryos other than Human

Arey, L. B., 1965. Developmental Anatomy. W. B. Saunders Company, Philadelphia. 7th ed. Chap. 29, The study of pig embryos.

Heuser, C. H., and G. L. Streeter, 1929. Early stages in the development of pig embryos, from the period of initial cleavage to the time of the appearance of the limb buds. Carnegie Contrib. Embryol., 20:1–29.

Thyng, F. W., 1911. The anatomy of a 7.8 mm pig embryo. Anat. Record, 5:17–45.

Lewis, F. T., 1902. The gross anatomy of a 12 mm pig. Am. J. Anat., 2:211–226.

Anatomy of Human Embryos

The following references to papers in the *Carnegie Contributions to Embryology* are listed according to the stage in development of the embryo.

Ingalls, N. W., 1920. (Beginning of segmentation) Carnegie Contrib. Embryol., **11**:61–90.

Payne, F., 1924. (7-somite embryo.) *Ibid.,* **16:**
117–124.

Corner, G. W., 1929. (10-somite embryo.) *Ibid.,*
20:81–102.

Heuser, C. H., 1930. (14-somite embryo.)
Ibid., **21**:135–154.

Atwell, W. J., 1930. (17-somite embryo.) *Ibid.,*
21:1–24.

Davis, C. L., 1923. (20-somite embryo.) *Ibid.,*
15:1–51.

chapter **12**

THE
CENTRAL
NERVOUS
SYSTEM

In its finer structure, the nervous system is exceedingly complex. No design of woven tapestry could be more intricate, yet more precise in detail, than the fabric of fibers which organizes the behavior of an animal in its complicated and ever-changing environment. The circulatory system shares its complexity but not its precision. It is indeed an astounding thought that the basic features of this faithful pattern of fibers are woven during the course of development, under the control of the genes, and in the interactions of part with part. But, although the finer detail of the nervous system is complex beyond imagination, yet its grosser pattern and organization are reasonably simple and quite subject to description.

THE DEVELOPMENT OF NERVOUS TISSUES

The nervous system is derived from the neural tube and neural crest, with some contributions from epidermal placodes. The neural tube becomes the central nervous system, i.e., the brain and spinal cord. The neural crest, together with cells from the epidermal placodes and fibers which grow out from the neural tube, give rise to the peripheral nerves and ganglia.

The Neural Tube

During neurulation, the neural plate folds inward and forms the neural tube. The surface which at first was exposed outwardly becomes the lining of the cavity of the neural tube, that is, the lining of the *neurocoel* (Fig. 8–3). The innermost cells of the neural tube have been termed the "germinal layer," for it is only in them that mitotic figures can be seen (Fig. 12–1). Furthermore, it was supposed that following each cell division, one of the daughter

cells remained in the germinal layer as a "stem cell," while the other migrated outwardly and ultimately became a nerve cell (neuroblast) or supportive cell (spongioblast).

Present evidence indicates, in contrast to this earlier view, that all the cells of a young neural tube are stem cells and germinal in character. Together they constitute the neuroepithelium. The outer cells are in interphase, but they remain connected with the lining of the neurocoel by slender cytoplasmic stalks. When they begin to divide (prophase of mitosis), they round up and by doing so pull inward into the innermost layer of the neural tube. Here they are seen in metaphase and anaphase (Fig. 12–1, *A*). Then, having divided (telophase), they migrate outwardly into the neuroepithelium.

The neuroepithelium remains as the inner or *ependymal layer* of the brain and spinal cord (Fig. 12–1, *B*). Very soon some of the ependymal cells break their connections with the lining of the neurocoel and cease to divide. They migrate outwardly and accumulate in a zone surrounding the ependymal layer, known as the *mantle layer*. Here they sprout nerve fibers and so transform into neurons. The term neuron includes both the nerve cell body and the fibers which grow out from it. In the simplest neurons there is just one long fiber, the *axon*, whose function it is to conduct nerve impulses away from the cell body. The shorter fibers which grow out from the cell body are called *dendrites*. They carry impulses toward the cell body.

The axons grow outwardly from the cell bodies of the mantle layer and accumulate in bundles around the mantle layer, forming the outer or *marginal layer* of the neural tube. Thus the neural tube consists of three layers:

1 An inner neuroepithelium composed of

stem cells, namely, the ependymal layer

2 An intermediate layer composed of nerve cell bodies and their shorter fibers, both dendrites and the end branches of axons —the mantle layer

3 An outer layer consisting of tracts of axons —the marginal layer

The mantle layer is often referred to as *gray matter*. Within it are located the nerve centers where impulses jump across *synapses,* places of near contact, from the end arborizations of the axons of one neuron to the dendrites and cell bodies of other neurons. In the spinal cord, the mantle layer is continuous from one end to the other. It forms the so-called *dorsal* and *ventral gray columns* (Fig. 12-2). In the brain, however, the mantle layer is broken up into

isolated masses of gray matter of larger or smaller size which are usually referred to as *nuclei* of the brain.

The tracts of axons which compose the marginal layer are termed *white matter*. This is because, in the later development of higher vertebrates, most of the axons become surrounded by sheaths of a glistening white substance known as *myelin*. Myelin serves to insulate the axons, or axis cylinders, and speed up the transmission of impulses. The term white matter, however, is not exact, for not all fibers of the marginal zone are myelinated.

The Neural Crest

The neural crest, with assistance from epidermal placodes, gives rise to the *cranial* and

FIGURE 12-1   *A.* Diagram which illustrates the proliferation of cells in the early neural tube (neuroepithelium). The cells shift centrally during metaphase and anaphase, and then return. *B.* Diagram which shows the multiplication, migration, and differentiation of neuroblasts in the later neural tube. The mantle layer is "gray matter"; the marginal layer is "white matter."

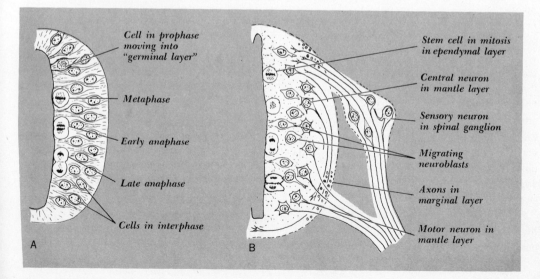

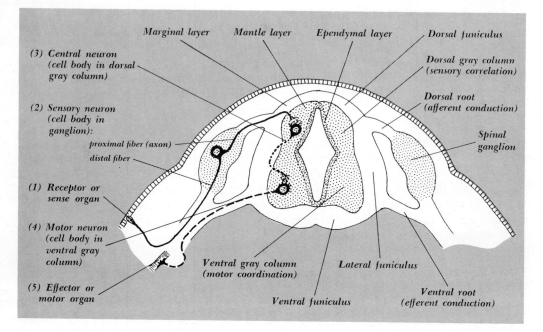

FIGURE 12-2 Schematic cross section of a developing spinal cord illustrating the elements of a three-neuron reflex arc.

spinal ganglia (Fig. 12-2). These are isolated masses of gray matter situated adjacent to the neural tube but entirely outside of it. Like other gray matter, they consist of supporting cells and neuroblasts. But unlike other gray matter, these neuroblasts of the ganglia are bipolar; that is, each of them sprouts not one but two fibers. One of them grows distally toward the periphery of the body until it makes contact with a sense organ. The other grows proximally and enters the neural tube. It may branch, but ultimately each branch terminates in an end-arborization at some nerve center in the mantle layer of the brain or cord.

Functionally the two fibers are unlike in that the distal one conducts impulses from the periphery toward the cell body in the ganglion,

while the proximal fiber, like axons generally, conducts impulses away from the cell body. Later the bases of the fibers, where they join the cell body, grow together so that each bipolar neuron may be said to become unipolar.

THE FUNCTIONAL PATTERN OF THE CENTRAL NERVOUS SYSTEM

The Cross Section of the Neural Tube

The functional pattern of the central nervous system is foreshadowed in the relations of the open neural plate to its surroundings. It will be recalled that the neural plate is a thickened area of ectoderm on the dorsal side of the embryo. It is bounded anteriorly and at the sides by the neural folds. It has a longitudinal neural groove at its center. The neural folds extend backward to the lateral lips of the blastopore (amphibians) or, what is the same thing, to the sides of the primitive streak (birds and mammals) (Fig. 12–3, *A*). The neural groove begins at the center of the forward region of the neural plate and ends posteriorly at the dorsal lip of the blastopore, or primitive node. As described by Kingsbury, the neural plate consists of three regions:

1 The central region of the neural plate, which corresponds to the neural groove, becomes the *floor plate*. It overlies, and for a time is in contact with, the notochord (Fig. 12–3, *B*). This lack of peripheral connections may be, at least in part, the reason that the floor plate undergoes relatively little nervous differentiation.

2 The zone which borders the floor plate anteriorly and laterally is known as the *basal plate* (Fig. 12–3, *A*). Its anterior regions overlie the prechordal mesoderm from which, among other things, the eye muscles are derived. Its lateral portions immediately overlie the axial mesoderm from which comes much of the body's musculature. We are not surprised, therefore, to learn that the basal plate gives rise to the *motor neurons* (efferent neurons), whose axons leave the central nervous system and enter the adjacent premuscle masses (myotomes, etc.) (Fig. 12–3, *B* and *C*). During subsequent development, the muscle masses may migrate far, but as they do so they tow the axons of their nerve supply after them. The nerve centers which develop within the mantle layer of the basal plate are all *motor-coordination centers*.

3 The outermost zone of the neural plate adjoins the basal plate anteriorly and laterally

FIGURE 12–3 The functional pattern and peripheral relations of the central nervous system. *A*. Open neural plate. *B*. Section through the neural plate. *C*. Section through the neural tube.

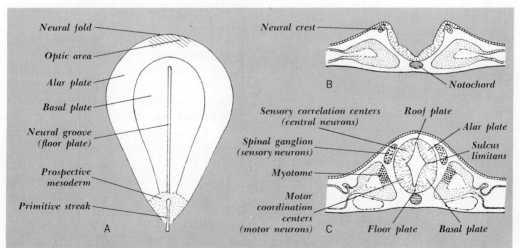

and is known as the *alar plate* (Fig. 12–3). It is located close to the neural-crest cells and to the epidermal placodes from which the cranial and spinal ganglia are formed. It receives the proximal axons of the *sensory neurons* (afferent neurons) whose cell bodies are in the ganglia (Fig. 12–2). Its own neurons are *central neurons* (internuncial neurons) whose fibers never leave the central nervous system. The centers which are derived from the alar plate are in all cases *sensory-correlation centers*. This is true even of the so-called motor areas of the cerebral cortex, whose axons carry orders to motor-coordination centers of the basal plate.

When the neural plate closes to form the neural tube, the dorsal suture (raphe or seam) becomes the *roof plate* (Fig. 12–3, *C*). The forward extension of the roof plate, formed by the union of the two sides of the anterior neural fold, is the *lamina terminalis* (Fig. 12–7). For the most part, the roof plate is not a nervous structure, but in places it is crossed by axons from the sides. For a time, a shallow groove along each sidewall of the neurocoel marks the division between the alar and basal plates. Each groove is a *sulcus limitans*.

Figure 12–2 shows the elements of a simple, three-neuron reflex arc superimposed on a cross section of a young embryo. Such an arc may be described, for pedagogical purposes, as consisting of five elements:

1 A sense organ (receptor) located outside the central nervous system.

2 A sensory neuron (afferent neuron), whose distal axon brings impulses inward from the sense organ via a cranial or spinal nerve. Its cell body is located in a cranial or spinal ganglion. Its proximal axon enters the central nervous system by the dorsal root of a spinal nerve or corresponding root of a cranial nerve

and terminates in a sensory-correlation center of the alar plate.

3 A central neuron (internuncial neuron), whose cell body is located in a sensory-correlation center of the alar plate, and whose axon travels usually in bundles with other axons to a motor-coordination center in the basal plate.

4 A motor neuron (efferent neuron), whose cell body is in a motor-coordination center of the basal plate, and whose axon promptly emerges from the brain or cord by way of a ventral root of a spinal nerve or comparable cranial nerve and proceeds by way of the nerve to a motor organ somewhere in the body.

5 A motor organ (effector), in this case a striated muscle cell or group of muscle cells. (The case of the reflex arcs which terminate in smooth muscle cells or in glands will be discussed in connection with the autonomic nervous system, pp. 301–303.)

A "Fate Map" of the Neural Tube

In order to visualize the functional organization of the central nervous system, it will be helpful to draw a "fate map" of the neural tube (Fig. 12–4). In its essential features, such a map is the same for all vertebrates; for the gross differences in the nervous systems of the different vertebrate classes arise during development as an expression of the relative functional importance of the several regions of the brain and cord. Thus, an animal which depends a great deal on the sense of smell develops a huge olfactory area in its brain, while an animal which is required to carefully maintain equilibrium has a large cerebellum.

The figure shows the brain of an early embryo as seen from the left side. It indicates that the forebrain, that is, the telencephalon and diencephalon, is wholly of roof plate and alar plate origin. Its neurons are all central neurons, i.e., neurons which are entirely within the central nervous system. The midbrain, or mesencephalon, is derived from the roof plate, alar plate, and basal plate. The hindbrain, namely, the metencephalon and myelencephalon, includes these plates and the floor plate as well.

The fate map further illustrates that the alar and basal plates are divisible into longitudinal zones or "cell columns." Beginning at the dorsal side, these are:

walls of the forebrain give rise to the optic vesicles. These bulge out laterally and then cave in to form the optic cups and, finally, the retinas of the eyes. It would seem reasonable to suppose that in view of this relation to the retinas the forebrain would produce the nervous centers for seeing. But originally this was not so. The primitive centers of reflex behavior based on sight are derived from the alar plate of the midbrain, a region which we have labeled the "eye brain" (Fig. 12-4).

The special somatic sensory zone includes

From the alar plate
 1 The special somatic sensory column ("nose, eye, and ear brains")
 2 The general somatic sensory column ("skin brain, etc.")
 3 The visceral sensory column (labeled "taste, etc.")

From the basal plate
 4 The general visceral motor column
 5 The special visceral motor column
 6 The somatic motor column

The *special somatic sensory column* includes those areas of the brain which are dominated by the special sense organs of the head, namely, the nose, eye, and ear. The latter includes both equilibrium and hearing.

The forebrain is close to the olfactory epithelium and was originally dominated by the sense of smell. The part which retains this function—a very important function in most vertebrates—is termed the *rhinencephalon,* literally the "nose brain" (Fig. 12-4). There is, however, an area of each side wall of the forebrain which escapes from the domination of the sense of smell and becomes a region of higher sensory correlation. It gives rise to the major portion of the *cerebral hemispheres* and to the *thalami.*

At a very early stage of development the

also the forward part of the hindbrain. The peripheral relation here is to the auditory vesicles, primarily in their role as organs of equilibrium. On the fate map this region is labeled "ear brain." The *cerebellum* is derived from it, as also are certain dorsal nuclei of the *medulla.*

The *general somatic sensory column* is ventral to the ear brain and extends the full length of the hindbrain and spinal cord. It is labeled "skin brain" on the fate map, although its functions include also the deeper body senses. It receives the sensory fibers of the great fifth cranial nerve or trigeminal and the dorsal roots of the spinal nerves. Beneath it lies the *visceral sensory column,* which on the diagram is labeled "taste, etc."

The most dorsal column of the basal plate is the *general visceral motor column.* It con-

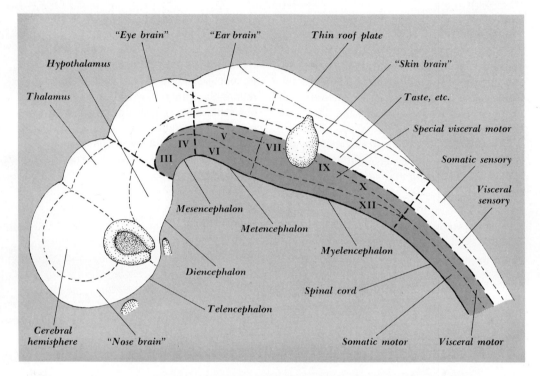

FIGURE 12–4 A "fate map" of the neural tube showing the prospective neural regions. The basal plate is shaded; the alar plate is not shaded. For an explanation see text.

sists of motor neurons whose axons leave the central nervous system and, after relaying in autonomic ganglia (see Fig. 13–9), supply smooth muscles, heart muscle, and glands throughout the body. It is present in the hindbrain and spinal column.

Below it is the *special visceral motor column,* whose motor neurons supply striated muscles derived from the branchial arches. As would be expected from the location of these arches, this column is practically restricted to the hindbrain.

The most ventral column of all, the *somatic motor column,* consists of the motor neurons which supply the axial and appendicular muscles of the body, namely, all the striated

muscles except those which are derived from the branchial arches. The somatic motor column is to be found in the midbrain, hindbrain, and spinal cord. Thus four of the six columns are present in both the brain and spinal cord, while two are found only in the brain.

The roman numerals in Fig. 12–4 give the approximate centers of origin of the motor nerves of the brain.

THE GROSS STRUCTURE OF THE EMBRYONIC BRAIN AND CORD

We now pick up the story of the development of the nervous system from where we left it in

Chapter 6, on the amphibian, and Chapter 8, on the chick. The emphasis will be placed on the mammal, and we shall carry the story farther than before. We shall start our account at the anterior end and work backward. Figure 12–5 illustrates the entire central nervous system of a 6-mm pig as seen from the left side. It might equally well represent the brain and cord of a 4-week human embryo.

What processes account for the gross form of the central nervous system? The answer involves, not only growth and proliferation of nerve cells, but also changes in the shape of cells, migration of cells, ingrowth of axons, and production of sheaths around the axons. In certain places, for example, in the cervical region of the spinal cord, an extensive degeneration of nerve cells takes place as a feature of normal development. Generally speaking, these various processes begin at the anterior end and progress posteriorly.

The Telencephalon

The telencephalon is of alar plate origin. Its function is sensory correlation. Its sidewalls expand outwardly and become the cerebral hemispheres. It will help to understand the telencephalon if we anticipate its future structure and function and describe it as consisting of the *rhinencephalon* or "nose brain" (Fig. 12–4), dominated by the sense of smell, and two lateral "higher correlation areas" which are not dominated by any one sense. The higher correlation areas give rise to the *striate bodies* and the *neopallia* (Figs. 12–6 and 12–7).

The rhinencephalon Phylogenetically this is the oldest region of the telencephalon. Its anterior end is invaded by axons from the neurons of the adjacent olfactory epithelia. It

responds by forming the *olfactory bulbs,* the primary sensory-correlation centers of smell (Figs. 12–6 and 12–8). The neurons of each bulb, in turn, send axons posteriorly by way of the *olfactory tracts* to secondary and tertiary centers of smell. Here the olfactory impulses are brought into relation to impulses ascending from other sense organs. Some of these centers are dorsal and form what is known as the *archipallium,* which includes the *hippocampus* (Figs. 12–6 and 12–10). Other centers are ventral and are termed the *paleopallium,* which includes the *piriform lobe.* The former presumably correlates olfaction with other somatic senses, while the latter is similarly related to impulses of visceral sensory origin. In embryos, and indeed in most adult mammals, a shallow groove known as the *rhinal fissure* marks the separation between the rhinen-

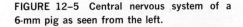

FIGURE 12–5 Central nervous system of a 6-mm pig as seen from the left.

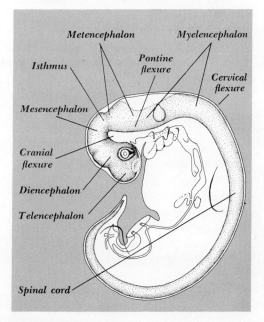

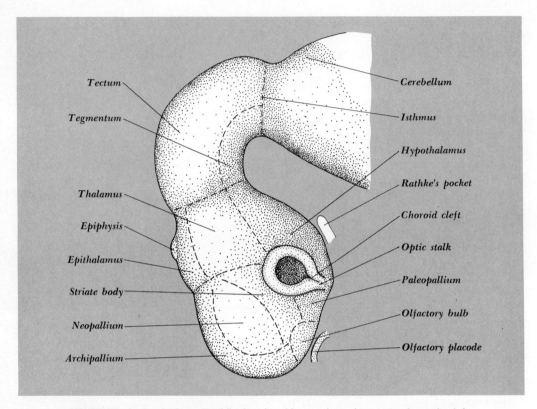

Tectum — *Cerebellum*

Tegmentum — *Isthmus*

Hypothalamus

Rathke's pocket

Thalamus — *Choroid cleft*

Epiphysis — *Optic stalk*

Epithalamus — *Paleopallium*

Striate body — *Olfactory bulb*

Neopallium — *Olfactory placode*

Archipallium

FIGURE 12–6 Forebrain and midbrain of a 12-mm pig embryo seen from the left. Regions are labeled according to their prospective fates.

cephalon and the higher correlation area (Fig. 12–13).

Smell is a primitive sense and remains, in most vertebrates, the principal discriminatory sense. It is not surprising therefore to find that its centers develop early even in the highest vertebrates including man. In man, however, smell is not emphasized, and the rhinencephalon is soon overshadowed by the greater growth of the higher correlation areas.

The striate bodies In lower vertebrates, the principal component of each higher correla-

tion area is a mass of gray matter which is known as the *striate body* (corpus striatum) (Figs. 12–6, 12–7, and 12–10). The corpora striata are the forward extension of the mantle layer of the neural tube into the telencephalon (Fig. 12–14).

The striate bodies are of special interest because they are large in birds and other animals with complicated innate patterns of behavior, that is, "instinct." It would seem that a three-dimensional structure, such as is possessed by the striate bodies, is favorable to stereotyped modes of behavior although experimental evi-

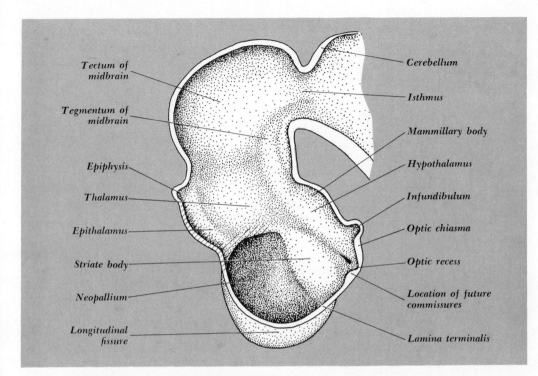

FIGURE 12-7   Median view of the right half of the forebrain and midbrain of a 12-mm
pig embryo. Some details have been inserted from a somewhat older embryo. The
diagonal stripes mark the location of the future choroid plexuses. *(Based in part on
M. Hines, 1922, J. Comp. Neurol., 34:73–171, Fig. 7.)*

FIGURE 12-8   Brain of a 20-mm pig embryo as seen from the left side.

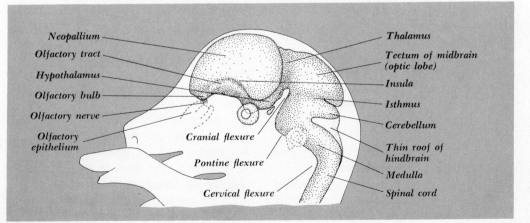

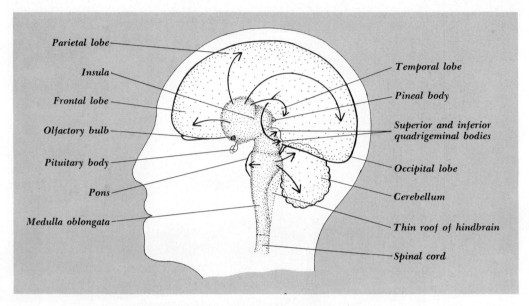

Parietal lobe

Insula

Frontal lobe

Olfactory bulb

Pituitary body

Pons

Medulla oblongata

Temporal lobe

Pineal body

Superior and inferior
quadrigeminal bodies

Occipital lobe

Cerebellum

Thin roof of hindbrain

Spinal cord

FIGURE 12–9 Diagram of a human brain. The arrows indicate the expansion of the cerebral hemispheres, cerebellum, and pons.

dence that this is true has not been forthcoming. In mammalian embryos the striate bodies are first recognized by the thicker ventral wall of each cerebral hemisphere.

The neopallium The dorsal wall of each cerebral hemisphere remains thin for a considerable time and is called the *pallium*. Some of it, namely, the archipallium and paleopallium, belongs to the rhinencephalon since they are principally concerned with smell; but most of it in the case of mammals is neopallium and is not dominated by any one sense (Figs. 12–6 to 12–10).

At first the pallium consists, as does the rest of the brain and cord, of an inner ependymal layer, a middle mantle layer of gray matter, and an outer marginal layer of white matter. But rather late in development, an unusual

process takes place: neuroblasts of the mantle layer migrate through the marginal layer and take a position next to the surface. Here they form a superficial layer of gray matter known as the *cerebral cortex* (Fig. 12–11). This migration occurs also to a limited extent in reptiles and birds, but it is characteristic of mammals.

The white matter, which now lies between the cortex and the striate bodies, affords an abundance of interconnections between the various regions of the cortex and between cortex, striate bodies, and lower centers of the brain and cord. Indeed, the cortex has often been compared to the front of a telephone switchboard, with the white matter behind it represented by the wires on the back of the switchboard. Spread out thus in two dimensions, the cortex is able to mediate the plastic functions of behavior such as learning, memory, thinking, conscious awareness, and voluntary action.

As the pallium expands, it overshadows the other regions of the forebrain. It flanks the thalamus. In man it even covers the midbrain and encroaches on the hindbrain (Fig. 12-9). Its forward expansions become the *frontal lobes* of the cerebral hemispheres; its upward expansions become the *parietal lobes;* the backward expansions form the *occipital lobes;* and, in man, it pushes downward and forward in a manner suggesting a ram's horns to form the *temporal lobes.* The great cleft between the cerebral hemispheres is the *longitudinal fissure* (Fig. 12-10). The central area of each side, around which expansion takes place, is the *insula* (island of Reil) (Figs. 12-8 and 12-9).

We return to the striate bodies. Each is located beneath the insula near the center of cerebral expansion. Nevertheless it shares to some extent in the expansion of the latter, for its forward region, known as the *caudate nucleus,* becomes drawn backward and downward as a sort of tail (Figs. 12-12 and 12-13). Its rear region, the *lenticular nucleus,* however, remains as a hub around which the expansion takes place. Between the caudate and lenticular nuclei there is a zone of white matter known as the *internal capsule* (Fig. 12-11).

White matter of the cerebral hemispheres
Most of each cerebral hemisphere is white matter and consists of axons of three sorts:

1 *Projection fibers:* These consist of axons descending and ascending between the cortex and lower levels of the nervous system (Fig. 12-12). After passing through the internal capsule, they fan out to all parts of the cortex by what is known as the *corona radiata.*

2 *Association fibers* (both long and short): These connect one region of a cortex with other regions of the same cortex (Fig. 12-12).

3 *Commissural fibers:* These cross the mid-

FIGURE 12-10 Oblique view of the brain of a 20-mm pig embryo. Enough of the neopallium has been removed to expose the median wall of the lateral ventricle.

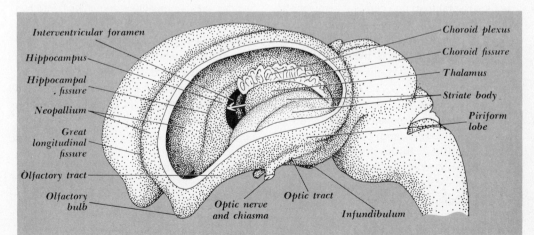

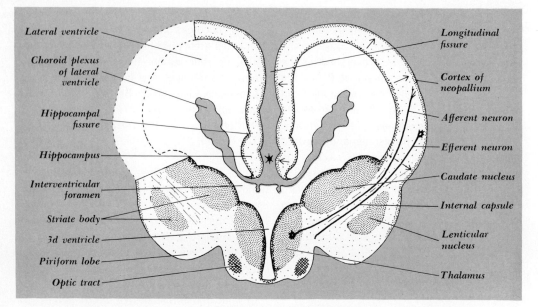

FIGURE 12-11 Vertical section through the forebrain which is illustrated in Fig. 12-10 (part of left neopallium removed). The section passes through the choroid plexus, thalamus, striate bodies, and optic tract (schematic). The asterisk marks the location of the future hippocampal commissure and corpus callosum. The arrows indicate the migration of neuroblasts from the ependymal layer through the marginal layer to form the cortex (superficial layer of gray matter).

line (decussate) from one cerebral hemisphere to the other.

Most of the commissural fibers make use of the lamina terminalis, that is, the suture formed by the coming together of the sides of the anterior neural fold (Fig. 12-7). Two of the commissures which are thus formed are primitive and serve the rhinencephalon. These are the *anterior commissure* and the *hippocampal commissure*. But, with the great development of the mammalian neopallium, there appears a new commissure, the *corpus callosum* (Fig. 12-13). It originates dorsal to the hippocampal commissure, and as the neopallium expands it also is drawn backward. In doing so it

invades each archipallium and splits it into two parts: a dorsal part which remains rudimentary (the so-called rudimentary hippocampus); and a ventral part which gives rise to a strange, nearly circular band of fibers, the *fornix*. Beginning at the rear of a hemisphere, the fibers of the fornix circle upward, forward, downward, and backward in a great bow. They carry impulses from the definitive hippocampus at the rear of each hemisphere to the ventral nuclei of the diencephalon, namely, the hypothalamus.

Ventricles of the telencephalon   As a result of the expansion of the cerebral hemispheres,

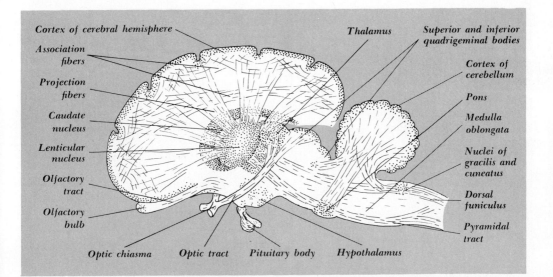

FIGURE 12-12 Diagram of a mammal brain from the left side illustrating some of
the tracts. The parts which are labeled can be easily found in an adult cat or sheep brain.

FIGURE 12-13 The development of the commissures of the human brain. The right
cerebral hemisphere is seen from the median surface after the brain stem has been
removed by a diagonal cut through the right thalamus.

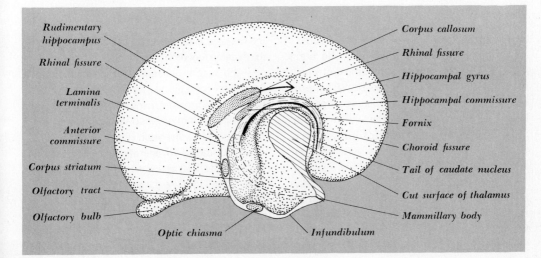

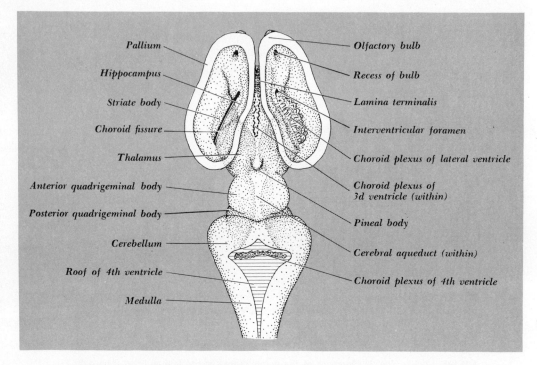

Pallium

Hippocampus

Striate body

Choroid fissure

Thalamus

Anterior quadrigeminal body

Posterior quadrigeminal body

Cerebellum

Roof of 4th ventricle

Medulla

Olfactory bulb

Recess of bulb

Lamina terminalis

Interventricular foramen

Choroid plexus of lateral ventricle

Choroid plexus of
3d ventricle (within)

Pineal body

Cerebral aqueduct (within)

Choroid plexus of 4th ventricle

FIGURE 12–14 Schematic dorsal view of the brain of a 25-mm pig. The pontine and cervical flexures are shown as straightened out. The roofs of the lateral ventricles have been cut away. The choroid plexus of the left lateral ventricle has been removed. The thin roof of the fourth ventricle is striated.

their cavities are drawn out into "horns" known as the *lateral ventricles* of the brain (Fig. 12–14). The central cavity, most of which is in the diencephalon, is the so-called *third ventricle* of the brain. The openings from the lateral ventricles into the third ventricle become constricted as a result of the growth of the striate bodies and are known as the *interventricular foramina* of Monro.

The roof of the third ventricle is thin and is termed the *tela choroidea*. At the junction between the telencephalon and the diencephalon, this thin region (the tela) extends laterally onto the sidewalls of the forebrain (Fig. 12–7). When

the hemispheres expand, each originally lateral thin region becomes stretched out on what is now the medial surface of each hemisphere. Soon the thin regions invade the cavities of the lateral and third ventricles as the membranous coverings of nets of blood vessels, the *choroid plexuses* of the forebrain (Figs. 12–10 and 12–11). The line of invagination is known as the *choroid fissure*. Because of the manner in which the thin regions are drawn out, the plexuses taken together have the form of a ♈, the symbol of the astronomical constellation Aries, the ram (Figs. 12–7 and 12–14). Throughout life, the choroid plexuses secrete cerebrospinal fluid into the ventricles of the brain.

The Diencephalon

The diencephalon, like the telencephalon, is of alar plate origin and consists of sensory-correlation centers. Its neurons are central neurons. It is possible that the diencephalon was originally the highest well-developed center of the brain and that the regions of the telencephalon expanded when the sense of smell, and later other senses, acquired increased importance as functions of discrimination in the life of the vertebrates. Some of the primitive functions of the diencephalon remain even in man, for impulses of temperature and pain are said to ascend no higher than the thalamus. Feelings of pleasure and displeasure are thought to enter consciousness here.

It will be recalled that early in development the walls of the diencephalon bulge laterally and give rise to the optic vesicles and that these cave in and form the optic cups and, finally, the retinas and optic nerves. In an embryological sense, therefore, the retinas are areas of the brain, and the optic nerves are tracts of the brain.

The roof of the diencephalon gives rise to a small bud which evaginates upward and becomes the epiphysis. This is usually considered to be the forerunner of the pineal body (Figs. 12–6 and 12–7). The complex relations of this organ are of interest to comparative anatomists but need not concern us here. In some of the lower vertebrates the pineal body possesses cones which may be compared to those of an eye. They are sensitive to light.

In mammals, the pineal body loses its nervous connections with the brain and becomes a gland of internal secretion. Its hormones, one of them known as melatonin, regulate the functions of the gonads. The pineal itself is controlled nervously by way of the sympathetic nervous system. Although in mammals the pineal body is not directly responsive to incident light, it does react when light falls on the retinas. Experiments on rats indicate that the mechanism is as follows: The light which falls on the retina indirectly *inhibits* the secretory activity of the pineal body. Its secretions have been found to *suppress* various reproductive functions, such as the growth of the ovary and the rate of the female cycle (estrous cycle). We thus have a double negative yielding a positive. Light on the retina tends to stimulate reproductive functions.

Each sidewall of the diencephalon develops three thickened areas (Figs. 12–6 and 12–7). The most dorsal of these is the *epithalamus,* a region of little consequence in mammals except that it receives fibers from the archipallium and presumably integrates somatic impulses such as those of sight with those of smell. The middle thickening is the *thalamus.* Its relations are to the higher sensory-correlation areas of the telencephalon; its principal function in the higher vertebrates is to relay impulses which come to it from all the senses except smell and project them forward to the cortex by way of the internal capsule and corona radiata and to the striate bodies.

The third and most ventral of the three swellings of the diencephalon is the *hypothalamus* (Figs. 12–8 and 12–12). Today we are discovering the vast importance of this organ in controlling the drives of the body, such as those of feeding, temperature control, sex behavior, fighting, and sleep. It consists of a complex of nerve centers called nuclei. These receive impulses from the rhinencephalon anterior to it and from visceral sensory centers posterior to it. They then correlate and integrate the impulses. In keeping with this primitive function, the hypothalamus develops early in all vertebrates.

On its ventral side, the diencephalon (it has no true floor plate) begins at the *optic recess.* Immediately posterior to this is the *optic chiasma,* where fibers from the optic nerve decussate to the opposite side of the brain (Fig. 12–12). In animals with stereoscopic vision, as is the case with man, the fibers of the inner half of each retina cross over to the other side, while those of the outer half do not. Thus it comes about that the right side of each visual field throws an image on the left side of each retina, which then is projected from both the left and right eyes to the left side of the brain.

Posterior to the optic chiasma, the floor of the diencephalon pockets downward as the *embryonic infundibulum* (Figs. 12–7 and 12–13). Its tip comes into contact with Rathke's pocket, and together they become the pituitary body or adult hypophysis (Fig. 10–1). Rathke's pocket, also called the *adenohypophysis,* gives rise to the anterior lobe and to additional tissue (pars intermedia) which joins with the embryonic infundibulum, or *neurohypophysis,* to form the posterior lobe. The floor of the diencephalon posterior to the infundibulum forms a swelling, the *mammillary body,* which is functionally a part of the hypothalamus (Fig. 12–7).

The Mesencephalon

The mesencephalon or midbrain possesses a roof plate, alar plates, and basal plates, but no floor plate. Its dorsal portion (alar plate) is known as the *tectum* and is sensory-correlation in function (Figs. 12–6 and 12–7). Its

ventral portion (basal plate), the gray matter of which is known as the *tegmentum,* consists of motor-coordination centers. The dorsal side of the midbrain elongates more rapidly than the ventral side, with the result that the brain bends sharply downward making the *cranial flexure.*

The midbrain is concerned mainly with reflex behavior based on sight and, to a less extent, on hearing. In vertebrates below the mammals, the axons of the *optic tract,* after leaving the optic chiasma, course across the lateral surface of each thalamus. Some terminate in the thalamus, but many go to sensory-correlation centers of the tectum (Fig. 12–12). In birds this function of reflex seeing is so important that the tectum develops into two large *optic lobes* (corpora bigemina), which may be even larger than the cerebral hemispheres. In mammals, on the other hand, most of the fibers of the optic tract terminate in the thalamus (lateral geniculate body), and from there impulses are relayed to the occipital lobes of the cerebral hemispheres. Yet even in mammals reflex functions of sight are retained by the midbrain, such as winking, the pupillary reflex, accommodation, and convergence. The optic lobes of mammals are reduced in size and are known as the anterior or *superior quadrigeminal bodies* (Fig. 12–12). The posterior or *inferior quadrigeminal bodies* are related in similar fashion to reflexes based on hearing.

The basal plates of the midbrain at first overlie the prechordal mesoderm from which the muscles of the eye are derived. In keeping with this fact, the gray matter of the basal plate (tegmentum) gives rise to the motor-coordination centers (nuclei of origin) of the nerves which supply the eye muscles. (We shall return to this topic in Chapter 13.)

The white matter of the midbrain increases

tremendously in volume as development proceeds and axons find their way to and from the forebrain. Most of the axons pass ventral to the tegmentum and form what is known as the *cerebral stalk,* or peduncle. Particularly important in mammals are the great tracts of fibers which descend from the cortex to lower motor-coordination centers in the hindbrain and cord. These are known as the pyramids or *pyramidal tracts* (Fig. 12–12).

The midbrain is separated from the hindbrain by a constricted region known as the *isthmus.* (Figs. 12–6 and 12–7).

The central cavity of the midbrain (neurocoel) becomes narrow and is known as the *cerebral aqueduct* of Sylvius (Fig. 12–14).

The Metencephalon

The hindbrain is generally considered as divisible into metencephalon and myelencephalon. Basically the two parts are so similar that they might be considered together. Both consist of roof plate, alar plates, basal plates, and a floor plate. In this respect, they are like the spinal cord; but they differ from the cord in that the roof plate of the hindbrain is broad, thin, and is entirely lacking in nervous tissue except where, as in the cerebellum, it has been invaded from the sides.

The cavity of the hindbrain, known as the *fourth ventricle,* has the shape of a diamond kite when viewed from above (Fig. 12–14). Its lateral recesses mark the boundary between the metencephalon and myelencephalon. In older mammal embryos, a reversed curvature takes place at this location, known as the *pontine flexure* (Figs. 12–5 and 12–8).

The alar plate of the hindbrain develops sensory-correlation centers (nuclei of termination) for all the sensory nerves of the head and visceral arches, except only those of smell and sight. Of special importance in the metencephalon are the special somatic sensory cell columns. These are related to the otocysts and to the eighth or auditory nerves. They push dorsally to the midline where they coalesce and form the beginning of the *cerebellum* (Figs. 12–8 and 12–14). Now, the cerebellum is a higher sensory-correlation region concerned with the smoothness of body motion and the maintenance of equilibrium. It is hugely expanded in birds and large also in active mammals. In those lower vertebrates, however, in which the maintenance of equilibrium is less of a problem, it remains small.

In mammals, certain of the neuroblasts of the alar plate which are some distance posterior to the primordium of the cerebellum migrate ventrally into the basal plate. Here they take up a position in the midst of the fibers which descend from the forebrain to lower levels, that is, in the midst of the pyramidal tract. They give rise to the nuclei of the *pons* (Fig. 12–12). Their function is to serve as collateral relay stations from which impulses go to the cerebellum, informing this organ of the orders which the cerebral hemispheres send forth.

The Myelencephalon

In most respects the myelencephalon is a continuation of the metencephalon. It begins at the level of the lateral recesses of the fourth ventricle, i.e., at the pontine flexure, and gradually tapers posteriorly into the spinal cord. It becomes the *medulla oblongata* (Figs. 12–12 and 12–14).

Its roof plate, as previously noted, is broad and thin and is known as the *tela choroidea* of the fourth ventricle (Fig. 12–15, *A*). Blood

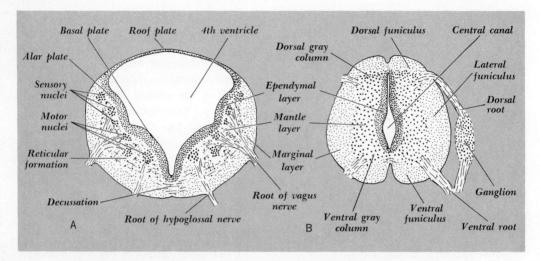

FIGURE 12–15 Diagrammatic sections through the hindbrain and spinal cord. A. Section through the medulla of a later embryo. B. Section through the spinal cord of a 25-mm embryo or later.

vessels above the tela form plexuses which, pushing folds of the tela before them, enter the cavity of the fourth ventricle as the tuftlike *posterior choroid plexus* (Fig. 12–14). Late in development, pores develop in the lateral recesses of the roof plate, which permit the cerebrospinal fluid secreted by the several choroid plexuses to escape from the neurocoel into spaces surrounding the brain and cord. These pores are known as the *foramina of Luschka*.

The demarcation between gray matter (mantle layer) and white matter (marginal layer) is less clear in the hindbrain than it is elsewhere in the brain and cord. Over a considerable region, the axons and the cell bodies with their dendrites tend to intermingle, forming what is known as the *reticular formation*. However, a number of more or less discrete nuclei of gray matter may be recognized within

it. These include the nuclei of termination and origin of the cranial nerves (Fig. 12–15, *A*).

Tracts of fibers connecting the brain and spinal cord make up a large part of the weight of the hindbrain of the adult mammal. In the lower vertebrates, these tracts are relatively of less importance, for much of the activity of these forms is carried on by the cord itself with little assistance from the brain. In the higher vertebrates, however, the brain exercises a large control. In mammals, nerve impulses, after coming forward in white matter of the dorsal side of the spinal cord, are relayed in nuclei (the nuclei of gracilis and cuneatus) on the dorsal side of the myelencephalon (Fig. 12–12). They are then carried forward by fresh neurons to the thalamus and cerebellum. From the thalamus they relay to the cortex. On the ventral side of the mammalian hindbrain, the great corticospinal or *pyramidal tracts* carry impulses all the way from the cortex to the motor-coordination centers in the ventral gray column of the spinal cord. Most of these fibers decussate in the medulla (Fig. 12–15, *A*); that is, they cross over to the opposite side. But in man, some do not decussate until they reach the cord.

The Spinal Cord

There is no sharp boundary separating the brain and spinal cord. Somewhat arbitrarily, either the place where the neural tube emerges from the future skull or the location of the first pair of spinal nerves is taken as the boundary. Except for swellings in those regions of the spinal cord which supply the limbs, the spinal cord tapers gradually to the tip of the tail bud.

The mantle layer of the spinal cord gives rise to the gray columns (Fig. 12–15, *B*). The alar plate becomes the *dorsal gray column*. It is sensory-correlation in function. The basal plate gives rise to the *ventral gray column* and is motor-coordination in function. These columns continue throughout the length of the cord and are not broken into separate nuclei as they are in the brain. In the embryo, a sulcus limitans marks the division between the alar and basal plates (Fig. 12–3, *C*).

The marginal layer, similarly, is divisible into white columns, or funiculi (Fig. 12–15, *B*). The *dorsal funiculi* are located above the dorsal gray columns and the dorsal roots of the spinal nerves. They consist of the ascending tracts which carry impulses forward to the brain. The *lateral funiculi* are situated between the dorsal and ventral roots. They include tracts from various sources and to various destinations, mostly short in length. The *ventral funiculi* are found below the gray

columns and ventral roots. In large part they consist of descending tracts. Most of them are short in the lower vertebrates, but they include the great corticospinal or pyramidal tracts in mammals.

The cavity of the spinal cord, as seen in cross sections, is at first diamond-shaped; but as its walls thicken, it becomes more and more compressed until only a small *central canal* remains. Above it, the alar plates of the two sides meet and fuse to form a dorsal septum.

Meninges of the Brain and Cord

Mesoderm shares in the development of the central nervous system by providing meninges, i.e., membranous coverings for the brain and cord. Primitive membranes can be seen in cross sections of 10-mm mammal embryos or demonstrated in hand-dissection of older embryos. The outer tough membrane is the *dura mater*. It is closely applied to the inside surface of the skull. The inner, more delicate membrane, the *pia mater,* hugs the outside surface of the brain and cord. It is richly supplied with blood vessels. In mammals the embryonic pia mater becomes invaded by cerebrospinal fluid and gradually splits into two layers: an outer *arachnoid,* and an inner pia mater proper. Connecting the two are numerous weblike threads of fibrous tissue.

The cerebrospinal fluid filters from the blood through the choroid plexuses of the fore- and hindbrain into the third and fourth brain ventricles. It then leaves the cavities of the brain through the foramina of Luschka in the thin roof of the hindbrain and enters the subarachnoid space. It finally reenters the bloodstream through villi of the arachnoid into venous sinuses of the skull and by way of the sheaths of cranial and spinal nerves. If, as sometimes happens, the foramina of the roof of the hindbrain fail to open, then the internal pressure of the cerebrospinal fluid within the brain causes the infant's head to swell. The condition is known as hydrocephaly.

EXPERIMENTS ON THE OUTGROWTH OF NERVE FIBERS

How do axons grow out from the nerve cell bodies? What guides them to their destinations?

In 1907 R. G. Harrison obtained an answer to the first of these questions in an experiment which marks the beginning of the culturing of tissues, that is, living tissues grown outside the body. He isolated small pieces of the neural tube of young frog embryos on clots of lymph taken from the lymph sacs of adult frogs. He then watched as axons sprouted from the neuroblasts and grew outward in much the manner that pseudopods push outward from the body of an amoeba. He thus was able to prove that axons grow all the way from nerve cell bodies to the periphery and are not formed enroute.

Paul Weiss, studying the regeneration of nerve fibers, demonstrated that an axon is a tube through which a stream of liquid "axoplasm" flows from the cell body toward the growing tip, driven along by "a sort of pumping action." The advancing tip may divide and start to flow in several directions; yet usually one tip dominates, and the other tips are withdrawn.

This, of course, is a description of how the nerve fiber grows out. It does not explain why a fiber takes one course rather than another and so contributes to the organized pattern of fibers in the nervous system.

Observations by Weiss and others have shown that the amoeboid tips of axons move along the surfaces of fibers and membranes. They never cross unbridged spaces. Furthermore, they tend to follow in the direction of lines of stress. These include the orientations of the macromolecules which take place when membranes are stretched. The axons also grow toward actively developing organs such as limb buds, possibly by following the oriented macromolecules in the surrounding region. Electrical and chemical factors do not seem to guide the growth of fibers, although such influences when applied experimentally may inhibit their growth.

According to Weiss's analysis, there is first a "pioneering phase" in which the outgrowths of the first fibers follow lines of stress and interfaces of contact. As a result, they reach their normal destination, namely, some peripheral sense organ or motor organ. Next there is an "application phase." In this, more fibers follow along in contact with the pioneer fibers. After reaching the peripheral organ, the fibers attach. Moreover, they retain their attachment as the peripheral organ (a premuscle mass, for example) migrates to a new location. Weiss calls this the "towing phase."

Principles of this sort still do not answer some basic questions: Why do the axons of a bipolar sensory neuron grow in opposite directions, one toward the periphery of the body, the other toward the central nervous system? Why do the axons of central neurons (alar plate) grow longitudinally and tangentially, while those of motor neurons (basal plate) take a radial course away from the brain and cord? Again, why do the tips of sensory neurons connect with sensory cells, while those of motor neurons join motor cells? Two views have been held in answer to these questions: One proposes that the nerve cells differentiate individually before the fibers grow out, that is, that the growing tips of the axons are chemically different from each other and each tip responds to the situation in which it grows in its own way. The other view is that the final functional specification of the nerve fibers is attained only *after* they have grown out. They become "modulated" by the peripheral cells to which they attach.

There is probably truth in both viewpoints. Much of the evidence comes from experiments on regeneration. Sperry performed operations in which he removed the eyes of salamanders or frogs, rotated them 180°, and then reimplanted them in the eye sockets of the same or opposite side. Recovery of sight took place but in a nonadaptive way (Fig. 12-16, *C* and *D*). The fibers which grew back from the retina to the brain became tangled, yet they reestablished connections with the tectum in an orderly manner and according to their original connections. For example, the original upper side of the retina—now the down side—became connected with that part of the tectum which originally receives fibers from the upper side

FIGURE 12-16 Experiments concerning the outgrowth of nerve fibers. *A.* In one experiment Weiss grafted the right forelimb of a salamander behind the left forelimb of another salamander. He arranged for it to become innervated by a branch of the brachial plexus. After recovery, the grafted limb moved as a mirror image of the normal limb. *B.* In another experiment he interchanged the right and left forelimbs. After recovery, the ingrafted limbs crawled backward when the animal sought to move forward. *C.* Sperry's experiment in which he grafted the right eye of a frog into the left eye socket of another frog but with the ventral side of the implanted eye up. After healing, when a fly (*) passed above and in front of the frog, the tongue shot forward and downward. *D.* Again, Sperry grafted a right eye into a left eye socket but with the anterior side toward the rear. The recovered frog responded to a fly passing at the rear by shooting its tongue forward. For explanation see text. *(Adapted from R. W. Sperry, 1956, Sci. Am., 194 (May): 50.)*

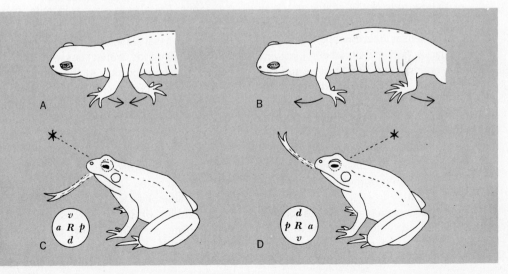

of the retina. As a result, when later a fly passed by, the tongue of the operated frog shot out reflexively; but it shot out downward when the fly passed overhead and upward when the fly was below.

It must be concluded that the growing tips of the regenerating axons are already specified and are able to find their proper destination. It has been suggested that the growing axon sends out multiple tips as "feelers" and then follows up the tip which discovers an appropriate connection.

In the case of motor neurons to the eye muscles, it has been found that the nerve fibers possess a considerable degree of specificity. In regeneration, the fibers establish connections most effectively with those muscle fibers with which they normally connect.

Chemical specificity of the growing tips of axons is probably not the whole answer to the patterning of the nervous system. If it were, what a tremendous variety of growing tips of axons there would have to be! Consider a motor-coordination center within the spinal cord which controls the movements of a limb. Some of the neurons of the center are flexor neurons and send impulses to flexor muscles. Others are extensor neurons and control extensor muscles, while still others innervate abductors, or adductors, etc. Shall we suppose that each axon, as it grows out during its development, "knows" what connections to make in order that the resulting limb movements will be orderly? Probably no. There is considerable experimental evidence that when a motor axon grows out from the basal plate, it makes functional connection with any muscle cell which is in its path.

Important observations by Weiss on the transplantation of limbs in urodele larvae throw light on this subject. He grafted a limb close behind the normal limb and arranged it so that it would be innervated by a branch (any branch) from the nerve supply of the normal limb. After the wound had fully healed, he observed that the second limb moved simultaneously with the normal limb and that it moved in the same manner (homologous response). It flexed when the normal limb flexed and extended when it extended. When the grafted limb came from the opposite side of the body, its motions were the mirror images of those of the normal limb (Fig. 12-16, A). In yet another experiment, he transplanted both forelimbs, right for left, each to the opposite side of the body (Fig. 12-16, B). When the animal, after having restored its nerve connections, sought to walk forward, its forelimbs moved in the wrong direction—opposite to the hindlimbs. It never learned to do otherwise.

How does one explain these findings? The answer seems to be that muscle cells differ from each other in a subtle way which is not as yet understood. They modify (the term Weiss uses is "modulate") the motor neurons which attach to them by imparting to them their own specificity. They, as it were, "infect" the axons. Then each axon transmits its acquired specificity backward to the motor cell body in the basal plate. The cell body responds by making dendritic connections with other cell bodies of like specificity. At the same time it develops a reciprocal relation to neurons of opposite specifications; that is, flexor neurons become antagonists to extensor neurons and vice versa, so that when one is excited the other is inhibited.

By way of summary we may say that the patterning of the nervous system passes through three phases: (1) a phase of regionalization, during which the various *areas* of the neural plate and neural crest become committed to their several fates; (2) a phase of guided outgrowth and selective attachment in the course of which *individual neurons* make functional connection with specific locations either inside or outside the central nervous system; and (3) a phase of modulation by which the peripheral organs, or centers within the central nervous system, exert a back influence on the neurons which supply them. The question of the relative importance of these three phases continues to be a subject of research and debate.

SELECTED READINGS

General References

Arey, L. B., 1965. Developmental Anatomy. W. B. Saunders Company, Philadelphia. 7th ed. Chap. 24, The histogenesis of nervous tissues. Chap. 25, The central nervous system.

Hamilton, W. J., J. D. Boyd, and H. W. Mossman, 1962. Human Embryology, The Williams & Wilkins Company, Baltimore. 3d ed. Chap. 12, Nervous system.

Romer, A. S., 1962. The Vertebrate Body. W. B. Saunders Company, Philadelphia. Chap. 16, The nervous system.

Bartelmez, G. W., and A. S. Dekaban, 1962. The early development of the human brain. Carnegie Contrib. Embryol., **37**:13–32. 30 plates.

Functional Pattern of the Nervous System

Kingsbury, B. F., 1922. The fundamental plan of the vertebrate brain. J. Comp. Neurol., **34**:461–491.

Herrick, C. J., 1924. Neurological Foundations of Animal Behavior. Holt, Rinehart and Winston, Inc., New York. Especially Chaps. 12 to 17.

Jacobson, C. O., 1959. The localization of presumptive cerebral regions in the neural plate of the axolotl larva. J. Embryol. Exptl. Morphol., **7**:1–21.

Kallen, B., 1965. Early morphogenesis and pattern formation in the central nervous system. In DeHaan and Ursprung (eds.), Organogenesis. Holt, Rinehart and Winston, Inc., New York. Pp. 107–128.

Watterson, R. L., 1965. Structure and mitotic behavior of the early neural tube. In DeHaan and Ursprung (eds.), Organogenesis. Holt, Rinehart and Winston, Inc., New York. Pp. 129–159.

The Outgrowth of Nerve Fibers

Harrison, R. G., 1907. Observations on the living developing nerve fiber. Anat. Record, **1**:116–118. Reprinted in Willier and Oppenheimer (eds.), 1964. Foundations of Experimental Embryology. Prentice-Hall, Inc., Englewood Cliffs, N.J. Pp. 98–103.

Weiss, P., 1939. Principles of Development. Holt, Rinehart and Winston, Inc., New York. Part IV, The development of the nervous system, pp. 491–557.

————, 1955. Organogenesis: Nervous system. In Willier, Weiss, and Hamburger (eds.), Analysis of Development. W. B. Saunders Company, Philadelphia. Pp. 346–401.

Sperry, R. W., 1965. Embryogenesis of behavioral nerve nets. In DeHaan and Ursprung (eds.), Organogenesis. Holt, Rinehart and Winston, Inc., New York. Pp. 161–186.

Szekely, G., 1966. Embryonic determination of neural connections. Advan. Morphogenesis, **5**:181–219.

chapter **13**

THE PERIPHERAL
NERVOUS
SYSTEM
AND SENSE
ORGANS

There is only one nervous system. The terms "central nervous system" and "peripheral nervous system" refer to parts of one functional whole. In fact the entire sensorineuromotor mechanism of the body performs as one unit. It is not surprising, therefore, to find that during development the several parts interact and are mutually interdependent.

The peripheral nervous system consists of nerves, that is, bundles of axons (white matter), and of ganglia, namely, aggregations of nerve cell bodies (gray matter) which are located entirely outside of the brain and spinal cord. The axons are of two types:

1 Sensory fibers (afferent fibers) carry impulses from the sense organs (receptors) to "nuclei of termination" (sensory-correlation centers) in the alar plate of the central nervous system (Fig. 12–2). They develop from neuroblasts of the neural crest. With the exception of the olfactory and optic nerves, their cell bodies are located in cranial and spinal ganglia.

2 Motor fibers (efferent fibers) carry impulses from "nuclei of origin" (motor-coordination centers) in the basal plate of the brain and cord to motor organs, namely, muscle and gland cells (effectors).

Most peripheral nerves consist of fibers of both types, but the sensory and motor fibers leave the spinal cord by different roots. They then combine to form the spinal nerve. The sensory fibers enter by the dorsal root in which the spinal ganglion is located. The motor fibers leave by the ventral root. In a modified way, this separation of sensory and motor fibers applies to the cranial nerves also, as we shall see.

It is customary to divide the components of the peripheral nervous system into six divisions which correspond to the six cell columns of the central nervous system (pp. 265, 266) (Fig. 13–1). These are:

1 The special somatic sensory division, which includes the olfactory, optic, and auditory nerves
2 The general somatic sensory division whose fibers come from the skin (exteroceptors) and body wall (proprioceptors)
3 The visceral sensory division, the fibers of which come mostly from the pharynx, branchial arches and gut wall
4 The general visceral motor division, which supplies smooth muscle and glands throughout the body—this constitutes the highly important autonomic nervous system (pp. 301–303)
5 The special visceral motor division, which sends fibers to striated muscles derived from the branchial arches
6 The somatic motor division, which goes to the striated muscles of the body wall, i.e., the axial and appendicular musculature

The peripheral nerves of the higher vertebrates include twelve pairs of cranial nerves and a varying number of spinal nerves (Fig. 13–2). The cranial nerves exit through foramina in the skull. Their functional components are as follows:

1. Olfactory	Special somatic and visceral sensory	Smell
2. Optic	Special somatic sensory	Vision
3. Oculomotor	Somatic motor	Four eye muscles
4. Trochlear	Somatic motor	One eye muscle

5. Trigeminal	Mixed	Skin senses of head and muscles of mastication
6. Abducens	Somatic motor	One eye muscle
7. Facial	Mixed	Taste (in part), salivary glands (in part), and muscles of facial expression
8. Auditory (acoustic)	Special somatic sensory	Equilibrium and hearing
9. Glossopharyn-geal	Mixed	Taste, salivary glands
10. Vagus	Mixed	Sensory and motor to larynx, lungs, heart, stomach, etc.
11. Spinal acces-sory	Special visceral motor	Trapezius
12. Hypoglossal	Somatic motor	Intrinsic muscle of tongue

The nerves referred to as mixed include general somatic sensory, visceral sensory, general visceral motor, and special visceral motor components.

A mnemonic device for remembering the order of the cranial nerves is the ditty: "On Old Olympus' Towering Top A Finn and German Viewed Some Hops." The first letter of each word is the same as that of the corresponding nerve.

The spinal nerves are pretty much alike in composition. Each originates from the spinal

FIGURE 13–1 Diagram of the components and central connections of the cranial nerves. The nerves are projected on a schematic cross section of the hindbrain.

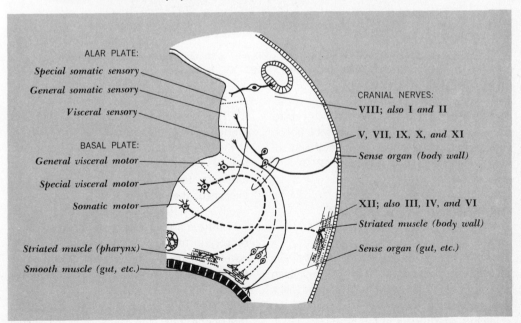

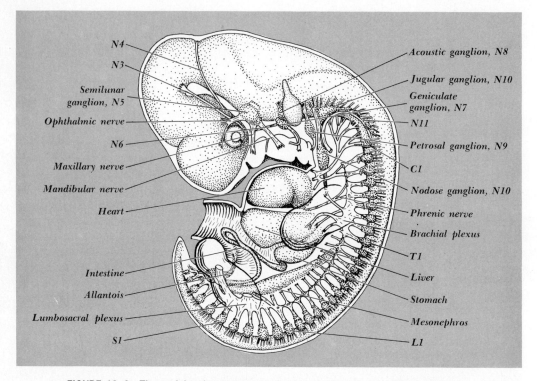

FIGURE 13–2 The peripheral nervous system of a 10-mm human embryo. (Redrawn from G. L. Streeter, 1908, Am. J. Anat., 8:285–301, Plate 1.)

cord by two roots: a dorsal root which carries somatic sensory and visceral sensory fibers, and a ventral root with visceral motor and somatic motor fibers (Figs. 12–1 and 12–2).

THE NOSE AND THE OLFACTORY NERVES (I)

The formation of the *olfactory placodes* was referred to in connection with the amphibian embryo (page 127) and the chick embryo (page 191). They arise as paired thickenings of the epidermis near the tip of the head (Fig. 13–3,

A). They develop in response to an induction exerted by the endoderm of the forward wall of the pharynx, but their normal development requires also the presence of neural-crest and forebrain tissue. After the placodes have thickened, they sink inward and become the *olfactory vesicles* or *nasal pits* (Fig. 13–3, *B*).

Fish have a pair of sacs of this kind, purely sensory in function. Air-breathing vertebrates develop snouts by the forward growth of the facial processes which border the nasal pits (page 291). The processes grow forward, the nasal cavities lengthen, and the original olfactory epithelia become deeply recessed. Also in air-breathing vertebrates, the nasal cavities break through the roof of the mouth cavity and form internal nares or *primitive choanae* (Fig. 11–3).

Our present concern is with the olfactory sensory epithelia. Many of the cells become bipolar neurons—the only instance in vertebrates where neurons retain the primitive character of epithelial cells. The peripheral pole of each neuron develops fine protoplasmic processes which project into the nasal cavity and serve as receptors of the sense of smell. The central pole sprouts an axon which grows toward the forebrain and, having entered it, terminates in an olfactory bulb (Fig. 13–3, *C*) (cranial nerves I). The bundles of axons thus formed are the olfactory nerves. In mammals, each olfactory nerve consists of several distinct strands instead of a single trunk.

THE EYES AND THE OPTIC NERVES (II)

The development of the eye is one of the most fascinating and instructive stories that embryology has to tell. Tissues from several sources contribute, yet the result is a single unified organ.

The Retina

The future retinal tissue is at first a single region situated at the anterior tip of the neural plate (Fig. 12–3, *A*). It becomes dumbbell

FIGURE 13–3   The development of the olfactory organ.

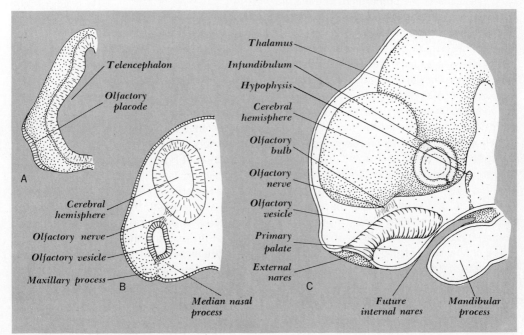

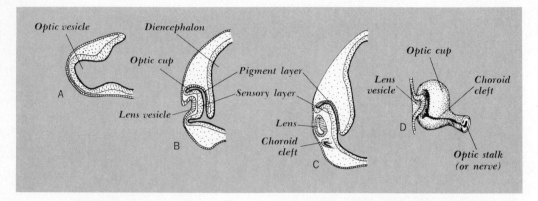

FIGURE 13–4 The development of the eye. The heavy line marks the original outer surface of the neural plate. It is this surface of the sensory layer of the retina which gives rise to the rods and cones.

shaped; each wing becomes the primordium of one of the retinas, and the middle connecting piece becomes the primordium of the *optic chiasma*. The latter is the most anterior tissue of the original neural plate.

After the closure of the neural tube, the retinal areas evaginate laterally as spherical *optic vesicles*. They push through the thin head mesenchyme until they reach the epidermis and temporarily adhere to it (Fig. 13–4, *A*). Both tissues then respond to the contact: the epidermis by forming *lens placodes;* the wall of the optic vesicle by thickening and invaginating to form the inside, or *sensory layer,* of the *optic cups* (Fig. 13–4, *B*). The outside layer of the optic vesicle, on the other hand, reacts to contact with the surrounding mesenchyme by becoming the thin *pigmented layer* of the retina.

The connections of the optic cups with the brain constrict and become the *optic stalks.* Later they are invaded by axons from the sensory layer of the retina and give rise to the optic nerves. Thus, embryology demonstrates that the retina is truly brain tissue and the optic nerves are tracts of the brain.

The inversion of the retina It will be instructive to trace the history of the original outer surface of the optic region of the neural plate. When the neural plate folds inward, its outer surface becomes the inner surface of the neural tube. When an optic vesicle invaginates and forms an optic cup, this original outer surface becomes the surfaces of contact between the sensory and pigmented layers of the cup (Fig. 13–4). It is from this same outer surface of the sensory layer that the *rods* and *cones* of the eye are formed (Fig. 13–5). In consequence of this arrangement, the light that enters the eye passes through the cornea, lens, and humors of the eye and then must pass through the entire thickness of the sensory layer before it reaches the receptors. This arrangement is spoken of as the "inversion of the retina."

Now the outer surface of the neural plate, like the outer surface of the epidermis, is potentially a ciliated epithelium. Hence, the inner surface of the neural tube of the embryo is potentially ciliated. Indeed, cilia may persist in certain of its areas even in the adult. It is therefore not surprising to learn that the rods and cones which are derived from the lining of the neural tube have been shown by the electron microscope to be modified cilia.

The choroid cleft   A problem now arises: How do the axons of the sensory layer of the retina find their way to the optic stalk? Must they pass around the margin of the optic cup? Or do they burst across the space that in the embryo separates the sensory and pigmented layers? They do neither. When an optic vesicle caves in and forms an optic cup, it does so, not only from its outer or distal surface, but also along its ven-

tral side (Figs. 8–14 and 13–4, *D*). Thus the optic cup is not a complete cup, but rather a cup with a cleft which extends to the optic stalk and even a little way along the stalk. The stalk, therefore, is a tube with one side folded inward. At the base of the optic cup both layers of the retina continue directly into the stalk. Thus the nerve fibers find a direct pathway to the brain. The place where nerve fibers leave the sensory layer is known as the *blind spot*. The choroid cleft also permits blood vessels to enter and leave the inner surface of the retina.

The Lens

Lens placodes form where the epidermis comes into contact with the optic vesicles. They thicken and then invaginate to form open

FIGURE 13–5 Longitudinal section through the human eye (diagrammatic).

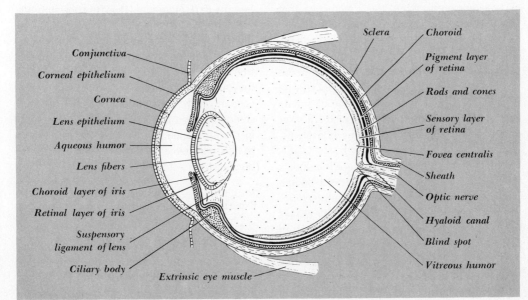

pits (Fig. 13–4). Shortly afterward they lose their connection with the epidermis and become hollow vesicles lying within the cavities of the optic cups. The outer wall of each sphere then thins and forms the lens epithelium, while the cells of the inner wall elongate and become columnar lens fibers. The central cavity is obliterated.

The Mesodermal Coats of the Eye

The outer coats of the eye are comparable to the meninges of the brain. The outer white, fibrous coat, the *sclera*, corresponds to the dura mater and serves to give strength to the eyeball (Fig. 13–5). It completely surrounds the eye, although its outermost part, where during development it comes into contact with the lens, becomes the transparent *cornea*. Actually the cornea consists of two layers: an outer epithelial layer which is derived from the outside epidermis (later the conjunctiva that lines the tear cavity), and an inner corneal layer of mesodermal origin. The inner coat of the eye, known as the *choroid*, corresponds to the pia mater. It is highly vascularized and, in the case of dark eyes, is abundantly supplied with pigment. The pigment of the choroid should not be confused with that of the pigment layer of the retina with which it becomes closely associated.

The choroid coat thickens where it contacts the periphery of the lens and forms a doughnut-shaped muscular ring known as the *ciliary body*. When in the functioning eye the muscles of this ring contract, they relieve the lens of outward tension, namely, the tension which results from the hydrostatic pressure within the eyeball. The result is that the lens, by reason of its own elasticity, thickens and so becomes optically a stronger lens. Thus nearby objects come into focus on the retina. At least this is true in the case of mammals. In birds, the ciliary body actually pushes on the periphery of the lens, forcing it to bulge. In bony fish, muscular contraction moves the lens backward so that it focuses on more distant objects. In amphibians it is the other way; contraction of muscle moves the lens forward as an adjustment to nearby objects. This process of focusing is known as accommodation.

Unlike the sclera, the choroid coat and the pigmented layer of the retina do not cover the front of the eyeball. Instead they extend forward only to the margin of the pupil. Both layers contain pigment which is seen through the transparent cornea as the colored *iris* of the eye. A dark iris is the result of the brown pigment of the choroid layer masking the dark purple of the pigmented layer which lies behind it. A blue iris is the result of a light-colored choroid scattering light and permitting the dark pigmented layer to show through.

A surprising feature is that the muscles of the iris by which it expands or contracts in adjustment to the amount of light are derived from the pigmented layer of the retina. Hence they are ectodermal in origin. This is the only case in which muscular tissue is derived from ectoderm. Also the ligamentous fibers, which in the developed eye connect the ciliary body to the perimeter of the lens, are said to be derived from ectoderm.

Experiments on the Eye

The ectoderm which is to form the retina undergoes induction during the course of gastrulation as a response to contact with the forward part of the archenteric roof (Fig. 5–17). This has been repeatedly demonstrated in amphibians and the chick. If the prechordal part of the archenteric roof fails to contact the ectoderm, no optic vesicle is formed.

The dependence of the eye on the archenteric roof does not end with gastrulation. Two eyes are the result of a process which takes place during neurulation, in which the prechordal mesoderm of the archenteric roof separates to the right and left and the overlying optic area is induced to do likewise. If, by chemical or other means, the separation of the prechordal mesoderm is suppressed, a single median cyclopic eye results. Moreover, the size of the eye depends upon the amount of prechordal mesodermal substrate. If, during neurulation, some of the mesoderm is removed, a small eye results. If the mesoderm of a species with a large eye is substituted, a larger eye will result.

During neurulation and for a considerable time thereafter, the optic areas continue to be harmonious equipotential systems; that is to say, any part may form a whole, or two optic vesicles superimposed may unite to form one single eye. Determination is progressive. As in the case of the limb bud, axiation takes place by three steps: first the anteroposterior axis becomes fixed, then the dorsoventral axis, and lastly the mediolateral axis. In fact, either the sensory or the pigmented layer of the retina is competent to give rise to a whole retina as late as the 72-hour stage of chick development. Eakin has found that if the sensory layer is reversed (turned inside out) almost up until the time the retina begins to function, the cells regulate. It seems likely that early contact of retinal tissue with the lens vesicle induces the formation of the sensory layer, that contact of retinal tissue with mesenchyme induces the pigmented layer, and that contact between the sensory and pigmented layer orients the cells of the sensory layer.

The lens of the eye is determined as a response of head epidermis to contact with an optic vesicle (Fig. 1-9). This was discovered by Spemann in 1901 and is a classic example of embryonic induction (page 12). If the optic vesicle is removed, ordinarily no lens forms. If foreign epidermis of a neurula, for example, epidermis from the flank, is grafted over the optic vesicle, or if the optic vesicle is implanted in the flank, the foreign epidermis is competent to form a lens. However, it was found that there are some species of amphibians in which a lens develops even though the optic vesicle is removed. This discovery led to much confusion and controversy.

The solution of the problem turns out to be that the determination of a lens takes place in two stages (pp. 130, 131). The first stage occurs during gastrulation when the region of the future lens comes under the influence of the archenteron. By this first induction it acquires lens-forming competence. In some species, this competence is sufficiently marked that the epidermis will produce ill-formed lenses, or lentoids, even though the optic vesicle never contacts it. The second stage we have already described. It takes place late in neurulation when the epidermis comes into contact with the optic vesicle. By this second induction the position of the lens is precisely fixed and its size determined. In some species, perhaps in most species, this second induction is so powerful that a lens will form even in strange epidermis, such as the epidermis of the flank.

THE MOTOR NERVES OF THE EYEBALL (III, IV, and VI)

Five nerves supply each eyeball (Figs. 13-2 and 13-6). The optic nerve (II) is the nerve of vis-

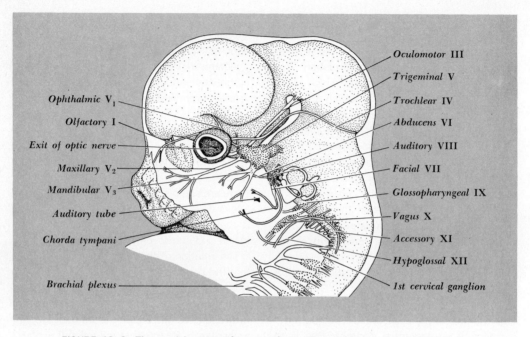

Ophthalmic V₁

Olfactory I

Exit of optic nerve

Maxillary V₂

Mandibular V₃

Auditory tube

Chorda tympani

Brachial plexus

Oculomotor III

Trigeminal V

Trochlear IV

Abducens VI

Auditory VIII

Facial VII

Glossopharyngeal IX

Vagus X

Accessory XI

Hypoglossal XII

1st cervical ganglion

FIGURE 13–6 The cranial nerves of a rat embryo. *(Drawn from J. A. Long and P. L. Burlingame, 1938, A Stereoscopic Atlas of the Rat Embryo.)*

ion; the trigeminal nerve (V) carries the sensory fibers of touch, muscle sense, and pain; while three purely motor nerves—the oculomotor (III), trochlear (IV), and abducens (VI)—innervate the muscles that move the eyeball. The oculomotor and trochlear arise from the midbrain. Abducens originates in the hindbrain. All three are somatic motor nerves whose cell bodies are located in the basal plate. Their axons leave the brain and enter the closely adjacent mesenchyme of the region from which the extrinsic eye muscles are developed. (The nerves to the smooth muscle and glands of the eye are referred to in connection with the autonomic nervous system.)

As the muscle masses migrate and insert on

the outer or sclerotic coat of the eyeball, they drag their nerves after them. The first premuscle mass gives rise to the anterior rectus, the dorsal rectus, the ventral rectus, and the ventral oblique. These, respectively, adduct, elevate, depress, and rotate the eyeball laterally. It is supplied by the *oculomotor nerve* (III). The second such mass produces the dorsal oblique which rotates the eyeball medially. It is served by the *trochlear nerve* (IV). This nerve is peculiar in that its fibers, instead of emerging from the floor of the midbrain directly, wander upward within the substance of the midbrain and leave the brain at the top of the isthmus (Fig. 13–6). The third premuscle mass gives origin to the posterior rectus. Since it abducts the eyeball, its nerve is known as the *abducens* (VI).

THE EARS AND
THE AUDITORY NERVES (VIII)

The ear of higher vertebrates consists of three parts: inner ear, middle ear, and outer ear. Of these, the inner ear develops from epidermis lateral to the hindbrain (Fig. 13–7). The middle ear is derived from the first pharyngeal pouch and is lined by endoderm. It encloses the little ear bones: three in mammals, one in other air-breathing vertebrates. These are derived from branchial cartilages, and hence are visceral mesodermal. The outer ear comes from the first branchial furrow, which is lined by ectoderm. Only the inner ear is present in water-breathing vertebrates, and its function is principally that of an organ of equilibrium. Hearing was added progressively during the evolution of higher vertebrates. The middle ear is present in amphibians; but an outer ear does not occur below reptiles, birds, and mammals. Indeed, a *pinna* projects from the side of the head only in mammals. The entire story of the ear is indeed a spectacular one.

The Inner Ear

The development of the otocysts has been described (page 130). At first these vesicles are open to the outside, but as development proceeds they become detached from their parent epidermis, except in primitive fish. In

FIGURE 13–7   The development of the ear.

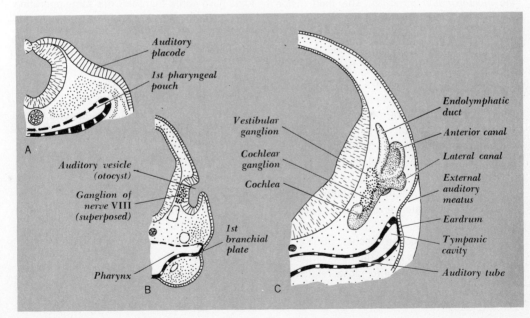

mammals, each consists of two regions: a dorsal region which gives rise to the *utricle* and to three *semicircular canals;* and a ventral region which becomes the *saccule* and *cochlea* (Fig. 13–8). The utricle, semicircular canals, and saccule are organs of equilibrium. The cochlea—it is represented by the so-called lagena in vertebrates lower than mammals—is concerned with hearing.

The otocyst and its derivatives become enclosed by a capsule of dense mesenchymatous tissue which in time becomes the periotic cartilage and finally, bone (Figs. 13–8 and 16–5). But this process does not take place before a liquid-filled space has hollowed out between the derivatives of the otocyst and the surrounding mesenchyme. The liquid-filled canals and chambers are called the *bony labyrinth*. The fluid which fills them is the *perilymph*. The parts derived from the otocyst constitute the *membranous labyrinth*. The fluid inside the membranous labyrinth is the *endolymph*.

What decides when and where an ear shall form? Experiments on amphibians indicate that an otocyst develops from head epidermis in response to two successive inductions (page 131). The first takes place at the end of gastrulation and the beginning of neurulation when the prospective auditory region comes under the influence of the mesenchyme of the head. It is an aspect of the general deuterencephalic induction of the posterior organs of the head (page 113). As a result, the auditory region acquires the competence to become a placode and then a vesicle.

The second induction occurs during the course of neurulation when the now-competent auditory region comes into close relation to the hindbrain. It then acquires the capacity to become a vesicle of a definite size, definite

form, and exact location. If at an early stage of neurulation brain tissue is removed, the otocysts, if they form at all, are small and imperfect. Later in neurulation, however, when the neural folds have come together, the same operation does not prevent normal otocysts from developing.

At first the otocyst, like so many other organs, is a harmonious equipotential system. If an auditory placode is cut away, the surrounding epidermis closes in, the wound heals, and an otocyst is regenerated. At this time any part will form a whole otocyst. Two otocysts implanted side by side will unite and form a single oversized vesicle. But the capacity to reorganize wanes as neurulation progresses. At first, a right auditory placode transplanted to the left side with its anteroposterior axis reversed will regulate and develop an otocyst conforming to the left side. But after the neural folds have come together, the same operation results in an otocyst of reversed polarity. This shows that the anteroposterior axis has been determined. Not until the tail-bud stages does the dorsoventral axis become fixed. The amount of regulation which takes place depends on the strength of the inductive influences of the environment. Tissues of a young host have more influence on a transplanted otocyst than those of an older host.

The relationship between the otocyst and the surrounding mesenchyme which forms the auditory capsule is one of inductor and induced. If an otocyst is removed, a cartilaginous capsule does not form. If, on the other hand, an otocyst is implanted in a strange location, for example, in the flank of a tail-bud embryo, cartilage-forming cells may gather around and generate a capsule—although not always a perfect one. The source of the cartilage cells in this case is sclerotome, or possibly somatic mesoderm.

The Middle Ear

The middle ear is derived from the first pharyngeal pouch and gives rise to the *tympanic cavity* and *auditory* (eustachian) *tube* (Fig. 13-7). It is homologous with the spiracular sac of the shark. The first branchial plate, formed where the endoderm of the first pouch comes into contact with the ectoderm of the side of the head, may or may not break through. It becomes the eardrum or, more precisely, it fills in and is later replaced by the eardrum. A structure of this sort is present for the first time in amphibians.

The first pharyngeal pouch is ventral to the otocyst, but it develops a wing which pushes dorsally into head mesenchyme until it comes to lie between the otocyst and the lateral epidermis. Rather late in development, this wing expands to form the *tympanic cavity*. As it does so, it enfolds the proximal ends of the cartilages of the first and second branchial arches. In the mammal the cartilages become the three ear bones: *malleus, incus,* and *stapes* (Fig. 13-8). Malleus and incus are from the first branchial arch and correspond to the proximal cartilages of the lower and upper jaws of lower vertebrates. Stapes is a cartilage of the second branchial arch and is homologous with the hyomandibular cartilage which suspends a shark's jaws from its skull. Stapes

FIGURE 13-8 Schematic diagram of the human ear. The relations of the outer, middle, and inner ear are shown as in the embryo, yet much of the detail is that of more advanced stages. (*The membranous labyrinth is from G. L. Streeter, 1906, Am. J. Anat., 6:139–166, Plate 2.*)

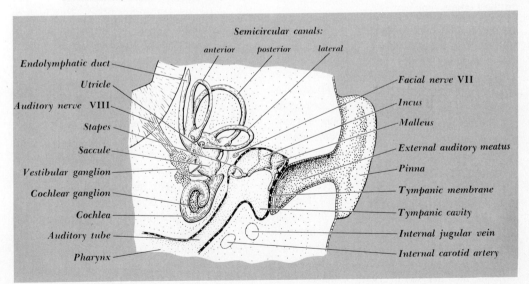

is the only ear bone present in amphibians, reptiles, and birds.

The Outer Ear

The outer ear forms from the ectodermal furrow which lies between the first and second branchial arches (Figs. 13-7 and 13-8). It is shallow in amphibians and does not persist. In reptiles, birds, and mammals, however, a pit develops from it which grows inward until it makes contact with the endoderm of the tympanic cavity. The pit is the *external auditory meatus*. The ectoderm of the pit, the endoderm of the cavity, plus a small amount of mesoderm give rise to the definitive ear drum, or *tympanic membrane*.

In mammals, the first branchial furrows are bordered by small hillocks which grow outward and mold into the *pinna* or visible "ear" (Figs. 11-2 and 13-8). (The word pinna is a general term; its synonym, "auricle," applies primarily to the human "ear.")

The Auditory Nerves (VIII)

The auditory or acoustic nerves (VIII) are listed as the eighth pair of cranial nerves. Like the olfactory and optic nerves they belong to the special somatic sensory division, but unlike these two nerves their cell bodies are located in ganglia. At first the ganglia of the auditory and facial nerves lie close together, just anterior to an otocyst (Figs. 13-2 and 13-6). The peripheral fibers of the bipolar neurons of the auditory nerve grow into small patches on the otocyst wall, namely, patches which become the sensory cristae of the semicircular canals, the maculae of the utricle and saccule, and the spiral organ of the cochlea

(Fig. 13-8). When the otocyst differentiates into a dorsal and a ventral part, the auditory nerve divides into a vestibular branch which functions in equilibrium and a cochlear branch which functions in hearing.

THE NERVES OF THE BRANCHIAL ARCHES (V, VII, IX, X, and XI)

The trigeminal nerve (V), facial nerve (VII), glossopharyngeal nerve (IX), vagus nerve (X), and the accessory (XI) supply sensory and motor organs derived from the branchial arches (Fig. XI is motor only). The cell bodies of their sensory neurons are in ganglia which are derived from the neural crest of the hindbrain and adjacent epidermal placodes. The motor neurons of the same nerves originate in the basal plate of the hindbrain. These branchial arch nerves resemble the dorsal roots of spinal nerves. They differ in that they carry visceral motor fibers. This feature is probably primitive, for the dorsal roots of the spinal nerves of primitive fish also carry visceral motor fibers.

The Trigeminal Nerve (V)

The trigeminal is the nerve of the first branchial arch. It also carries sensory fibers from the entire face, as well as from the optic, nasal, and oral cavities. As its name suggests, it has three branches: (1) the *ophthalmic nerve* from the region of the eye and forehead, (2) the *maxillary nerve* from the upper jaw, and (3) the *mandibular nerve* from the lower jaw. Its great semilunar or Gasserian ganglion is an outstanding feature of even young embryos (Figs. 8-14, 8-19, and 13-6). Its three divisions can be recognized in sections of a 6-mm pig.

Sensory fibers are carried by each of the three divisions of the trigeminal nerve. For example, the ophthalmic carries impulses of touch and pain from the eyeball, the maxillary carries touch and pain from the tooth sockets of the upper jaw, and the mandibular carries like impulses from the lower jaw. Motor fibers are present only as an accompanying part of the mandibular branch. They supply the muscles derived from the first branchial arch, namely, the muscles of mastication and also a small muscle which inserts on the malleus, the tensor tympani.

The Facial Nerve (VII)

The facial nerve supplies the second branchial arch. In fish and larval amphibians it is a sensory nerve of the lateral-line organs of the head, but this function has been lost in higher vertebrates. It does carry visceral sensory fibers, however, from the walls of the pharynx, including some from the taste buds of the tongue. It carries general visceral motor fibers to the lacrimal, sublingual, and submaxillary glands via a branch which crosses the eardrum. The branch is known as the chorda tympani (Fig. 13–6). Its principal function in mammals, however, is that of a special visceral motor nerve to the muscles of expression (page 350). It is this function, and not its sensory functions, which gives it its name, the facial nerve. Its ganglion, known as the geniculate ganglion, can be seen in young embryos closely associated with the ganglion of the eighth nerve and immediately anterior to the otocyst (Fig. 13–2).

The Glossopharyngeal Nerve (IX)

This is the proper nerve of the third branchial arch. It emerges from the hindbrain a short distance posterior to the otocyst. It has two ganglia: one of them dorsal and close to the brain (the superior ganglion), and one of them ventral and more distant from the brain (the petrosal ganglion) (Fig. 13–6). Its main sensory function is to serve, along with the facial nerve, the taste buds of the tongue. It is also the main motor nerve to the parotid gland, the salivary gland which is ventral to the ear.

The Vagus Nerve (X)

The vagus nerve serves the remaining branchial arches and organs of the splanchnopleure as far posterior as the intestine. Its sensory functions include supplying the receptors of the pharynx, larynx, trachea, lungs, heart, esophagus, and stomach. It carries visceral motor fibers to cardiac muscle and to the smooth muscles and glands of the lungs, esophagus, stomach, and intestine.

In the young embryo, the vagus nerve can be seen arising by a series of roots from the side of the hindbrain (Figs. 12–15, A, 13–2, and 13–6). The several roots converge to form a trunk which circles forward and ventrally and then turns posteriorly. The trunk has two ganglia: one close to the brain (the jugular ganglion), and one located a short distance posterior to it (the nodose ganglion).

The Spinal Accessory Nerve (XI)

The spinal accessory nerve arises from the spinal cord and courses forward to join the vagus nerve (Figs. 13–2 and 13–6). In reptiles, birds, and mammals it enters the skull along with the spinal cord. It is listed as a cranial nerve because it emerges along with the vagus through a foramen of the skull. Its counterparts in fish and amphibians do not exit through the skull and hence are not considered to be cranial nerves. It is separate from the vagus, however, and supplies the trapezius group of shoulder muscles (page 349). Since these muscles are historically of branchial origin, the accessory nerve is listed as a special visceral motor nerve.

THE HYPOGLOSSAL NERVE (XII), THE MOTOR NERVE OF THE TONGUE

Four cranial nerves supply the tongue. The trigeminal (V) is the nerve of touch and pain. The facial (VII) and glossopharyngeal (IX) carry taste fibers. The hypoglossal (XII), which is the last of the cranial nerves, supplies the intrinsic muscles of the tongue. Its fibers emerge from the somatic motor column of the hindbrain ventral to those of the vagus nerve (Figs. 12–15, A, 13–2, and 13–6).

The so-called tongue of gill-breathing vertebrates is not a muscular organ, and the muscles which in them correspond to the tongue muscles of higher vertebrates are hypobranchial muscles located ventral to the pharynx. These, like the tongue muscles of higher vertebrates are derived from occipital somites (page 349) and reach their destination by detouring around the posterior border of the pharyngeal arches and then growing forward on the ventral side of the head. As they do so they tow their nerve supply after them. These are the nerves which correspond to the hypoglossal nerve of higher vertebrates. They

are not cranial nerves in lower vertebrates because they do not exit through the skull.

In mammals, a migration of the muscles of the tongue has not been directly demonstrated; but the circuitous course taken by the hypoglossal nerve as it loops downward behind the pharynx and then forward to the floor of the mouth leaves little doubt that in some manner a migration of this sort has taken place (Fig. 13–6).

THE SPINAL NERVES

Although the segmental nature of the cranial nerves is obscure, that of the spinal nerves is definite and clear. Roots and ganglia repeat themselves at regular intervals through the full length of the spinal cord (Fig. 13–2). Now, this segmentation of the nervous system is secondary to that of the mesodermal somites, for if axial mesoderm is experimentally removed, the neural crest does not subdivide into ganglia. Moreover, the number of nerve fibers which a spinal nerve carries is regulated by the size of the area, sensory or motor, which the nerve supplies.

The spinal nerves are fundamentally all alike. Each arises by two roots: a *dorsal root* within which the spinal ganglion is located and which is made up of sensory axons produced by neuroblasts of the ganglion; and a *ventral root* of motor fibers from motor neurons of the basal plate of the spinal cord (Fig. 13–9). The dorsal and ventral roots join to form the trunk of a spinal nerve. Then almost immediately the trunk divides into (1) a *dorsal ramus* (actually dorsolateral), which goes to the skin, the deep muscles of the back, and the spinal column, and (2) a *ventral ramus* which supplies the lateral and ventral body wall. The

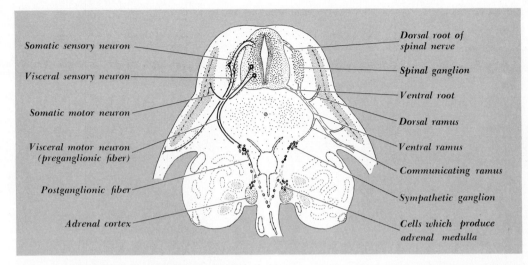

FIGURE 13-9 Diagrammatic section of a mammal embryo showing a spinal nerve and its functional components.

relation of the spinal nerves to the autonomic nervous system will be described presently.

In the region of the fore- and hindlimb buds, the ventral rami join with one another by connecting branches to form plexuses: the brachial plexus of the forelimb and the lumbosacral plexus of the hindlimb (Fig. 13-2). It is of interest that the nerves of these plexuses do not send branches to the ventral body wall as do the spinal nerves which are not included in the plexuses. This fact may have some bearing on the origin of the limbs.

THE AUTONOMIC NERVOUS SYSTEM

The nerves which supply smooth muscle, heart muscles, and glands are to a large extent "a law to themselves"—hence the name *autonomic nervous system*. They constitute the general visceral motor division of the nervous system.

Among the peculiarities which distinguish them from the nerves which go to striated muscle (that is, the somatic motor and special visceral motor nerves), is the fact that the primary visceral motor neurons relay their impulses to a second set of motor neurons located outside of the central nervous system (Fig. 13-9). The primary motor neurons which leave the brain or cord, whose cell bodies are in the basal plate, are known as *preganglionic neurons*. The motor neurons to which they relay impulses and whose cell bodies are located in autonomic ganglia or in the walls of the organs which they supply are called *postganglionic neurons*. This arrangement permits one preganglionic neuron to activate many postganglionic fibers, each of which, in turn, stimulates a large number of smooth muscle or gland cells into action.

From the functional standpoint there are two sets of autonomic fibers: parasympathetic fibers and sympathetic fibers (Fig. 13-10):

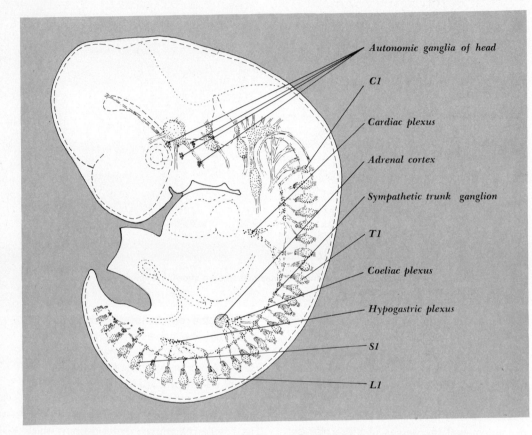

FIGURE 13–10 Diagram of the autonomic nervous system. See Fig. 12–2 for the names of the cranial nerves. Nerves V, VII, IX, and X, and also the sacral nerves carry autonomic fibers of the parasympathetic division. The trunk ganglia, and the cardiac, coeliac, and hypogastric plexuses belong to the sympathetic system. (Based on G. L. Streeter, 1912, in Keibel and Mall (eds.), Manual of Human Embryology, Lippincott, Philadelphia, Vol. 2, p. 151.)

1 The *parasympathetic fibers* arise in the hindbrain and lower spinal cord and travel by way of the trigeminal, facial, glossopharyngeal, vagus, and sacral nerves to smooth muscles and glands throughout the body. Because of their origin the parasympathetic nerves are also called the cranio-sacral nervous system.

They act primarily in the normal control of the organs which they innervate.

2 The *sympathetic fibers* leave the thoracic and lumbar regions of the spinal cord by way of the ventral roots of the spinal nerves. They travel forward and backward in the sympathetic nerve trunks to the same smooth muscles and glands that the parasympathetic nerves supplied. Because of their origin, the sympathetic nerves are often called the thoracico-lumbar nervous system. They come into action mainly at times of emotion and stress.

Thus, it comes about that each smooth muscle and gland is controlled by two nerves. Moreover, these nerves usually have opposite effects. For example, in the case of the salivary glands and stomach, the parasympathetic nerves stimulate the smooth muscles and glands into normal activity, while the sympathetic nerves inhibit the functioning of these same organs. In the case of the heart, the parasympathetic nerves hold the organ in check, while the sympathetic nerves cause it to speed up.

The postganglionic motor neurons originate from neuroblasts which are derived from the neural crest; but they migrate farther from their origin than do the sensory neuroblasts which give rise to the cranial and spinal ganglia. Those of the parasympathetic system follow the branchial nerves (V, VII, IX, and X) to the walls of the organs (salivary glands, heart, stomach, etc.) which they supply. They relay the impulses which come to them, to smooth muscle and glands.

The neuroblasts which give rise to the sympathetic nervous system, on the other hand, migrate to the sides of the aorta. Here they consolidate into chains of autonomic ganglia, the sympathetic trunks. Some neuroblasts migrate even farther and become ganglionic masses near the bases of the coeliac and mesenteric arteries. But wherever they go, they drag their nerve supply, namely, the pre-ganglionic fibers, after them (Figs. 13-9 and 13-10).

THE ADRENAL GLANDS

There is an important exception to the principle that preganglionic neurons relay their impulses to postganglionic neurons. This is the *adrenal gland* (Figs. 13-9 and 13-10). An adrenal gland has two parts: an outer layer, or cortex, and an inner "pith" or medulla.

1 The *cortex* is derived from the mesodermal epithelium of the surface of the mesonephros (Fig. 18-8). It is comparable in its origin to the secretory cells of the ovary and testis. It is comparable to them also with respect to the chemical nature of the hormones which it secretes. They are steroids.

2 The *medulla* of the adrenal gland, on the contrary, is of the same origin as the autonomic ganglia. Its secretory cells are derived from the neural crest and hence are to be thought of as modified postganglionic neurons. The hormone epinephrin (adrenalin) which they secrete is chemically and functionally similar to the norepinephrin (noradrenalin) liberated at the nerve endings of the sympathetic nervous system.

At the beginning of development these two groups of cells, cortex and medulla, are separate, and indeed they remain separate in some lower vertebrates; but in birds and mammals, the cells of the future medulla invade the mesodermal rudiment which gives rise to the adrenal cortex, and together they form the adrenal gland.

SELECTED READINGS

Arey, L. B., 1965. Developmental Anatomy. W. B. Saunders Company, Philadelphia, 7th ed. Chap. 26, The peripheral nervous system. Chap. 27, The sense organs.

Goodrich, E. S., 1930. Studies on the Structure and Development of Vertebrates. The Macmillan Company, London. Republished by Dover Publications, Inc., New York. Chap. 14, Peripheral nervous system and sense organs.

Hamburger, V., and Rita Levi-Montalcini, 1949. Proliferation, differentiation, and degeneration in the spinal ganglia of the chick embryo under normal and experimental conditions. J. Exptl. Zool., **111**:457–502.

Hörstadius, S., 1950. The Neural Crest. Oxford University Press, London.

The Eye

Twitty, V., 1955. Organogenesis: the eye. In Willier, Weiss, and Hamburger (eds.), Analysis of Development. W. B. Saunders Company, Philadelphia. Pp. 402–414.

Lopashov, G. V., and O. G. Stroeva, 1961. Morphogenesis of the vertebrate eye. Advan. Morphogenesis, **1**:331–377.

Coulombre, A. J., 1965. The eye. In DeHaan and Ursprung (eds.), Organogenesis. Holt, Rinehart and Winston, Inc., New York. Pp. 219–251.

Eakin, R., 1947. Determination and regulation of polarity in the retina of Hyla regilla. Univ. Calif. Publ. Zool., **51**:245–287.

The Nose and Ear

Yntema, C. L., 1955. Organogenesis: Ear and nose. In Willier, Weiss, and Hamburger (eds.). Analysis of Development. W. B. Saunders Company, Philadelphia. Pp. 415–428.

Streeter, G. L., 1906. On the development of the membranous labyrinth and the acoustic and facial nerves of the human embryo. Am. J. Anat., **6**:139–166.

The Autonomic Nervous System

Nawar, G., 1956. Experimental analysis of the origin of the autonomic ganglia in the chick embryo. Am. J. Anat., **99**:473–506.

Levi-Montalcini, Rita, and P. V. Angeletti, 1965. The action of nerve growth factor on sensory and sympathetic cells. In DeHaan and Ursprung (eds.). Organogenesis. Holt, Rinehart and Winston, Inc., New York. Pp. 187–198.

Kuntz, A., 1953. The Autonomic Nervous System. Lea & Febiger, Philadelphia, 4th ed.

chapter 14

THE
DIGESTIVE
AND
RESPIRATORY
SYSTEMS

Generally speaking, all many-celled animals higher in the zoological scale than coelenterates and flatworms have complete alimentary tracts; i.e., they have digestive systems which begin with a mouth and end in an anus. Typically, there is a *stomodaeum,* or cavity lined with ectoderm, at the anterior end; a midregion, or *enteron* (gut proper) lined with endoderm; and lastly a *proctodaeum* lined by ectoderm at the posterior end (Fig. 14–1). (The terms foregut, midgut, and hindgut are commonly used in invertebrate zoology

to designate these three regions; but, as we have seen, they have other meanings in vertebrate embryology.)

THE STOMODAEUM

The stomodaeum was described in Chapter 10 in connection with the development of external form. Recall that at first it is shallow and remains so in the lower vertebrates. In the higher vertebrates it becomes deeper through the

FIGURE 14–1 Digestive and respiratory systems of a 6-mm pig embryo. The ectoderm of the stomodaeum and proctodaeum are shown by fine stippling; the endoderm of the gut proper is indicated by coarser stippling. An asterisk marks the partition which invades the pharynx and divides it into a ventral larynx and trachea and a dorsal-laryngeal pharynx. Another asterisk indicates the partition which divides the cloaca into the ventral urogenital sinus and the dorsal rectal sinus.

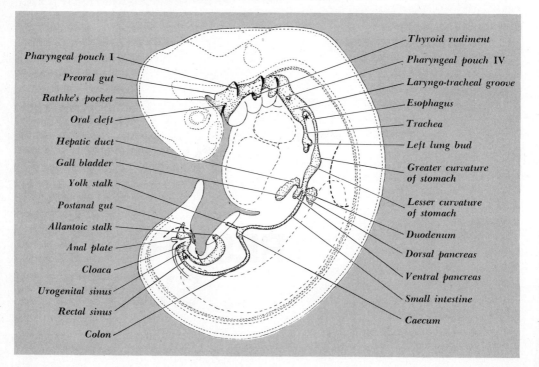

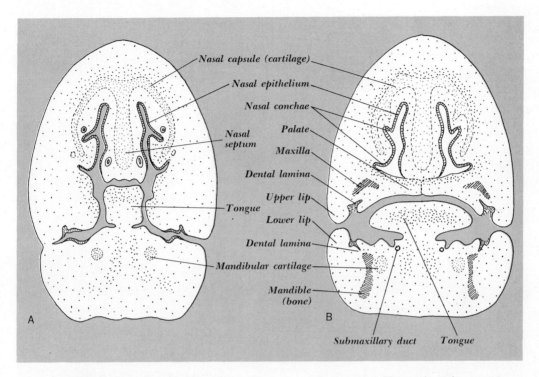

FIGURE 14–2 Sections through the snout of an 18-mm and a 25-mm pig embryo (somewhat schematic).

forward growth of the nasal, maxillary, and mandibular processes which form the face. The palatine shelves and nasal septum divide the stomodaeum into two nasal cavities above and the oral cavity below (Figs. 11–3 and 14–2).

The oral plate at first lies between the stomodaeum and enteron. Later it disappears and leaves no indication of its location. Nevertheless, the larger part of the oral cavity of birds and mammals is of stomodaeal origin. The lips, teeth, and salivary glands, as well as most of the surface of the body of the tongue, are derived from ectoderm rather than endoderm (Fig. 14–3). The place where Rathke's

pocket (the hypophysis) evaginated marks the posterior limit of ectoderm along the roof of the stomodaeum.

The Teeth

Vertebrate teeth are derived from the ectoderm and adjacent mesoderm of the stomodaeum. (In some lower vertebrates teeth are derived also from endoderm.) They are therefore not endoskeletal structures in the usual sense. Their mode of origin suggests the scales of sharks; that is, there is a layer of epidermis which surrounds a central dermal papilla. The inner cells of the epidermis se-

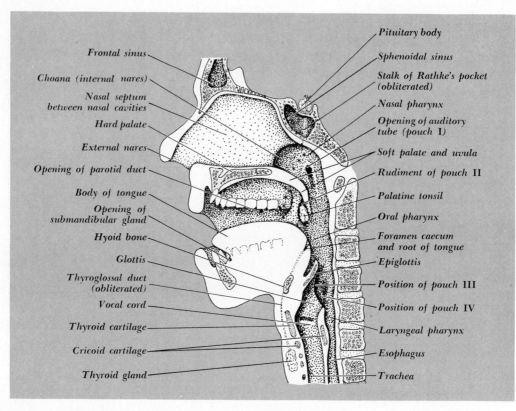

FIGURE 14–3 Median view of the right half of a human head, showing the derivatives of the stomodaeum and pharynx. The approximate boundary between the ectoderm and endoderm is marked by a row of asterisks. *(Adapted from B. M. Patten, 1964, Foundations of Embryology, McGraw-Hill Book Company, New York, 2d ed.)*

crete a stony layer, the *enamel.* The outer cells of the dermal papilla deposit a bonelike substance known as *dentine.* The central cells of the papilla remain as a soft *pulp.*

Tooth development in mammals begins near the borders of the upper and lower jaws, just inside the *labial groove* which separates the lips from the gums (Fig. 14–2). It begins when a ridge of ectoderm folds inward, invading the mesenchyme of the gum. The ridge, known as

the *dental lamina,* is at first continuous, but soon *tooth germs* bud off from it (Fig. 14–4, *A*). Then each germ invaginates on the side away from the mouth cavity and forms a sort of cap or bell which encloses a papilla of mesenchyme. The inner lining of the cap is termed the *enamel organ;* the papilla is the *dental papilla.* The shape of the surface of contact between the ectoderm of the enamel organ and the mesenchyme of the papilla determines the form of the crown of the future tooth. Once calcification has begun, no further change of pattern is possible.

The process of calcification begins at what will be the apex of the tooth (it begins at the tip of each cusp in the case of molar teeth) and progresses toward the root of the tooth (Fig. 14-4, *D*). The inner cells of the enamel organ, called *ameloblasts,* elongate and secrete prisms of crystalline calcium phosphate (apatite), the mineral constituent of *enamel.* The outermost cells of the dental papilla, known as *odontoblasts,* also deposit calcium phosphate; but they do so in a softer form, namely, dentine, which differs little from bone except that cells do not become trapped within it, as the osteoblasts do in bone. The center of each dental papilla remains as a *pulp cavity* richly supplied with a nerve and blood vessels.

The dental laminae degenerate and disappear but not before they have given rise to a second series of tooth germs, the primordia of the permanent dentition.

The ameloblasts and odontoblasts deposit their respective secretions layer upon layer with the result that the crowns thicken and the roots grow in length. When the enamel of the crown is complete, the ameloblasts cease to secrete. For a time, they remain as a sheath around the root of the tooth; yet, even this sheath disappears, and the mesenchymal cells which surround the root organize as a *dental sac* comparable to the periostium of a bone. The sac deposits a bonelike material around the root, known as *cementum.* It also produces a connective tissue membrane, the *periodontium,* which anchors the tooth in its bony socket (Fig. 14-4, *E*).

One by one, beginning at the front of the

FIGURE 14-4   The development of human teeth. A to C. The development of enamel organs. (Adapted from L. B. Arey, 1965, Developmental Anatomy, W. B. Saunders Company, Philadelphia, 7th ed. After Eidmann.) D and E. Differentiation of a human incisor. (Adapted from Arey, 1965.)

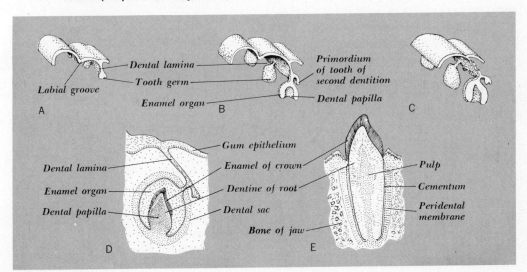

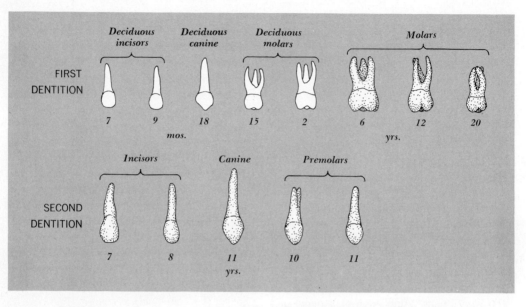

FIGURE 14–5 The sequence of the human dentition (upper jaw). Numerals indicate the approximate age at which each tooth erupts. The permanent teeth are stippled. (× ¾).

jaws, the teeth erupt through the gums. In man, the sequence is as indicated in Fig. 14–5. The first dentition consists of two incisors, one canine, two deciduous molars, and three molars on each side of each jaw. These teeth, with the exception of the molars, are replaced by a second dentition. The molars plus the teeth of the second dentition are permanent teeth.

The teeth of mammals furnish clear evidence against the Lamarkian concept that evolution has been accomplished through the cumulative effects of use and disuse. Although mammalian teeth are remarkably adapted to each animal's manner of life, yet they are fully patterned before they erupt through the gums and come into use. Indeed, teeth of characteristic pattern are occasionally found in tumors deep within an animal's body.

Experiments on the Determination of Teeth

What decides when and where a tooth shall develop? What determines its pattern? A complete answer to these questions is not as yet forthcoming, but experiments on salamander embryos have yielded some information.

It has been demonstrated that tissues from three sources are involved in the production of teeth: neural-crest cells of the head, ectoderm of the stomodaeum, and endoderm of the floor of the pharynx. The neural-crest cells provide the material of the dental papillae. If the neural crest is removed, no teeth are formed. But neural crest alone will not form teeth. When explanted, neural-crest cells produce pigment cells, fibroblast-like cells, neuroblasts, and mesenchyme but never tooth structures.

The ectoderm of the stomodaeum forms the dental ridge and the tooth germs, but, here again, by itself it does not form teeth. Wilde has found that when neural-crest cells and ectoderm of the stomodaeum are explanted together, they still do not form tooth structures. They give rise to epidermis, brain tissue, mesenchyme, and occasionally even to nasal placodes but never to teeth. He found, however, that by adding endoderm of the floor of the pharynx to the explants of neural crest and stomodaeal ectoderm, he obtained a variety of head tissues, including teeth. This takes place even when endoderm itself does not enter into the composition of teeth.

Holtfreter found that ectoderm taken from elsewhere on the body and grafted to the ventral side of the head gives rise to teeth. Even more surprising, Balinsky discovered that if the ectoderm of the stomodaeum is removed, endoderm may move in and take its place and produce enamel organs.

It seems that some factor related to the underside of the head is instrumental in the determination of teeth. What is this factor? It may be that Wilde uncovered the answer when he observed that tooth germs always arise in close relation to oral-plate tissue. When oral-plate material appeared in his explants, tooth germs were apt to be present. Otherwise they were not present.

Tooth germs, once they have been determined, are self-differentiating. When transplanted to new locations, they produce enamel organs and papillae. However, the epithelium and the papillae are interdependent. If the inner epithelium is removed, a papilla does not form odontoblasts or give rise to dentine. Reciprocally, if the odontoblasts are taken away, the ameloblasts of the epithelium do not produce enamel, but instead, revert to a squamous epithelium.

The Salivary Glands

Well-developed salivary glands are characteristic of mammals but less so of other vertebrates. Their function is to produce a mucus which lubricates the food while it is chewed and swallowed. In some cases—and this includes man—they secrete an enzyme, ptyalin, which initiates starch digestion.

There are typically three pairs of salivary glands: (1) the *parotid glands,* located ventral to the ear, whose ducts open in the cheek (Fig. 14-3); (2) the *submaxillary glands,* ventral to the lower jaw, with ducts which open beneath the tongue; and (3) the *sublingual glands* beneath the tongue. The glands develop as buds from the epithelium which lines the oral cavity. Presumably, this epithelium is of ectodermal (stomodaeal) origin.

The buds grow into the adjacent mesenchyme where they branch repeatedly in the manner characteristic of compound glands. The final "twigs" terminate in vesicles of secretory cells, known as *acini* (Latin for "grapes"). The mesenchyme responds to the epithelial buds by forming dense capsules around the buds and their branches.

It has been demonstrated by explantation experiments that both the epithelium and the mesenchyme are necessary if the differentiation of tubules and acini is to take place (page 319).

THE PHARYNX

The bird and mammal pharynx is a triangular structure, broadest at the anterior border, and tapering to an apex behind (Fig. 14-6). Anteriorly it is thin from roof to floor. Posteriorly it is thin from side to side. Its most anterior tip, the preoral gut, or Seessel's pocket, is

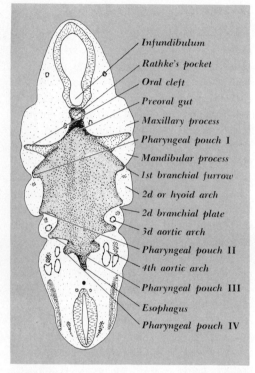

Infundibulum

Rathke's pocket

Oral cleft

Preoral gut

Maxillary process

Pharyngeal pouch I

Mandibular process

1st branchial furrow

2d or hyoid arch

2d branchial plate

3d aortic arch

Pharyngeal pouch II

4th aortic arch

Pharyngeal pouch III

Esophagus

Pharyngeal pouch IV

FIGURE 14–6 Dorsal view of the stomodaeum and pharynx of a 6-mm pig embryo. Reconstructed from serial sections. The endoderm of the pharynx is darkly shaded.

temporary and soon disappears. The endoderm of the floor of the pharynx near its anterior end downpockets and comes into contact with the ectoderm of the stomodaeum. Together, they form the oral plate.

The Pharyngeal Pouches

A series of five pairs of pharyngeal pouches bulge laterally from the sides of the pharynx. In mammals, however, the fifth pouches are attached to the fourth pouches and do not amount to much (Fig. 14–7).

The first pouches, known also as the *hyo-*

mandibular pouches, push laterally between the mesenchyme of the first (mandibular) and second (hyoid) branchial arches. When they come into contact with the epidermis, they adhere and form closing plates (branchial plates). These anterior plates may (birds) or may not (mammals) break through and become open *branchial clefts,* or "gill slits." Even if they do break through, the openings close again, and plates are reformed which become the *tympanic membranes,* or eardrums (Fig. 13–7). Details of how the first ectodermal furrows become the external ear and how the first pharyngeal pouches give rise to the auditory tubes (eustachian tubes) and middle ears were given in Chapter 13.

When one examines his open mouth in a mirror, he will note two folds on each side separating the oral cavity in front from the oral part of the pharynx (the "throat") behind. These

FIGURE 14–7 The derivatives of the mammalian pharynx. *(After Keibel and Mall, 1912, Manual of Human Embryology, J. B. Lippincott, Philadelphia.)*

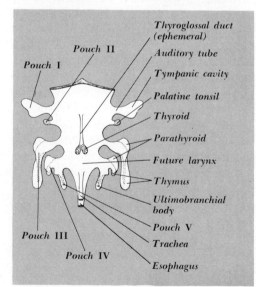

Thyroglossal duct (ephemeral)

Pouch II

Pouch I

Auditory tube

Tympanic cavity

Palatine tonsil

Thyroid

Parathyroid

Future larynx

Thymus

Ultimobranchial body

Pouch V

Pouch III

Trachea

Pouch IV

Esophagus

folds are the palatine arches and represent parts of the second and third branchial arches. The space between them is occupied by the *palatine tonsils* and what little remains of the second pharyngeal pouches (Fig. 14–3).

The other pharyngeal pouches disappear except for certain glandular derivatives. From the epithelium of the third and fourth pouches come the *parathyroid glands*. They secrete parathormone, the hormone which is important in regulating the calcium of the body (Fig. 14–7). From the third pouches (and sometimes from the fourth) are derived the *thymus glands*. These are now known to play a part in the development of immunological response. They are well developed and important in young animals but decline as maturity is approached (see pp. 379, 380). Other derivatives of the pharyngeal pouches are of less importance.

The Tongue

The floor of the oral cavity and forward part of the pharynx gives rise to the tongue (Fig. 14–3). Its body, that is, its forward part, is covered mostly with ectoderm. The body arises as swellings on the inner faces of the mandibular processes. The root of the tongue (ventral to the throat) is covered entirely with endoderm. It is derived mainly from the bases of the second branchial arches. The intrinsic muscles of the tongue are of somite origin (page 349). The complicated nerve supply to the tongue is witness to its compound origin (page 300).

The Thyroid Rudiment

The floor of the pharynx midway between the first and second pair of pharyngeal pouches evaginates downward as the primordium of the *thyroid gland* (Figs. 8–12 and 14–1). Now,

in primitive chordates such as tunicates, amphioxus, and the ammocoetes larvae of lampreys, there is an open groove on the floor of the pharynx known as the endostyle. It secretes a colloidal mucus that entraps microscopic food particles. It has been interpreted as the forerunner of the thyroid gland, although this interpretation has been disputed. At any rate, in all vertebrates above the agnathans (lampreys and their kin), the thyroid gland loses its connection with the floor of the pharynx and becomes a gland of internal secretion. Its hormone, thyroxin, serves in the regulation of the body's metabolism. In mammals, the spot where the thyroid rudiment budded off from the floor of the pharynx remains as a small pit between the body and root of the tongue. It is known as the *foramen caecum* (Fig. 14–3).

The Trachea and Lung Buds

A longitudinal groove on the floor of the pharynx toward its posterior end is known as the *laryngotracheal groove* (Figs. 8–20 and 14–1). It gives rise to the *larynx* and *trachea*. From its posterior end the right and left *lung buds* arise. As the buds grow out, they branch repeatedly, producing several generations of bronchi. Finally they give rise to the entire bronchial tree.

Beginning at the posterior end of the laryngotracheal groove and progressing gradually forward, the sides of the groove come together and separate the dorsal part of the pharynx, which leads to the esophagus, from the ventral larynx and trachea (Fig. 14–1). The pinching, however, is not completed, for an opening remains at the anterior end of the groove, namely, the *glottis* (Fig. 14–3). At first the glottis is a longitudinal slit on the floor of the pharynx between the fourth pouches. Later it becomes T-shaped because of the backward growth of the *epiglottis* from the floor of the pharynx anterior to it.

The Division of the Pharynx

The palatine shelves were described (pp. 242, 243) as dividing the stomodaeum into the nasal cavities above and the oral cavity below. These shelves continue a short way posteriorly at the sides of the pharynx and partially divide the nasal pharynx from the oral pharynx (Fig. 14–3). That part of the palate which lies between the nasal and oral cavities is known as the *bony palate*. The posterior part of the palate (beneath the nasal pharynx) lacks bone and is known as the *soft palate*. It ends posteriorly in a hanging, fingerlike process, the *uvula*.

The embryonic mammalian pharynx thus becomes divided into four parts: (1) the nasal pharynx dorsal to the soft palate; (2) the oral pharynx ventral to the soft palate and dorsal to the root of the tongue; (3) the "laryngeal" pharynx, often thought of as the anterior end of the esophagus, although it is of pharyngeal origin; and (4) the larynx and trachea. The auditory tubes (eustachian tubes), derivatives of the first pharyngeal pouches, open into the nasal pharynx. The third and fourth pharyngeal pouches (and the fifth, if indeed there is a fifth) disappear in the region of the larynx (Fig. 14–3).

ESOPHAGUS, STOMACH, AND INTESTINE

The Esophagus

The esophagus of the early embryo is short and narrow. In the chick it even closes for a time, but later reopens. In amphibians and in embryos of mammals it is lined by a ciliated epithelium. Later, in the case of mammals the ciliated epithelium is replaced by a stratified epithelium.

The esophagus continues to be short in the lower vertebrates which have no neck, but becomes long in the vertebrates which possess a neck and large thorax. It runs in the mediastinum, namely, in the partition between the right and left pleural cavities (Fig. 11–13, *B*).

It is not easy to decide just where the pharynx ends and the esophagus begins. The muscle of the pharynx is striated muscle, while the esophagus is composed mostly of smooth muscle; but the two types of muscle overlap. Usually the esophagus is a simple tube, but in birds its posterior portion is swollen into a *crop* for the storage of food.

The Stomach

In the young embryo, the stomach is short and narrow and marked by only a slight expansion of the gut. Soon, however, it deepens dorsoventrally. Then its original dorsal border swings to the left and becomes its so-called *greater curvature* (Figs. 11–14, *A*, and 15–7). Its original ventral border remains near the midline as the *lesser curvature*. A bulge to the left near the anterior end of the greater curvature is the future *fundus*, or storage part of the stomach. The fundus is comparable functionally but not morphologically to the crop of a bird. This rotation of the stomach to the left is characteristic of vertebrates generally.

The posterior region of the stomach remains narrow and is known as the *pyloric limb*. Its function is to churn the food. In this respect it is comparable to the *gizzard* of a bird. The pyloric limb tapers to the *pyloris,* which is the outlet from the stomach into the duodenum.

The Intestine

The intestine of amphibian embryos is richly supplied with yolk and, as such, is commonly referred to as the midgut. Its lumen is narrow. It is even closed for a time in salamanders. After much of the yolk has been digested and absorbed, the intestine reopens, but the new lumen is at a more ventral level than it was originally.

In birds, the intestine is at first widely open to the surface of the yolk. But the anterior intestinal portal migrates posteriorly, and the posterior intestinal portal migrates anteriorly until finally the attachment of the intestine to the yolk sac is reduced to the narrow yolk stalk (Fig. 8–28). Similar remarks apply to the mammal, except that in their case the yolk sac contains essentially no yolk.

At first the intestine is a nearly straight tube located in the midplane of the body. It remains just that in primitive fish. But in other vertebrates it elongates more rapidly than the surrounding body wall and is thrown into loops and coils. The bending does not take place haphazardly. The *duodenum* (the first part of the intestine) charactristically bows to the right in somewhat the manner in which the stomach bows to the left (Figs. 14–9 and 15–7). Indeed, the stomach and duodenum taken together form a clockwise spiral: The stomach begins near the dorsal side of the body cavity, curves to the left, then turns ventrally to where it joins the duodenum. The duodenum, in turn, swings to the right and curves dorsally. Within the axis of this spiral, the dorsal pancreas develops as an outgrowth from the duodenum. Near the same location, but originally from the

ventral side of the duodenum, the hepatic diverticulum evaginates.

In birds and mammals, the intestine posterior to the duodenum becomes drawn into the umbilical stalk as a ventral hairpin loop. The yolk stalk is attached at its tip (Fig. 14–1). The anterior prong of the loop gives rise to the proximal two-thirds or so of the small intestine. The posterior prong of the loop becomes the rest of the small intestine and also the entire large intestine, or *colon.* A swelling near the beginning of the colon is the future *caecum* (Fig. 15–7).

In mammals, the loop of the intestine twists counterclockwise on its long axis in much the way one might twist an old-fashioned wire hairpin. The torsion has already begun in a 6-mm pig. The posterior prong of the loop comes to be anterior to the anterior prong. Ultimately the twisting brings the caecum (in the posterior prong) around to the right of the original anterior prong—a total rotation of some 270° (Fig. 15–7). This explains why, in the adult body, the caecum and appendix (which is attached to the caecum) are on the right side and why the colon crosses from the right side of the body to the left, passing ventral to the small intestine.

For a time the intestine protrudes into the coelomic cavity of the umbilical cord (Fig. 13–2). Later, this "normal umbilical hernia" of the embryo is withdrawn and the opening from the body cavity into the umbilical cord closes. Occasionally a vestige of the yolk stalk persists. It is known as Meckel's diverticulum.

The Liver

The liver has its origin in a pocket of endoderm (two pockets in the chick) (Figs. 8–14 and 8–20) which evaginates from the floor of

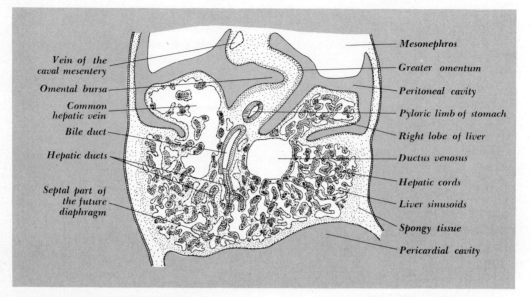

Vein of the
caval mesentery

Omental bursa

Common
hepatic vein

Bile duct

Hepatic ducts

Septal part of
the future
diaphragm

Mesonephros

Greater omentum

Peritoneal cavity

Pyloric limb of stomach

Right lobe of liver

Ductus venosus

Hepatic cords

Liver sinusoids

Spongy tissue

Pericardial cavity

FIGURE 14–8 Section through the liver of a 6-mm pig embryo. Note the interlacing of the endodermal cords (trabeculae) with the endothelia of the sinusoids. From a serial section, but modified in that it shows the fine structures of the liver much coarser than they actually are.

the gut (future duodenum) immediately posterior to the stomach. The pocket is the *liver diverticulum*. The pocket grows forward within tissue which is morphologically the ventral mesentery of the stomach. It proliferates rapidly and in so doing broadens the mesentery until it fuses on each side with the lateral mesocardia (Fig. 15–3, *B*). Together, these three structures (the ventral mesentery and the two lateral mesocardia) constitute the *septum transversum* which separates the pericardial cavity in front from the peritoneal cavity behind (pp. 145, 146, 202).

As the liver grows, its right and left lobes bulge into the peritoneal cavity on each side of the gut, but its main mass expands within the septum transversum (Figs. 11–14, *A*, and

14–8). Later in development, the peritoneum (lining of the peritoneal cavity) pushes between the liver and the body wall until only a small "bare area" remains where the liver is attached to the persisting septum, or diaphragm.

As the hepatic diverticulum invades the ventral mesentery it divides into two regions (corresponding to the anterior and posterior hepatic diverticula of the chick). The anterior region gives rise to the hepatic duct from which *cords* of liver cells sprout and proliferate. Actually, these so-called cords are plates of cells, one to several cells thick. Each cell remains attached to a tiny *bile capillary* which is tributary to the hepatic duct.

The posterior region of the liver diverticulum expands into the saclike *gall bladder* and its duct, the *cystic duct* (Fig. 14–9). The stump of the hepatic diverticulum is the *common bile duct*. It discharges into the duodenum.

As the hepatic diverticulum grows forward in the ventral mesentery, it comes into contact with the vitelline veins. The hepatic cords which sprout from it invade the vitelline veins and break them up into a net of small blood vessels, the *liver sinusoids*. Thus, the liver tissue becomes a sort of sponge consisting of cords (or plates) of liver cells intermingled with a net of sinusoids (Fig. 14–8). This structure is seen when sections of a 6- or 10-mm pig embryo are viewed under the high power of the microscope. One has to look carefully to see the thin endothelia of the sinusoids, and the bile capillaries will probably be missed.

The afferent vessels which are derived from the vitelline veins, that is, the branches of the portal vein in liver tissue which carry blood toward the sinusoids, are known as *hepatoportal veins*. The efferent vessels, namely, those which collect blood from the sinusoids, are the *hepatic veins*. As a result of this arrangement all the blood which comes from the stomach and intestines must pass through the sinusoids before it proceeds onward toward the heart.

Late in development, mesenchyme migrates into the liver by way of the sheaths of the hepatoportal veins. It is accompanied by bile ducts and branches of the hepatic artery (a branch of the coeliac artery). The spongy tissue of the liver then organizes into a sort of honeycomb, each multicellular unit of which is a *liver lobule*. Thus it comes about that each lobule of the liver consists of a central vein (a tributary hepatic vein) surrounded by a sponge of liver cords and sinusoids. Separating one lobule from another are sheaths of connective tissue in which run three vessels:

branches of the hepatoportal veins, the hepatic artery, and the bile ducts.

The growth of liver tissue is accompanied by the proliferation of liver cells and the splitting and budding of the central veins of the lobules. At first there are only a few central veins (about six in a 10-mm pig); but ultimately the number of central veins and lobules becomes innumerable.

As the liver develops, it takes over functions previously performed by the endoderm and vessels of the yolk sac. For example, certain of the endothelial cells which line the sinusoids give rise to red blood corpuscles, while others, known as Kupfer cells, become scavengers

FIGURE 14–9 The development of the gall bladder and pancreas. (Adapted from F. W. Thyng, 1908, Am. J. Anat., 7:489–503.)

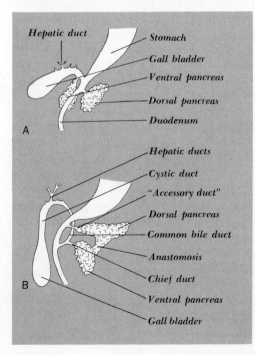

(macrophages) which devour particles that are loose in the bloodstream.

In many ways the liver is a unique organ: It is the only organ of the body which is predominantly endodermal in its origin. Compared with most other organs, it is relatively homogeneous; that is, its cells are structurally and functionally almost alike. Yet it has many functions, some of them glandular, some of them metabolic, while others are concerned with transforming and storing foodstuffs. Still other functions are excretory. All these services to the body are performed by the same type of liver cells!

The gross form of the liver is not the result of self-differentiation such as is usually the case with embryonic structures. On the contrary, its form is impressed upon it by the surrounding organs. As it grows, it occupies whatever space is available. If some of the liver is cut away, the cells which remain begin to grow, and they continue to grow until the surrounding space is again occupied. It is said that one-sixth of a human liver is capable of regenerating a whole liver. In the process of its regeneration, there is no dedifferentiation of cells followed by redifferentiation, such as occurs in most regeneration. Instead, the remaining liver tissue proliferates directly.

The Pancreas

The pancreas develops from two (or three) clusters of cells which evaginate from the duodenum in the region of the hepatic diverticulum (Figs. 11–14, *B,* and 14–9). Each cluster gives rise to ducts and vesicles (acini). The cells of the acini secrete important digestive enzymes.

One of the evaginations grows into the dorsal mesentery of the duodenum at a point almost exactly opposite the liver diverticulum. This is the *dorsal pancreas.* As it expands, it not only occupies the mesoduodenum but pushes anteriorly into the mesogastrium (Fig. 15–7). The other evagination (two in the chick) arises on the ventral side of the duodenum close to the hepatic diverticulum itself. It becomes the *ventral pancreas.* In man, the latter is actually an appendage of the liver diverticulum. Later, when the duodenum swings to the right, the dorsal pancreas and the ventral pancreas come together and unite into a single organ. In man, but not in mammals generally, the duct of the ventral pancreas taps the duct of the dorsal pancreas and becomes the duct of both. In turn, it joins the common bile duct and enters the duodenum by a single pore.

At a late stage in development, cords of cells bud off from the pancreatic ducts, break loose, and become "islands" of cells embedded in the pancreas known as the *islets of Langerhans.* They secrete insulin into the bloodstream, the hormone which makes it possible for the body to store and utilize sugar. Thus the pancreas is two sorts of a gland: It is a gland of external secretion (an exocrine gland) which secretes a digestive juice containing several enzymes; and it is a gland of internal secretion (an endocrine gland) which manufactures and releases the hormone insulin into the blood.

Experiments on the Derivatives of the Gut

The prospective endoderm of the amphibian egg possesses considerable capacity for self-differentiation from the late blastula stage on. This expresses itself first in the characteristic mass movements of gastrulation. There are histogenetic tendencies as well. But this early regional differentiation is not final, for there remains considerable power of regulation with respect to the adjacent mesoderm. The final pattern of the gut, therefore, is brought into being through the interaction of mesoderm and endoderm (page 109). It would seem that the mesoderm and the endoderm carry forward their regionalization together.

The capacity to produce a liver is present and localized in the blastoderm of the chick as early as the 3-somite stage. Dorothea Rudnick has found that when pieces of endoderm and splanchnic mesoderm are taken from the area lateral to the anterior end of the primitive streak and are transplanted to the chorioallantoic membrane of an older chick, they differentiate into liver and heart tissue. Le Douarin finds, however, that prospective liver endoderm does not actually possess within itself the competence to form liver tissue until it comes into contact with heart mesoderm at the anterior intestinal portal. It will be recalled that heart mesoderm migrates forward and medially from the sides toward the folds of the anterior intestinal portal (page 182). Here it forms the heart tube, and, coming into contact with the endoderm of the floor of the portal, it evokes in the endoderm the competence to become liver buds. No other tissue has been found which has the capacity to do this. If prospective liver endoderm is separated from the heart-forming tissue before the 4-somite stage and is transplanted posteriorly, even into the region of future liver mesenchyme, it does not form liver. Instead, it becomes incorporated into the endoderm of the intestine.

But this initial determination of competence is not the final determination. It is only when the endodermal bud (hepatic diverticulum) comes into contact with prospective liver mesenchyme that it sprouts hepatic cords which invade the mesenchyme and produce typical liver tissue. Mesenchyme of the lung region also has the capacity to stimulate the liver bud in this manner. The mesoderms of the head, limb buds, and somites do not have the power.

Prospective liver mesenchyme has a greater capacity to self-differentiate than has liver endoderm. If it is prevented from coming into contact with the endodermal bud, it nevertheless forms a sponge of strands of cells. These intersperse with blood sinusoids which surround and open into the vitelline veins.

The interdependence of mesoderm and epithelium in the development of the diverticula of the gut has been demonstrated in the mammal. Clifford Grobstein removed submaxillary glands of 13-day mouse embryos and exposed them briefly to calcium- and magnesium-free trypsin solutions. He was then able to separate the ectodermal epithelia from the surrounding mesenchyme capsules. When he cultivated the two tissues separately, neither usually differentiated. The epithelium spread out as a thin sheet of cells. The mesenchyme gave rise to homogeneous masses of fibroblast-like cells. But when he placed the two in contact with each other, they interacted by producing the ducts and connective tissue of a gland. Grobstein found that this inductive capacity is specific for the mesenchyme of the gland primordium. Mesenchyme of the somites, lateral plate, or lung region does not give this response.

Similar experiments have been performed by Auerbach on the thymus primordium and by Golosow and Grobstein on the pancreas, with similar results. The development of the pancreas went as far as the production of zymogen granules, the precursors of digestive enzymes. In the case of the thymus and pancreas, mesenchyme other than the gland's own mesenchyme is at least partially effective in facilitating the growth of the epithelium.

The structural differentiation of a gland primordium passes through two phases: During the first phase, the ducts and acini form as a result of cell divisions and morphogenetic movements. Also during this phase DNA is synthesized, and the cells are sensitive to those substances (nucleotide analogues) which produce abnormal nucleic acids. During the second phase the cells differentiate functionally and in the case of the pancreas produce zymogen granules. By this time cell division has largely ceased, DNA is no longer synthesized, and the cells are less sensitive to nucleotide analogues. This second phase, commonly termed *cytodifferentiation,* is to be distinguished from the regional differentiation and morphogenetic movements which preceded it. Both are differentiation processes.

THE ORGANS OF RESPIRATION

Originally, the ancestors of the vertebrates depended on the surface of their bodies to exchange gases with the environment. This was adequate when they were small, elongated creatures with soft, moist skins. But when the animals became larger, and especially when their skins became thick and tough, a more efficient mechanism for respiration was needed. The gill arches and clefts, which were evolved originally as an adaptation for filter feeding, took over this added function. Filaments grew out from the branchial arches, producing external gills, or from the walls of the pharyngeal pouches, producing internal gills. These contained capillary loops from the aortic arches. Thus the blood was aerated as it traveled from the heart to the tissues of the body.

When vertebrates became air breathers and invaded the land, new problems of respiration arose. Gill filaments, extending outward from the body, were no longer suitable for this function. They dried out and became impermeable to the respiratory gases. Lungs were the answer, that is, internal cavities with walls which could be kept moist and soft.

There is reason to believe that the first lungs were a modified posterior pair of pharyngeal pouches (Fig. 6–9). Instead of pushing laterally through the body wall, as gill pouches do, they expanded backward as sacs in the body cavity. Lungs of this sort are found in amphibians and reptiles today. The fact that lungs receive their blood supply from the last pair of aortic arches lends support to this theory.

The Larynx and Trachea

Earlier in this chapter we noted that the posterior part of the pharynx becomes pinched into a dorsal part and a ventral part by a partition which begins at the posterior end of the pharynx and works progressively forward. The dorsal part is the "laryngeal pharynx" (Figs. 14–1 and 14–3). The ventral part is the larynx and trachea. This pinching between the two divisions is never complete, for an opening, the *glottis,* remains at the anterior end connecting the larynx with the pharynx. The lung buds push out laterally from the posterior end of the trachea.

Figure 16–5 names some of the cartilages of the larynx which are derived from the posterior branchial arches. Cartilages continue to form in the walls of the trachea and are present even in the larger bronchi. The muscles of the larynx, like other muscles derived from the walls of the pharynx, are visceral striated muscle. They are supplied by the vagus nerve.

The Lungs

The growth of the lung buds is preceded by the development of a thick, dense layer of mesenchyme around the posterior end of the pharynx. The lung buds grow out into this mass, pushing mesenchyme ahead of them. As they do so, they branch repeatedly by a process of lateral budding. The largest branches are the *primary bronchi*. Then there are secondary bronchi and tertiary bronchi, and so on. Ultimately they form a bronchial tree, the final twigs of which (the *bronchioles*) terminate in lobed air sacs. The lobes are known as *alveoli*. The blood vessels of the pulmonary circulation invade and ramify within this mesenchymal mass.

At first, the mesenchyme which covers the branching bronchi is thick, but as functional maturity approaches the mesenchyme over the bronchioles and air sacs becomes thin. Even the endodermal lining of the bronchioles and air sacs become reduced and flattened until finally the lungs may be described as a sponge of thin-walled air passages surrounded by innumerable capillaries. The process is not completed until after birth. In this respect the newborn rat is less advanced than the human child, and the child is less advanced than the lamb.

While these processes are going on, the pleural cavities expand around the lungs. The pleurae (linings of the pleural cavities) invade the loose mesenchyme lateral and ventral to the heart until the pleural cavities almost completely surround the pericardial cavity containing the heart. This expansion of the pleural cavities is independent of the growth of the lungs, for it takes place before the lungs are fully expanded and will occur even if the lungs have been cut away.

Experiments on the Development of the Lungs

When a lung bud, consisting of both endoderm and mesenchyme, is isolated and cultivated on a plasma clot, it self-differentiates in a remarkably normal manner. The epithelium branches and rebranches to form a typical bronchial "tree." But if the epithelium is deprived of its mesenchymal cells, it fails to branch. If mesenchymal cells of a lung bud are placed next to the epithelium, they migrate toward it, invest it, and bring about its differentiation. Other mesenchyme of the embryo will not do this. (Mesoderm of the metanephros produces some effect.)

SUGGESTED READINGS

General References

Arey, L. B., 1965. Developmental Anatomy. W. B. Saunders Company, Philadelphia. 7th ed. Chap. 13, The mouth and pharynx. Chap. 14, The digestive tube and associated glands. Chap. 15, The respiratory system.

Hamilton, W. J., J. D. Boyd, and H. W. Mossman, 1962. Human Embryology. The Williams & Wilkins Company, Baltimore. 3d ed.

Chap. 10, Alimentary and respiratory systems, pleural and peritoneal cavities.

Copenhaver, W. M., 1955. Vertebrate organogenesis: Heart, blood vessels, blood, and entodermal derivatives. In Willier, Weiss, and Hamburger (eds.), Analysis of Development. W. B. Saunders Company, Philadelphia. Pp. 440–461.

The Teeth

Patten, B. M., 1964. Foundations of Embryology. McGraw-Hill Book Company, New York. 2d ed. Pp. 433–447.

Schour, I., 1955. Vertebrate organogenesis: Teeth. In Willier, Weiss, and Hamburger (eds.), Analysis of Development. W. B. Saunders Company, Philadelphia. Pp. 492–498.

Wilde, C. E., Jr., 1955. The urodele neuroepithelium. I. The differentiation in vitro of the cranial neural crest. J. Exptl. Zool., **130**:573–595.

The Liver

Croisille, Y., and N. M. Le Douarin, 1965. Development and regeneration of the liver. In DeHaan and Ursprung (eds.), Organogenesis. Holt, Rinehart and Winston, Inc., New York. Pp. 421–466.

Poole, B., 1966. The stimulus to hypertrophic growth. Advan. Morphogenesis, **5**:93–129.

Experiments on Gut Derivatives

Grobstein, C., 1953. Inductive epithelio-mesenchyme interaction in cultured organ rudiments of the mouse. Science, **118**:52–55.

Borghese, E., 1958. Inductive interactions of embryonic tissues after dissociation and reaggregation. In McElroy and Glass (eds.), The Chemical Basis of Development. The Johns Hopkins Press, Baltimore. Pp. 704–773.

Auerbach, R., 1960. Morphogenetic interaction in the development of the mouse thymus gland. Develop. Biol., **2**:271–284.

Golosow, N., and C. Grobstein, 1962. Epitheliomesenchymal interaction in pancreatic morphogenesis. Develop. Biol., **4**:242–255.

Grobstein, C., 1964. Cytodifferentiation and its control. Science, **143**:643–650.

———, 1964. Interactions among cells in relation to cytodifferentiation: A general survey. J. Exptl. Zool., **157**:121–125.

Wessells, N. K., 1964. Tissue interaction and cytodifferentiation. J. Exptl. Zool., **157**:139–152.

Rutter, W. J., N. K. Wessells, and C. Grobstein, 1964. Control of specific synthesis in the developing pancreas. Natl. Cancer Inst. Monogr. No. 13. Reprinted in Bell (ed.), 1965. Molecular and Cellular Aspects of Development. Harper & Row, Publishers, Incorporated, New York.

The Lungs

Goodrich, E. S., 1930. Studies on the Structure and Development of Vertebrates. The Macmillan Company, London. Reprinted by Dover Publications, Inc., New York. Chap. 11, Air bladder and lung.

Sorokin, S., 1965. Recent work on developing lungs. In DeHaan and Ursprung (eds.), Organogenesis. Holt, Rinehart and Winston, Inc., New York. Pp. 467–491.

chapter **15**

THE
COELOM
AND
MESENTERIES

This chapter and the ones which follow concern the mesoderm. The mesoderm is well protected. It is surrounded by barriers of ectodermal and endodermal epithelia, and its cells are bathed by a uniform body fluid, or internal medium. Shielded thus from outside influences, it pursues a relatively independent pattern of morphogenetic change.

The Body Cavity

A true *coelom,* or body cavity, is a cleft in mesoderm which separates an outer body wall, or *somatopleure,* from an inner viscera wall, or *splanchnopleure.* The outer wall is formed of ectoderm and somatic (parietal) mesoderm; the inner wall consists of splanchnic (visceral) mesoderm and endoderm.

The function of a coelom is to permit the internal organs to move without involving the body wall. Thus, the heart beats, the lungs expand and contract, and the stomach and intestines are mobile within their respective parts of the body cavity. The same principles apply to development. The heart tube elongates and becomes S-shaped, the stomach swings to the left, and the intestine coils freely within the body cavity. It is possible that originally the vertebrate coelom had also a distributive function, as do the various body cavities of the invertebrates. An indication of this is the fact that the first excretory tubules, which are formed by the embryo, open into the body cavity, and an egg, after it leaves the ovary, lies temporarily within the body cavity.

Originally, cavities were present in all three regions of the mesoderm: axial mesoderm, intermediate mesoderm, and lateral mesoderm. The cavities (myocoels) of young somites open, by way of the intermediate mesoderm, into the coelomic cavities of the lateral mesoderm (Figs. 8-8, *D,* and 15-1). A definitive

coelom, however, is present only in the lateral mesoderm.

At first the coelom on each side extends the full length of the gut from the pharynx to the cloaca. In amniotes, each body coelom is open laterally to the extra-embryonic coelom (exocoel) which occupies the space between the chorion and the yolk sac. The folding downward of the lateral body folds, however, gradually cuts off the extra-embryonic cavities from the coeloms of the embryo proper. The last communication with the extra-embryonic coelom is a cavity within the umbilical cord, which persists for a short time and then closes.

The Dorsal and Ventral Mesenteries

The sheets of splanchnic mesoderm, which form the inner linings of the coelomic cavities, push toward the midline both above and below the gut and thus enfold the gut. Where they meet they produce *dorsal* and *ventral mesenteries* (Fig. 15-1, *C*). A mesentery, therefore, consists of two layers of mesoderm with a varying amount of mesenchyme between. The regions of the dorsal mesentery are named according to the organs to which they are attached. Thus we speak of the mesoesophagus, mesogastrium, mesoduodenum, mesentery proper (of the small intestine), mesocolon, and mesorectum (Fig. 15-2). There is no mesentery dorsal to the pharynx.

The stomach rotates so that its original dorsal side comes to be its left or greater curvature. As a result of this rotation the dorsal mesogastrium is pulled out on the left side (Fig. 15-3, *C*). Two organs then make their appearance within its folds: the *spleen,* which is an organ of purely mesodermal origin, and the anterior part of the *dorsal pancreas* (Fig. 15-7).

The small intestine elongates and becomes

convoluted, and its mesentery spreads out like a fan.

The ventral mesentery has a varied fate. Beneath the posterior part of the pharynx it enfolds the endothelial tubes of the heart (Fig. 15-3, *A*). The part which lies between the heart and the underside of the pharynx is the *dorsal mesocardium*. It disappears except for small remnants at its anterior and posterior ends. The part of the ventral mesentery which is ventral to the heart is the *ventral mesocardium*. It is even more ephemeral. It may be said to originate, in birds and mammals, when

the lateral body folds close in beneath the heart; but it immediately breaks through and disappears. As a result the heart lies free within its part of the body cavity except, of course, at its two ends where the great blood vessels enter and leave.

Beneath the stomach, the ventral mesentery carries the two vitelline veins. It is here invaded by the endoderm of the hepatic diverticulum. Cords of liver cells from the diverticulum ramify in the midst of the endothelia of the veins. Together, the endodermal cords and the endothelia become the liver (Fig. 15-3, *C*).

FIGURE 15-1 Schematic sections of a bird or mammal embryo to illustrate the origin of mesenteries and body cavity. A. Before; B, during; and C, after the closing in of the lateral body folds to form the ventral body wall.

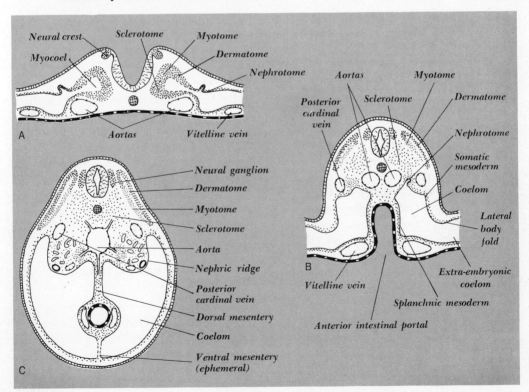

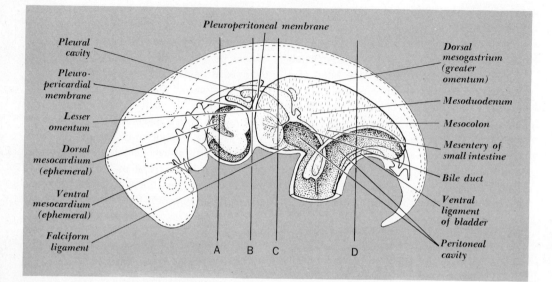

FIGURE 15-2 Diagrammatic lateral view of the body cavities of a bird or mammal embryo. The vertical lines indicate the locations of the schematic cross sections shown in Fig. 15-3.

Thus, in principle at any rate, the ventral mesentery of the stomach encloses the liver. That part which lies between the liver and the lesser curvature of the stomach is the *gastrohepatic ligament* (lesser omentum). The part which is ventral to the liver has been given the name of the *falciform ligament*.

Beneath the intestine, the ventral mesentery breaks through so that the right and left coelomic cavities (peritoneal cavities) become confluent with each other (Fig. 15-3, *D*). The posterior margin of the lesser omentum is the *hepatoduodenal ligament*. It carries the bile duct from the liver to the duodenum (Fig. 15-2). It also carries the hepatic portal vein and the hepatic artery. The posterior margin of the falciform ligament carries the umbilical veins. The right umbilical vein closes early. At birth, the left becomes a cord in the margin of the falciform ligament.

A ventral mesentery is potentially present ventral to the allantoic stalk (Fig. 15-3, *D*). In the embryo it is represented by the broad attachment of the allantoic stalk to the ventral body wall. It ultimately becomes the *ventral ligament of the bladder*. Its anterior margin carries the intra-embryonic part of the allantoic stalk (urachus) and the right and left umbilical arteries. These close at birth and become a cord.

The Lateral Mesocardia and Septum Transversum

Quite early in development the vitelline veins in the region which will be the sinus venosus bulge laterally and come into contact with the body wall. Fusion then takes place, and bridges are formed across the coelom between the mesoderm of the body wall (somatopleure) and the mesoderm of the sinus venosus of the heart (splanchnopleure) (Figs. 8–25, *A*, and 15–3, *B*). The bridges, known as the *lateral mesocardia,* carry the common cardinal veins from the body wall to the sinus venosus.

The lateral mesocardia partially divide the coelomic cavity into an anterior portion, namely, the *pericardial cavity,* and a posterior portion, the *peritoneal cavity.* (Compare Fig. 6–17.) The pericardial and peritoneal cavities communicate for a time by a pair of canals which are dorsal to the lateral mesocardia. In most vertebrate embryos they also communicate ventrally. Since the dorsal canals lie lateral to the lung buds and are later occupied by the

expanding lungs, we shall call them the *pleural canals.*

The ventral pericardio-peritoneal communications quickly close by the expansion of the liver within the ventral mesentery (Fig. 15–3, *B*). Indeed, in mammals they seem never to exist. The partition thus formed by the union of the lateral mesocardia and the expanded ventral mesentery is the *septum transversum* (Fig. 15–4). In principle, it is bounded anteriorly by the pericardial cavity, dorsally by the right and left pleural canals, and posteriorly by the peritoneal cavity. Laterally and ventrally, it adheres to the body wall.

In the young embryo, the septum transver-

FIGURE 15–3 Diagrammatic cross sections of a mammal embryo at the levels indicated in Fig. 15–2. A. Section through the pharynx and heart. B. Through the lateral mesocardia and ventral mesentery. These unite to form the septum transversum. C. Through the stomach and liver. D. Through the intestine and allantoic stalk.

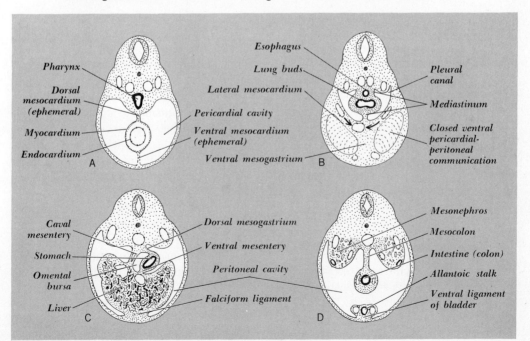

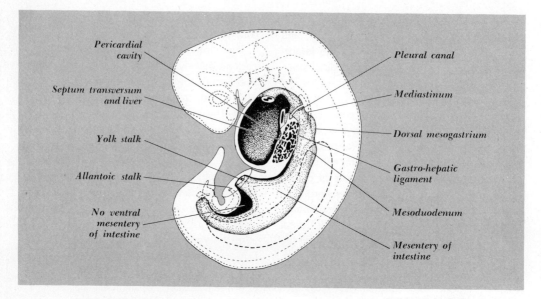

FIGURE 15–4   The body cavities and mesenteries of a 6-mm pig embryo. The location of the left lateral mesocardium is indicated by a white arrow. Compare Figs. 8–18 and 15–6.

sum is occupied by the liver. Only its anterior face (the face next to the pericardial cavity) is future septum (diaphragm). Later the peritoneum (lining of the peritoneal cavity) pushes forward from the rear between liver tissue and the body wall (page 316). It even partially separates the liver from the forward septal tissue until, finally, only a small area of attachment to the septum remains, the so-called "bare area" of the liver (Fig. 15–5).

The Pleural Cavities and Diaphragm

At first, each pleural canal communicates anteriorly with the pericardial cavity and posteriorly with the peritoneal cavity. Then oblique ridges develop on the body wall (Fig. 15–6), which narrow the communications and finally entirely isolate the pleural cavities both front and behind.

Each anterior communication becomes closed by the formation of a ridge medial to the common cardinal vein where the vein turns ventrally and enters the lateral mesocardium. The ridges become the *pleuropericardial membranes*. Ultimately they form the pericardium or sac which surrounds the heart.

The posterior communications, namely, those between the pleural canals and peritoneal cavity, remain open in amphibians and reptiles. As a result, in them the lungs expand backward at the sides of the abdominal viscera. In mammals, however, the nephric folds of each side continue forward along the dorsal body wall until they approach the lung buds. They then turn ventrally as oblique ridges along the posterior borders of the lateral meso-cardia. They expand and become the *pleuro-peritoneal membranes* which close the open-ings between the pleural and peritoneal cavities.

The pleural cavities, having been thus cut off before and behind, now expand as though to accommodate the future growth of the lungs. They invade the loose mesenchyme of the body wall lateral and posterior to the pericardial cavity and finally nearly surround this cavity. Both the pleuropericardial and the pleuroperi-toneal membranes receive additions from the body wall.

A muscular diaphragm used in breathing is a distinguishing characteristic of the class Mammalia. It is derived from the forward tissue of the septum transversum, the pleuroperi-toneal membranes, and additions from the body wall. The latter include premuscular tis-sue. At first, the diaphragm is located well for-

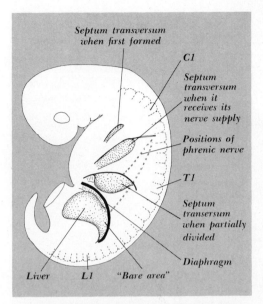

FIGURE 15–5 Schema to illustrate the migra-tion posteriorly of the septum transversum, and the almost complete separation of the embryonic septum transversum into diaphragm and liver.

FIGURE 15–6 Detailed reconstruction of the region of the left lateral mesocardium of a 6-mm pig embryo. The larger arrow passes through the left pleural canal. The shorter arrow points into the right pleural canal.

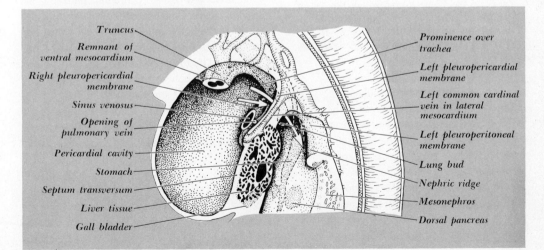

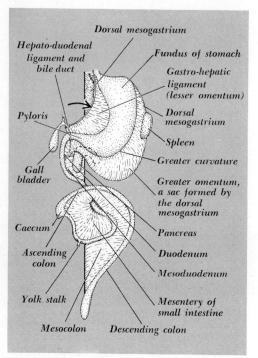

Dorsal mesogastrium

Hepato-duodenal
ligament and
bile duct

Fundus of stomach

Gastro-hepatic
ligament
(lesser omentum)

Pyloris

Dorsal
mesogastrium

Spleen

Greater curvature

Gall
bladder

Greater omentum,
a sac formed by
the dorsal
mesogastrium

Caecum

Pancreas

Ascending
colon

Duodenum

Mesoduodenum

Yolk stalk

Mesentery of
small intestine

Mesocolon   Descending colon

FIGURE 15-7 Diagram showing the rotation
of the alimentary tract and its mesenteries.
The arrow points into the cavity of the omental
bursa which lies dorsal to the lesser omentum,
dorsal to the stomach, and within the sac of
the greater omentum. *(Adapted from L. B.
Arey, 1965, Developmental Anatomy, 7th ed.,
W. B. Saunders Company, Philadelphia,
Fig. 242.)*

ward in the future neck region (Fig. 15-5).
While in this location, it receives its nerve supply from the fourth pair of spinal nerves. Then,
as the neck forms and the pleural and pericardial cavities expand, the diaphragm migrates posteriorly. As it does so it drags its
nerve supply (phrenic nerves) after it. Finally,
it is located ventral to approximately the 20th
pair of spinal nerves.

The Omental Bursa

The early embryo ceases to be bilaterally
symmetrical when the heart tube first bows to
the right. Somewhat later, the stomach swings
to the left in the manner which has already
been described. Its original dorsal border, the
greater curvature, to which the dorsal mesogastrium is attached, becomes its left border.
A pocket of the right coelomic cavity now lies
between the dorsal mesogastrium and the original right side of the stomach. This is the *omental bursa.* Much later, the posterior part of the
dorsal mesogastrium expands to form a bag
which hangs down ventral to the intestine (Fig.
15-7). The bag is the *great omentum.*

As a result of the rotation of the stomach, its
original ventral border, or *lesser curvature,*
faces toward the right. The gastrohepatic ligament (lesser omentum), instead of occupying
a sagittal plane, is now frontal in its orientation.

In the meantime, recesses of the right and
left coelomic cavities push forward from the
rear into the thick splanchnic mesoderm which
surrounds the esophagus. Each recess separates a lateral *pulmonary fold* (into which the
lung buds grow) from a median partition, the
mediastinum. Both recesses are to be found in
birds. In mammals, the left recess is transitory,
but the right recess remains. The right pulmonary fold grows posteriorly and forms a sort
of mesentery between the dorsal body wall and
the right lobe of the liver (Fig. 15-3, *C*). We
shall call it the *caval mesentery* (plica vena
cava) because the posterior vena cava finds its
way across this mesentery from the dorsal
body wall to the right lobe of the liver.

The caval mesentery and right pulmonary
recess add to the omental bursa, which now is
bounded on the left and posteriorly by the
great omentum (dorsal mesogastrium), and

ventrally by the stomach and lesser omentum. It is partially cut off on the right by the caval mesentery. The constricted opening into the right coelom which remains is known as the *epiploic foramen.*

The Mesentery of the Intestine

The first division of the small intestine is the duodenum. We have already noted that it forms a loop to the right in somewhat the way that the stomach swings to the left (Fig. 15–7). The dorsal pancreas forms in the mesentery of its loop. As the dorsal pancreas enlarges, it pushes forward a short distance into the dorsal mesogastrium.

The rest of the intestine develops a hairpin loop which, for a brief time, protrudes into the umbilical cord (Figs. 13–2 and 15–2). The superior mesenteric artery runs in the mesentery within this loop. The anterior prong of the loop elongates and becomes the major portion of the small intestine. As it lengthens, it becomes convoluted, and its mesentery becomes spread out like a fan (Fig. 15–7).

The posterior prong of the loop gives rise to the posterior part of the small intestine and to the first part of the large intestine (ascending colon). The loop of the intestine then twists counterclockwise in the manner described in Chapter 14 (page 315). The posterior prong finally comes to lie to the right of the anterior prong. Its mesentery (the last part of the mesentery of the small intestine plus the ascending mesocolon) is carried over to the right with it. The rest of the large intestine (descending colon and rectum) remains in approximately the midline of the trunk, attached to the dorsal body wall by the descending mesocolon and mesorectum.

In man, several adhesions take place between the dorsal mesentery and the body wall. Apparently, they serve to support the viscera when the body is standing upright. The duodenum adheres to the dorsal body wall. The ascending and descending mesocolons adhere to the body wall of the right and left sides, respectively. Adhesions of this sort are not characteristic of mammals generally. (It is of interest that the adhesions are always on the morphological left side of the mesentery.)

The development of the body cavities has not been extensively studied by experiments. Clefts frequently form in explants of mesenchyme and have sometimes been interpreted as "coelomic cavities." The movements of mesenteries and of the lining of the coelom during development are active and predictable. They should make a suitable subject for experimental analysis.

SELECTED READINGS

Arey, L. B., 1965. Developmental Anatomy. W. B. Saunders Company, Philadelphia. 7th ed. Chap. 16, The mesenteries and coelom.

Goodrich, E. S., 1930. Studies on the Structure and Development of Vertebrates. The Macmillan Company, London. Reprinted by Dover Publications, Inc., New York. Chap. 12, Subdivisions of the coelom and diaphragm.

Wells, L. J., 1954. Development of the human diaphragm and pleural sacs. Carnegie Contrib. Embryol., **35**:107–134.

chapter **16**

SKIN, SKELETON, AND MUSCLE

Details concerning the skin, skeleton, and musculature are best studied in courses in comparative anatomy. This is because these three systems are late developing, as compared with other systems, and their early rudiments are delicate and ill-defined. Except for the epidermis of the skin, they arise as condensations in loose mesenchyme rather than by obvious foldings and moldings of layers of cells.

At first the mesoderm is small in amount, compared with the ectoderm and endoderm, but it steadily increases until it far exceeds the other tissues. Only the central nervous system (ectodermal) and the liver (endodermal) compare with it in volume. The human body is over 90 percent mesodermal in origin.

The early development of the mesoderm was described in the chapters on the amphibian and chick. It will be recalled that most of the mesoderm breaks down sooner or later into loose mesenchymal cells which fill in the spaces between the other tissues: (1) The mesoderm of the head is mesenchymal from the start. (2) The sclerotomes promptly break into separated cells. (3) The dermatomes retain their epithelial nature for a time, but they also become a loose tissue. (4) Even the myotomes, nephrotomes, and lateral mesoderm (with the exception of the coelomic lining) produce mesenchyme. Yet mesenchyme, diffuse as it seems to be, has the capacity to organize into highly complex three-dimensional structures, namely, muscles, skeleton, and the dermis of the skin. Much of this capacity is retained even after the mesenchymal cells have been experimentally dispersed and then reaggregated.

THE SKIN

The *integument,* that is, the skin and its derivatives, constitutes an organ system with many functions. It protects the body against physical, chemical, and biological injury. When the skin is soft and moist, it aids in respiration. Its glands share in excretion and in the regulation of temperature. Its accessory structures, such as scales, feathers, hair, beaks, claws, and horns, serve numerous special purposes.

The skin of vertebrates consists of two layers of unlike origin. The outer *epidermis* is an epithelium derived from ectoderm. The inner *dermis* (corium) is a supportive tissue which originates in mesenchyme. The forerunners of the two tissues come together for the first time during gastrulation. They then strongly adhere and interact, and together they produce a single integrated system.

The Epidermis

When fully developed, the epidermis is a stratified epithelium; that is, it is a layer several cells thick. In fish and amphibians all the cells are alive. As they grow old, the outer cells slough off either singly or in a layer at the time of a molt. The lost or dying cells are continually replaced from beneath by the growth and multiplication of the cells of the inner or *basal layer* of the epidermis. The outer epidermal cells of reptiles, birds, and mammals, after they die, are retained as a protective horny covering, the so-called *stratum corneum.*

The epidermis of a bird or mammal develops in the following manner (Fig. 16–1): First it is a single layer of cuboidal cells. Then, beginning about the fifth day in the chick, or when the embryo is 10 mm long in the case of the pig or man, it becomes two layers. The outer layer, or *periderm,* serves temporarily for protection. The inner or basal layer becomes the actively proliferating *stratum germinativum* of the mature skin. An *intermediate layer* arises by the mitotic division of the cells of the basal

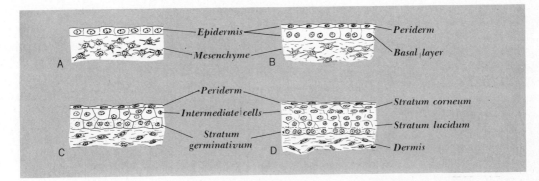

FIGURE 16–1 Sections through the developing skin. *A.* One-layered epithelium. *B.* Two-layered epithelium. *C.* Three-layered epithelium. *D.* Stratified epithelium.

layer. As the cells move outward, they gradually become cornified through the deposition of a horny substance (keratin) in their cytoplasm. They flatten and in time die. After birth or hatching, the periderm is lost, and the stratified intermediate cells take over as the protective outer layer (stratum corneum).

The stratum corneum is the source of the horny derivatives of the skin, such as claws, horny scales of reptiles, feathers and beaks of birds, and hair and horns of mammals. Buds from the germinative layer sink into the underlying dermis and give rise to various glands of the skin.

The Dermis

The dermis is a leathery sheet of connective tissue derived from mesenchyme. In fish it forms bony plates and scales (not to be confused with the horny scales of reptiles). The shell of turtles includes both outer horny plates of epidermal origin and underlying bony plates of dermal origin.

The mesenchyme which gives rise to the dermis comes from several sources. In the head, some of it is derived from the neural

crest and is known as mesectoderm. Some may come from the forward end of the pharynx and hence is mesendoderm. But most of it is from prechordal mesoderm and from the forward continuation of the axial mesoderm. The dermis of the trunk is derived in part from the breakdown of the dermatomes (outer layer) of the somites. Some embryologists have assumed that cells from the dermatomes migrate ventrally and give rise to the entire dermis, but careful observation does not bear this out. The dermis of the limbs and probably most of the dermis of the flanks and ventral side of the body are derived from the outer cells of the somatic layer of the lateral mesoderm.

Experiments on the Skin

The relationship between the dermis and the epidermis, during the development of the skin, has been made the subject of investigation. If the prospective ectoderm of an amphibian early gastrula is isolated before it comes into contact with mesoderm, it becomes nothing more than a wrinkled, undifferentiated epithelium which soon degenerates. (See discussion of exogastrulation, pp. 106–108.) But if it comes under the inductive influence of chorda-mesoderm, it differentiates. What it produces —whether neural plate, neural crest, or dorsal, lateral, or ventral epidermis—depends upon the mesoderm with which it is in contact. If the mesoderm is dorsal chorda-mesoderm, it becomes nervous tissue. If it is ventral mesoderm, it gives rise to epidermis.

Normally, the epidermis of an amphibian neurula develops cilia which beat in a polarized manner, mainly from anterior to posterior. As a result, a young embryo which has been removed from its membranes glides forward as it lies on its side on the bottom of a dish. Isolated ectoderm, lacking a mesodermal substrate, also develops cilia; but the cilia in this case beat in an uncoordinated fashion. The polarity of the epidermis, therefore, depends upon the mesoderm which lies beneath it.

Prospective ectoderm of an early gastrula must contact mesoderm on one side only if it is to differentiate into normal epidermis. If it is completely enclosed in mesenchyme its cells may lose their epithelial character and join with the surrounding mesenchyme cells.

Mesoderm which has been deprived of contact with ectoderm fares somewhat better than ectoderm free of mesoderm. It may differentiate into various mesodermal tissues. It will not, however, produce skin structures. To accomplish this, the cooperation of ectoderm is required.

During gastrulation and the neurulation which follows, the epidermis, having come under the influence of endoderm and mesoderm, acquires regional characteristics. The epidermis of the head acquires the competence to form sensory placodes, the hypophysis, and gill furrows, and the epidermis of the limb areas acquires the capacity to respond to the mesoderm of a limb blastema by forming an apical ectodermal ridge or cap (pp. 250–251). Similarly the epidermis of other areas of the body acquires capacities to produce hair, feather tracts, claws, horns, etc. For a time the differentiation of the epidermis is labile; that is, it will "regulate" when transplanted into changed relations in a new location. It then gives rise to structures suitable to the new location. But by the end of neurulation most of this lability is lost; the epidermis is now strongly inclined to differentiate according to its origin.

McLoughlin experimented with the epidermis of a 5-day chick (Fig. 16–2). When she isolated the epidermis on a plasma clot in a nutrient solution, it rounded up into epithelial nodules, and no further cell division took place. The cells keratinized and soon died. But when she placed the epidermis in contact with mesenchyme, it differentiated. What it became depended on the source of the mesenchyme. If the mesenchyme was from a limb bud, the epidermis underwent mitotic cell divisions and formed a stratum corneum such as is typical of skin. If the epidermis was brought into contact with mesenchyme of the gizzard, it became a mucous membrane which secreted mucus and sometimes even bore cilia. Myoblasts of the heart (future muscle cells) stimulated the epidermis to spread and become a thin squamous epithelium.

From these experiments we conclude that the embryonic epidermis has a large repertoire of possible fates. In older embryos, even in the adult, when epidermis is transplanted to new locations and comes under the influence of other mesoderm, it may modulate and differentiate in new directions. But note this: The epidermis always produces epidermal structures, never structures of a nonepidermal nature.

Other examples of regional specificity in the

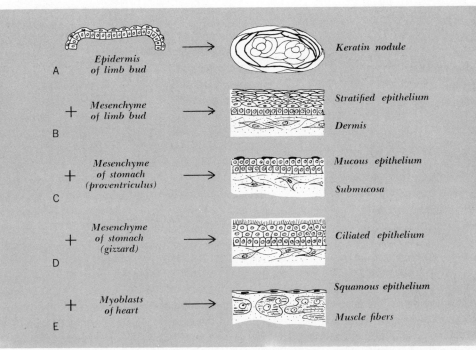

FIGURE 16–2 The influence of mesenchyme on epithelial differentiation. A. When isolated, the epidermis of a limb bud of a 5-day chick rounds up and keratinizes. *B* to *E*. When cultured with mesenchyme, epidermis differentiates according to the source of the mesenchyme. *(Adapted from C. B. McLoughlin, 1963, in Cell Differentiation, 17th Symposium of the Society for Experimental Biology, pp. 359–388.)*

epidermis are to be found in the response of epidermis to stimulation by hormones. At the time of metamorphosis, the epidermis of a tadpole's tail undergoes autolysis; that is, its cells digest away. At the same time, the apparently similar epidermis of the trunk thickens and strengthens. Now, if a patch of epidermis of a tadpole's tail is grafted to the trunk, it performs according to its original location. It does so even though it is now a part of the trunk; that is, it regresses when the host's tail regresses. Moreover, it will do this even when it is younger than the host tissue into which it is grafted and

normally would not yet be old enough to metamorphose.

The Development of Feathers

The development of feathers and feather tracts has been studied extensively by Lillie, Willier, Rawles, Saunders and others. Much that has been learned applies also to the pelage of mammals.

A feather is one of nature's masterpieces. The structure of feathers and the manner in which they are placed on a bird's body are out-

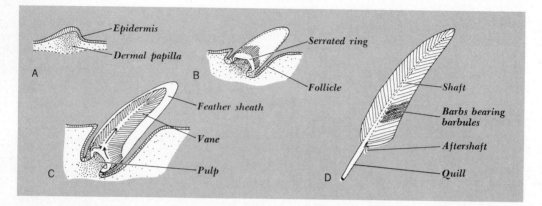

FIGURE 16–3 The development of feathers. A. Stage of the dermal papillae. B. The young follicle with a serrated epidermal ring. C. Developing vane. D. A mature feather. (Adapted from A. S. Romer, 1962, The Vertebrate Body, W. B. Saunders Company, Philadelphia, 3d ed. After Lillie and Juhn.)

standing examples of adaptation. Feathers serve variously for insulation, smoothness of body contour, social display, and flight. Yet all this precision of structure and arrangement is predetermined while the feather germs are yet in their follicles, long before they are ready to function as feathers. A functioning feather is a dead feather.

The development of a feather takes place in the following fashion. About the fifth or sixth day of incubation (chick), condensations of mesenchyme appear beneath the epidermis (Fig. 16–3, A) due to a local increase in the rate of cell division. These grow and project outwardly as *dermal papillae,* each covered with a layer of epidermis. Then the epidermis sinks inward into the dermis, pushing the dermal papilla before it. The result is a tubular *feather follicle.*

The epidermis at the bottom of the follicle thickens to form a ring around the dermal papilla as a core. (The papilla is the pulp of the future feather.) The peripheral border of the ring becomes serrated (Fig. 16–3, B). The ring then elongates and produces a tube, the quill. In the case of contour and flight feathers, this elongation is mainly on one side, namely, the side which is to become the shaft (rachis) of the feather (Fig. 16–3, C). (A smaller aftershaft may develop on the opposite side of the ring.) The serrations of the border gather toward the side of the shaft as it grows outward and become the barbs which make up the vane of the feather. Each barb in turn has smaller branches, the barbules, which are bordered by hooks and ridges. These catch together in such a way that the vane is strong and durable. Moreover, it is capable of being restored by preening, in case the feather becomes ruffled.

The color of the feather is produced mostly by pigment cells derived from the neural crest. These migrate into the epidermis from the dermal papilla and deposit pigment granules. This entire development takes place while the dermal papilla is yet alive and supplied with blood. When the development is complete, the blood supply is withdrawn, the papilla dries up, and the feather is ready to function.

The determination of the feathers of the chick begins about two days before the first mesenchymal condensations appear. After it has begun, a piece of skin removed from the embryo and replaced upside down will develop with its feathers disarranged. Although the substance of the future feather is ectodermal, the initiative for its production and the seat of its early determination are in the mesoderm. Yet the relation between the ectoderm and mesoderm is mutual. If the epidermis is removed, dermal papillae are not formed. If they are already present when the epidermis is removed they disappear.

THE SKELETON

The Notochord

The earliest skeleton of a vertebrate is the notochord. In Chapter 4 the notochord was referred to as the distinctive fact of chordate structure. It made the elongated form and darting locomotion of the vertebrate ancestors possible. Most primitive creatures swim or glide by means of cilia. The ancestral arthropod presumably rowed through the water by means of bristly lateral appendages which were moved by internal muscle fibers. But the original chordates, so we believe, had lithe bodies and sped through the water by undulatory movements just as larval fish and tadpoles do today.

The notochord is a basic feature of chordate development. This is so because it is the center around which the dorsal structures of the vertebrate embryo are organized. Yet, in all vertebrates except the lowest, the original skeletal function of the notochord has been largely or wholly taken over by the bodies of the vertebrae which form around it, and the notochord itself is reduced or eliminated.

Skeletal Tissues

Skeletal tissues usually pass through four stages during their development: (1) loose mesenchyme. (2) dense mesenchyme ("membrane"), (3) cartilage, and (4) bone (Fig. 16–4). (There are exceptions, of course. In sharks and some other groups, the development of the skeleton does not go beyond the cartilage stage.)

Mesenchyme, as we have noted several times, is the loose tissue which fills the spaces between the other tissues of the embryonic body. Its cells are star-shaped, and their processes are in contact with each other. The result is a three-dimensional net or sponge (reticulum) with a thin, jelly-like fluid filling the interstices between the cells.

Dense mesenchyme originates by condensation within the loose mesenchyme. It is the result of cell division and, to a lesser extent, cell migration. Some of the condensations, for example, the dermis of the skin, take the form of layered membranes. Their cells become fibroblasts which weave connective tissue fibers. But since the membranes do not line cavities, they are not classed as epithelia.

Cartilages form by chondrification within the dense mesenchyme (Fig. 16–4, *A*). Starting at definite centers, the mesenchyme cells (known as chondrocytes) round up and secrete a gelatinous matrix around themselves. The matrix is a conjugated protein known as chondrin. Its principal constituent is a mucopolysaccharide termed chondroitin sulfate. Cartilages grow by interstitial growth; that is, the chondrocytes divide and then continue to secrete matrix between each other. Sooner or later a definite cellular boundary layer, the *perichondrium,* forms around each cartilage as a whole. It also secretes matrix, and in doing so adds to the dimensions of the cartilage.

Bones originate in centers of ossification

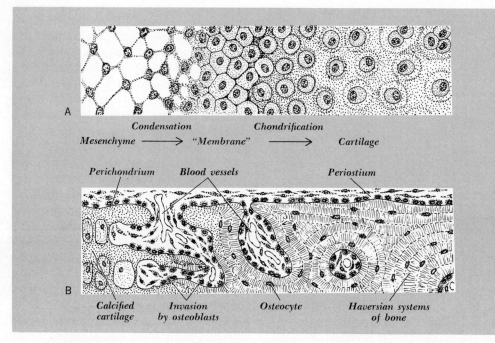

Condensation   Chondrification
Mesenchyme ——————→ "Membrane" ——————→ Cartilage

Perichondrium   Blood vessels   Periostium

Calcified   Invasion   Osteocyte   Haversian systems
cartilage   by osteoblasts   of bone

FIGURE 16–4 Histogenesis of cartilage and bone. *A.* Formation of cartilage by condensation and chondrification. *B.* Replacement of cartilage by bone.

within the cartilages. The so-called dermal bones (membrane bones) of the skull are exceptions. These ossify directly from dense mesenchyme without passing through a cartilaginous stage. In both cases ossification is not just a deposition of calcium salts within a previous tissue. It is a process of replacement in which the original tissue is progressively eaten away and a new matrix, consisting of a mineral and a protein, is deposited (Fig. 16–4, *B*). The mineral is chiefly calcium phosphate. The protein is collagen, a substance which converts to gelatin when boiled.

The cells which deposit the bony matrix are known as *osteoblasts*. Some become trapped in microscopic lacunae within the matrix. This does not, however, completely isolate them

from each other and from the blood vessels which are present in the bone, for fine canals (canaliculi) remain and radiate throughout the matrix. The perichondrium transforms into the periosteum and continues to deposit layer upon layer—in this case bone.

The Skull or Cranium

The skull is produced by the mesenchyme of the head. There are several steps in its formation, the details of which must be left to comparative anatomy. These are (1) the membranous cranium, (2) the cartilaginous cranium proper, (3) the visceral cranium, (4) the dermal cranium, and finally (5) the completed bony cranium.

1 The *membranous cranium* is produced by condensations within the loose mesenchyme which surrounds the brain and sense organs of the head. Mesenchyme also extends ventrally into the branchial arches.

2 The *cartilaginous cranium proper* (chondrocranium) is the result of chondrification at centers within the membranous cranium. Typically, there are a pair of cartilages lateral to the

hypophysis and ventral to the forebrain known as the *trabeculae* (Fig. 16–5, *A*). There are also cartilages lateral to the notochord and ventral to the hindbrain termed the *parachordal cartilages*. In mammals these pairs of cartilages tend to be united together from the first appearance. Cartilaginous capsules also form around the nose and ear. These several elements expand and join together until they be-

FIGURE 16–5 The development of the mammalian skull (highly diagrammatic). *A*. Early cartilaginous stage. *B*. The beginning of ossification. The chondrocranium is stippled; dermal bones are shown by hachures; visceral cartilage in solid black.

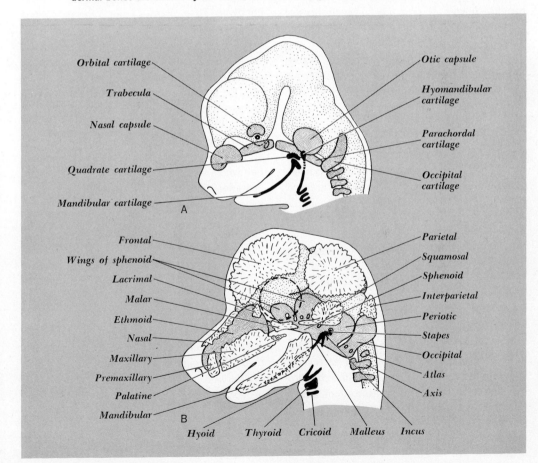

come a sort of cradle (neurocranium) in which the brain lies, supported beneath and at the sides. Foramina (holes) remain through the cartilage and permit the passage of nerves and blood vessels (Fig. 16–5, *B*).

3 The *visceral cranium* (splanchnocranium) consists of the cartilages which form in the membranous tissue of the branchial arches. (1) The first branchial arch gives rise to the quadrate cartilage of the upper jaw, and the mandibular cartilage (Meckel's cartilage) of the lower jaw. (2) The second branchial arch produces the hyomandibular cartilage and the hyoid cartilages. The manner in which three of these elements—the mandibular cartilage, the quadrate cartilage, and the hyomandibular cartilage—become the malleus, incus, and stapes of the middle ear was referred to in Chapter 13. (3) The remaining branchial arches give rise to cartilages which contribute to the hyoid apparatus and the larynx (Fig. 16–5).

4 The *dermal cranium*. There is no dermal cranium in sharks. Instead, the chondrocranium itself grows upward until it all but encloses the brain. In most vertebrates, however, dermal bones, that is, bones which ossify directly from membrane, roof over the brain and so complete the cranium (Fig. 16–5, *B*). They also form sheaths around the cartilaginous capsules of the nose and ear and provide a protecting rim around each eye socket and a roof above the mouth cavity (bony palate). In addition they provide the bones of the jaws, which bear the teeth.

5 The *bony cranium*. The bones of the dermal cranium spread and join with each other and with the bones which ossify in the chondrocranium to produce the mature skull. Sometimes the bony elements fuse together to form single bones of compound origin. Sometimes they knit together but do not fuse, and so produce immovable joints, or sutures. Figure 16–5, *B*, gives the names of some of the bones of the skull.

The Vertebral Column

It will be recalled that each somite has three regions: dermatome, myotome, and sclerotome. It is with the sclerotome that we are now concerned. The sclerotome is that part of a somite which is adjacent to the notochord and the ventral part of the neural tube. Its cells quickly lose their epithelial character and become mesenchyme. They migrate around the notochord and give rise to the bodies of the vertebrae. Some spread upward at the sides of the neural tube and become neural arches. Others push laterally into the somatopleure and produce ribs (or costal processes) and the transverse processes of vertebrae (Fig. 16–11).

Vertebrae originate in the following manner: First, each sclerotome differentiates into an anterior half which is less dense and a posterior half which is more dense (Fig. 16–6). Then the posterior more dense half of each sclerotome joins with the anterior less dense half of the sclerotome next posterior to it, and together, by chondrification and later ossification, they become the body of a definitive vertebra. The more dense half also gives rise to a cartilaginous intervertebral disc. In other words, each vertebra originates by the union of the adjacent halves of two consecutive sclerotomes.

As a consequence of this manner of development, the definitive vertebrae alternate with the myotomes from which the muscle segments are derived. When dorsal and lateral processes grow out from the less dense halves of the sclerotomes, they push into the septa between the successive muscle segments and thus serve for the attachment of muscle fibers.

The Ribs and Sternum

Ribs arise by chondrification and later ossification in the mesenchyme of sclerotome origin which has pushed laterally between the myotomes. All the vertebrae bear costal processes, but only in the thoracic region do the coastal processes become freely movable ribs.

The sternum originates independently of the ribs from condensations in the mesenchyme of the body wall. At first the condensations are paired right and left. But they move toward the midventral line and unite into a single ventral cartilage. Secondarily they become connected with the extremities of the ribs. In mammals, ossifications within the sternal cartilage give rise to a series of segments or sternebrae. (Fusion of sternebrae forms the body of the sternum in man.) In birds, the sternum is the huge breastbone bearing a keel to which the muscles of flight are attached.

Experiments have established that the mesenchyme of the sternal condensations is capable of self-differentiation; but the segmentation of the sternum into sternebrae is controlled by the ribs.

The Appendicular Skeleton

It is customary to divide the skeleton into the axial skeleton and the appendicular skeleton. The former includes the skull, vertebral column, ribs, and sternum. These have been discussed. The appendicular skeleton consists of the cartilages and bones of the anterior and posterior limbs. Each limb includes a free appendage and the girdle to which the free appendage is attached (Figs. 16–7 and 16–11). The appendicular skeleton is derived from the mesenchyme of the limb buds. This in turn is derived from the outer layer of the somatic

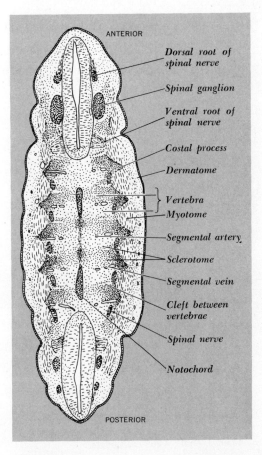

FIGURE 16–6 Frontal section through posterior trunk somites of a 10-mm pig.

mesoderm of the body wall. The condensations which take place in the mesenchyme become the skeletal, muscular, and connective tissue elements of the appendage.

Chondrification in the mammal embryo begins about the time that the limb buds acquire the shape of paddles (about a 12-mm embryo). In general it proceeds in a proximodistal direction. Ossification begins when the embryo is 18 to 25 mm long. It takes place first at the

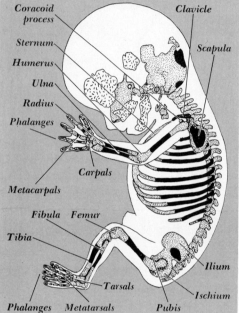

Coracoid process
Clavicle
Sternum
Scapula
Humerus
Ulna
Radius
Phalanges
Carpals
Metacarpals
Fibula Femur
Tibia
Ilium
Tarsals
Ischium
Phalanges Metatarsals Pubis

FIGURE 16–7 Centers of ossification in a young human embryo of about 40 mm. (The skull is adapted from C. C. Macklin, 1921, Carnegie Contrib. Embryol., 10:57–103, Plate 2. The limbs are from C. R. Bardeen, 1905, Amer. J. Anat., 4:265–302, Plate 5.)

length ceases. Increase in the cross section of a bone continues, however, by the laying down of new bone layer upon layer by the periosteum and the hollowing out of the interior of the bone by cells known as osteoclasts.

Experiments on Cartilage

What decides that a mesenchyme cell will give rise to cartilage instead of to some other tissue? A comparatively simple case is the differentiation of a somite as studied by Holtzer and his associates. They found that if a fragment of notochord is inserted into the somite region of a salamander embryo, it induces the production of cartilage by cells which otherwise would have formed muscle. A single cartilaginous mass is produced which is closely applied to the notochordal sheath.

If, instead of notochord, a fragment of the ventral half of the neural tube is implanted in the somite region, cartilage is also formed. But in this case several cartilages are produced and are separated from the nervous tissue by a narrow space. The authors interpret these results to mean that the body of a vertebra is induced by the sheath of the notochord, while the neural arch is induced by the ventral cells of the neural tube.

Like results are obtained when the experiments are performed in vitro, that is, when the somite is removed and cultured on a plasma clot. If no other tissue is present, the cells of the somite spread out as a flat disc and do not differentiate (except for a few fibroblasts). If, however, a fragment of notochord is placed in their midst, they move toward the fragment, round up, and produce a single mass of cartilage. A fragment of the ventral side of the neural tube also induces the production of cartilage; but in this case several cartilages are

centers of the shafts of the longer cartilages and proceeds toward the two ends of the cartilage. About the time of birth, secondary centers of ossification appear in the cartilaginous tissue at the ends of the long bones (Fig. 16–8). These produce caps of bone known as *epiphyses*. As long as zones of cartilage remain between the epiphyses and the shaft, increase in length of the long bone is accomplished through the growth of the intervening cartilage. But finally, late in adolescence, the bony epiphyses unite with the shaft, and growth in formed and are separated from each other and from the neural fragment.

Induction can take place when the inductor and the induced are not in immediate contact with each other. Neural tissue will induce when it is separated from the somite tissue by a thin millipore filter 20 μ thick. Notochord will not accomplish this. Moreover, an extract of neural tube or notochord added to the nutrient medium will induce somite cells to become cartilage.

Holtzer found that it may take as long as four days from the time the induction occurs until the first cartilage is produced. During this time some sort of chemical activity is going on

even though it cannot be directly demonstrated. Possibly the inducing chemical is an enzyme which starts a chain of enzymatic reactions.

A different explanation is needed to account for the differentiation of the cartilages of the limb buds. According to Holtzer, notochord and neural tube do not induce the production of cartilage in limb-bud mesenchyme or lateral-plate mesoderm. Yet these tissues do form cartilage, namely the cartilages of the appendicular skeleton.

Lash, working with material from 3-day chicks, made the unexpected discovery that mesenchyme taken from the mesonephric re-

FIGURE 16–8 Diagram to illustrate the growth of a long bone. Hyaline cartilage is shown with light stippling; calcified cartilage with heavier stippling; bone is solid black. A. Mesenchyme and blood vessels have invaded the shaft (diaphysis) and ossification is well under way. The periosteum has deposited a collar of compact bone around the shaft. B. Centers of ossification (epiphyses) have appeared near the ends of the forming bone. Throughout the period of growth in length, a plate of cartilage remains between each epiphysis and the shaft. C. At maturity (about age 19 in the case of the human femur) the epiphyseal plates ossify and growth in length ceases. Hyaline cartilage remains at the surfaces of joints.

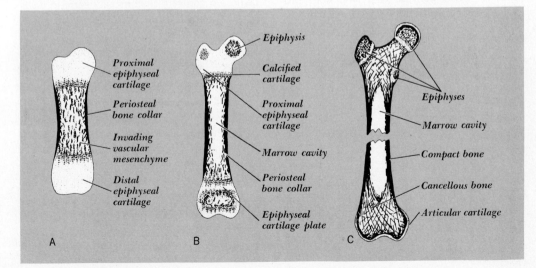

gion adjacent to the limb buds and then explanted and grown in culture will produce cartilage of limb-bud type. Other regions of the mesonephros-forming mesenchyme do not do so. The explanation seems to be that the "field" of limb-producing competence is larger than the actual limb bud and extends into adjacent nephrogenic tissue. It may even extend into nearby somites, for Muchmore has found that when several somites from the environment of the limb bud are explanted and grown in culture, they sometimes form limb structures.

Axial and appendicular cartilage, then, although they are alike histologically, are yet different morphogenetically. Moscona has demonstrated that limb-bud mesenchyme of young chick embryos, when dissociated in a suitable culture medium, tends to reaggregate in clusters. The same is true in the case of sclerotome mesenchyme. When the clumps are cultivated on plasma clots, they both produce cartilage; but the cartilages are different in size and shape. The cartilages of sclerotome origin form nodules. Those derived from limb buds produce flattened plates. When clusters from limb buds and sclerotomes are placed together on the same plasma clot, they form cartilages which grow into contact with each other; but the two types of cartilage do not fuse. Is it possible that incompatibility of this sort is the origin of joints between cartilages?

MUSCLE

Muscle fibers are derived from mesenchyme cells (myoblasts) which elongate and organize bundles of delicate myofibrils within their cytoplasms. They are of three kinds: smooth muscle, striated muscle, and cardiac muscle. These differ from one another both structurally and functionally. (1) The myoblasts which become smooth muscle remain spindle-shaped. Each fiber has a single nucleus, and their myofibrils are simple and threadlike. (2) The myoblasts which produce striated muscle become elongated and cylindrical. They acquire many nuclei in each fiber, and their myofibrils are beaded with alternating dark (anisotropic) and light (isotropic) bands. Since the bands of adjacent myofibrils line up crosswise, the entire fiber has a cross-striated appearance. (3) Cardiac muscle is a specialized type of visceral muscle which is present only in the heart. Like striated muscle, its fibers are crossbanded; but unlike striated muscle, its cells join with one another by protoplasmic connections to make what appears to be a continuous syncytial net. This is true at least functionally, for the net conducts impulses in every direction. If any place in the net is stimulated, the whole net contracts. This is known as the "all-or-none law" of the heart. Cardiac muscle differs also in that it contracts spontaneously. Indeed, it begins to contract rhythmically early in development, long before any nerve fibers have come near it (page 186).

How do striated muscle fibers become multinucleate? At first the myoblasts of striated muscle possess one nucleus per fiber. By the fifth day of incubation (chick), muscle fibers (myotubes) begin to appear which have several nuclei per fiber (Fig. 16–9). Is this the result of mitotic division of the nuclei? No, for mitotic figures are few. Rather, it results from the fusion of myoblasts. Wilde mingled chick and mouse myoblasts in tissue culture and obtained multinucleate fibers containing nuclei of both species. Holtzer mixed chick myoblasts, some of which had been labeled with a radioactive nutrient (tritiated thymidine) and some

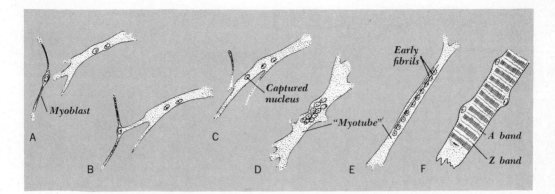

FIGURE 16–9 Histogenesis of a striated muscle fiber. *A to C.* A myoblast is incorporated into a growing muscle fiber. *D and E.* A young muscle fiber or "myotube" with nuclei central and arranged in a column. *F.* Portion of an embryonic muscle fiber in which the nuclei have been displaced to the side by the multiplication of fibrils. The latter are now striated. *(Based on photomicrographs by C. R. Capers, 1960, J. Biophys. Biochem. Cytol., 7:559–565.)*

of which had not. He obtained myotubes in which some of the nuclei were labeled and others were not.

Morphogenesis of Muscles

It will be convenient to describe the development of the muscles of the body under three heads: (1) axial muscles, (2) appendicular muscles, and (3) visceral muscles. Axial muscles and appendicular muscles consist of striated muscle fibers only. The term visceral muscle, on the other hand, includes muscle tissue of all three types: striated, cardiac, and smooth.

The Axial Muscles

The axial muscles of the head, trunk, and tail are derived from the myotomes of the somites with additions from the somatic layer of the lateral mesoderm. At first the myotomes

are small. After the sclerotomes have become mesenchyme, the myotomes swing to a position beneath the dermatomes (Fig. 15–1). Then together the myotomes and dermatomes expand dorsally and ventrally. For a time, both retain the character of layered tissues, or epithelia. There may even be a cavity, the myocoel, between them. But later the dermatomes break down into loose mesenchymal cells. However, the myotomes tend to remain as segments.

Typically—this is true of the lower vertebrates especially—the myoblasts which are derived from the myotomes elongate parallel to the axis of the embryo, each fiber spanning the distance from one myoseptum (partition between myotomes) to the next (Fig. 16–6). In the higher vertebrates the arrangement is more complicated. Successive myotomes tend to unite with each other and form premuscle masses which more or less lose their original segmental character and migrate often to con-

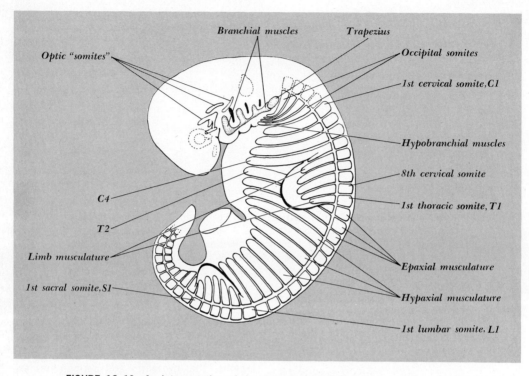

FIGURE 16–10 An interpretation of the skeletal musculature. *(Adapted in part from B. M. Patten, 1964, Foundations of Embryology, McGraw-Hill Book Company, New York, 2d ed., Fig. 16–19.)*

siderable distances. Also, the directions of their fibers may change, and they frequently subdivide into separate muscles. Some of the myoblasts become fibroblasts and give rise to connective tissues which surround the muscles (fascias) or attach the muscle to the skeleton (tendons).

In all vertebrates above the cyclostomes, the myotomes divide into a dorsal part, the *epaxial muscles,* and a ventral part, the *hypaxial muscles* (Fig. 16–10). Between the two parts there is a septum which is called, in fish at any rate, the *horizontal septum.* The epaxial muscles

extend the full length of the vertebral column and become the deep muscles of the back. Their function is to extend the spine (that is, bend it backward). The hypaxial muscles are derived in part from mesoderm of the somatopleure. They give rise to muscles of the lateral and ventral body wall and to muscles of the appendages.

Extrinsic muscles of the eye   Theoretically, at least, the most anterior axial muscles are those which move the eyeball. In the development of the shark, these arise from somites of the head anterior to the otocysts. Even in the chick and in man, transient cavities (myocoels) have been described in the walls of which the eye muscles appear. Moreover, the nerves which supply the eye muscles (III, IV, and VI) are somatic motor nerves homologous with those of the myotomes of the trunk.

There are no head somites in the region of the auditory capsules which surround the otocysts. The auditory (otic) cartilage takes up the available space.

Muscles of the tongue   We again encounter somites in the occipital region (Fig. 16–10). The first four somites of a chick or mammal embryo belong to the head. Their sclerotomes share in forming the occipital bone. The epaxial parts of their myotomes disappear, but the hypaxial parts migrate ventrally posterior to the branchial arches. They then move forward to a position beneath the pharynx. Here they give rise to the intrinsic muscles of the tongue. As they migrate, they tow their nerve supply after them, namely, fibers of the twelfth or hypoglossal nerve. The circuitous route taken by this nerve leaves no doubt but that in some fashion a migration of this sort actually takes place.

Muscles of the trunk and tail   The somites, beginning with the fifth pair, give rise to the muscles of the trunk and tail (Fig. 16–10). Their epaxial parts retain something of their original segmental character as they become the deep muscles of the back. Their hypaxial parts in the neck and tail become flexors of the spine. In the trunk, the hypaxial parts are augmented by contributions of the somatopleure and give rise to the lateral and ventral musculature of the body wall. The muscles of the mammalian diaphragm are derived from hypaxial muscles of the neck, mainly from the 4th cervical somites.

The Appendicular Muscles

Limb muscles are derived predominantly from the limb buds, and these in turn originate from the somatic layer of the lateral mesoderm. In the shark embryo, slips of premuscle tissue can be seen to bud off from the ventral margins of adjacent myotomes and enter the fin folds. A comparable migration has not been shown to occur in birds and mammals; yet, in principle, something of the sort must take place, for the nerves to the muscles of the limbs are comparable to those which go to the hypaxial muscles of the flank.

The mesenchyme of a limb bud differentiates into three layers (Fig. 16–11): a dorsal layer which gives rise to the premuscle masses from which the extensor muscles of the appendage are developed; an intermediate layer which becomes the skeleton of the limb; and a ventral layer which produces the premuscle masses from which the flexor muscles of the limbs are produced. The premuscle masses migrate, subdivide, and finally differentiate into the various muscles of the limb. As they migrate they trail their nerve supply after them. Hence the study of the nerve supply of a muscle reveals its source, history, and homologies.

An especially interesting muscle in this respect is trapezius, the great superficial muscle of the back. Historically, it is a muscle of the branchial region which has migrated posteriorly and inserted on the scapula (Fig. 16–10). It is therefore to be classed as a visceral striated muscle and as such is supplied by a visceral motor nerve, the spinal accessory (XI).

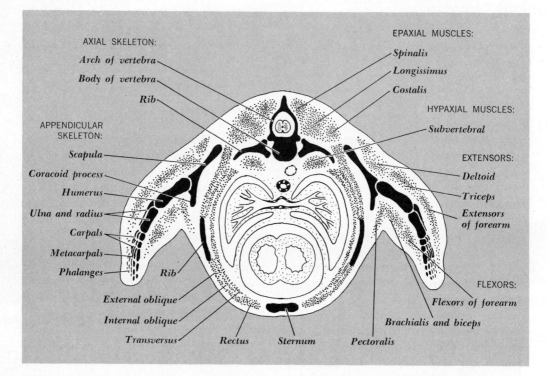

AXIAL SKELETON:
Arch of vertebra
Body of vertebra
Rib

APPENDICULAR
SKELETON:

Scapula
Coracoid process
Humerus
Ulna and radius
Carpals
Metacarpals
Phalanges   Rib

External oblique
Internal oblique
Transversus   Rectus   Sternum   Pectoralis

EPAXIAL MUSCLES:
Spinalis
Longissimus
Costalis

HYPAXIAL MUSCLES:
Subvertebral

EXTENSORS:
Deltoid
Triceps
Extensors
of forearm

FLEXORS:
Flexors of forearm
Brachialis and biceps

FIGURE 16–11 Diagrammatic section through the thorax to show the elements of the skeleton and musculature. *(Modified from B. M. Patten, 1964, Foundations of Embryology, McGraw-Hill Book Company, New York, 2d ed., Fig. 16–21.)*

The Visceral Muscles

The term visceral muscle, as used in embryology and comparative anatomy, refers to a muscle derived from the splanchnopleure, whether that muscle histologically is smooth, cardiac, or striated. (1) The muscles derived from the branchial arches are visceral striated muscles and are supplied by nerves of the special visceral motor column. The first branchial arches give rise to the *muscles of mastication* and are supplied by the mandibular division of the trigeminal nerve (V). The second branchial arches give rise to the *muscles of facial expression* and are served by the facial nerve (VII). They reach their greatest development and diversification in man. The other branchial arches give rise to muscles of the pharynx and larynx. They are innervated by the glossopharyngeal and vagus nerves (IX and X). (2) The muscles of the myocardium are cardiac muscle. We have already noted some of the peculiarities of this muscle (page 346). (3) The muscles of the alimentary tract and its derivatives posterior to the pharynx are smooth muscle. Their fibers are usually arranged as sheets around each endodermal organ: an outer sheet in which the fibers run longitudinally, and an inner sheet composed of circular fibers.

Experiments on Muscle Tissue

When is it decided that a given mesenchyme cell shall become a muscle cell? According to present concepts it takes place when the genes which lead to the synthesis of contractile (and no doubt other) proteins are activated. The genes transcribe their information to messenger RNA. The mRNA then migrates into the cytoplasm and, in association with ribosomes, begins the synthesis of myosin and actin. These in turn array themselves as myofibrils.

Electron micrographs of the somites of chicks of less than 60 hours of incubation show some cells which appear to be engaged in synthesis. They are no longer dividing. Polysomes are abundant, and associated with them are filaments of what well may be myosin molecules. A little later (after 60 hours) light microscopes show spindle-shaped cells, each with a single nucleus and a few cross-striated myofibrils.

How is it decided that certain cells will give rise to muscle? Muchmore explanted somites of amphibian embryos into ectodermal jackets and found that the cells have little tendency to produce striated muscle. Some differentiated and gave rise to nephric tubules, heart tissue, even limb buds—or they remained undifferentiated. But when he added tissue of the spinal cord, muscle tissue appeared. Holtzer cultivated somite tissue of the chick on plasma clots and obtained similar results. Not only is the presence of neural tissue necessary to initiate the differentiation of muscle, but its continuing presence is needed to keep muscle

tissue already produced from deteriorating. This property is peculiar to the cord and is not shared by peripheral nerves and ganglia. Brain tissue is said to inhibit rather than stimulate the differentiation of muscle.

What decides the specific nature of striated muscle? Why does one premuscle mass produce epaxial muscles and another produce hypaxial muscles? Why does one mass give rise to forelimb muscles and another to hindlimb muscles? What accounts for the differentiation of flexors, extensors, etc. Apparently the spinal cord does not control these differentiations. Muchmore finds that the inductive influence of the spinal cord is nonspecific. It favors the production of muscle, but it does not specify which muscles are produced.

The work of Weiss, discussed in Chapter 12, suggests that specificity originates in the muscle tissue itself. The muscle tissue then transfers its specificity to the nerves which supply it (pp. 282–283).

Another line of evidence which points to the specific nature of muscle tissue comes from experiments on the regeneration of amphibian limbs. When a fore- or hindlimb of a salamander larva is transected, the wound becomes covered with a layer of epidermis, and then a growth of new mesenchymal cells (derived from muscle and supportive tissues) takes place at the cut surface. In time, the "blastema" thus produced differentiates and restores the skeleton and musculature of the part of the appendage which was removed. Now, it has been demonstrated that the cut ends of the muscles of the stump have much to do with controlling the specific character (whether forelimb or hindlimb) and orientation of the part of the limb which is restored. The skeleton of the stump may have some influence; but a missing segment of a limb will regenerate, including its skeleton, even if the bone or cartilage of the stump has been removed and only softer tissue remains.

What then determines the specificity of muscle? No satisfactory answer has as yet been given. We may suppose that at an early stage in development local signs provided by the position of the mesenchyme in the gradient fields of the embryo decide what the mesenchyme will become.

SUGGESTED READINGS

The Skin

Rawles, Mary E., 1955. Vertebrate organogenesis: The skin and its derivatives. In Willier, Weiss, and Hamburger (eds.), Analysis of Development. W. B. Saunders Company, Philadelphia. Pp. 499–519.

Romer, A. S., 1962. The Vertebrate Body. W. B. Saunders Company, Philadelphia. 3d ed. Chapter 6, The skin.

Wessells, N. K., 1962. Tissue interactions during skin histodifferentiation. Develop. Biol., 4:87–107.

McLoughlin, C. B., 1963. Mesenchymal influences on epithelial differentiation. Soc. Exptl. Biol., Cell Differentiation (17th Symp.), pp. 359–388.

Bell, E., 1965. The skin. In DeHaan and Ursprung (eds.), Organogenesis. Holt, Rinehart and Winston, Inc., New York. Pp. 361–374.

Feathers

Lillie, F. R., 1942. On the development of feathers. Biol. Rev., 17:247–266.

Wang, H., 1943. The morphogenetic functions of the epidermal and dermal components of the papilla in feather regeneration. Physiol. Zool., 16:325–350.

Saunders, J. W., Jr., and M. T. Gasseling, 1957. The origin of pattern and feather germ tract specificity. J. Exptl. Zool., 135:503–527.

———, 1958. Inductive specificity in the origin of integumentary derivatives in the fowl. In McElroy and Glass (eds.), The Chemical Basis of Development. The Johns Hopkins Press, Baltimore. Pp. 239–253.

Cohen, J., 1966. Feathers and patterns. Advan. Morphogenesis, 5:1–38.

Cartilage and Muscle Tissue

Holtzer, H., 1961. Aspects of chondrogenesis and myogenesis. In Rudnick (ed.), Synthesis of Molecular and Cellular Structure. The Ronald Press Company, New York. Pp. 35–88.

Gross, J., 1956. The behavior of collagen units as a model in morphogenesis. J. Biophys. Biochem. Cytol., 2: Suppl., 261–274. Reprinted in Bell (ed.), 1965. Molecular and Cellular Aspects of Development. Harper & Row, Publishers, Incorporated, New York. Pp. 416–427.

———, 1961. Collagen. Sci. Am., 204 (May): 120–130.

Lash, J. W., F. A. Hommes, and F. Zilliken, 1962. The in vitro induction of vertebral cartilage with a low-molecular-weight tissue component. Biochim. Biophys. Acta, 56: 313–319. Reprinted in Bell (ed.), 1965. Molecular and Cellular Aspects of Development. Harper & Row, Publishers, Incorporated, New York. Pp. 95–100.

Konigsberg, I. R., 1963. Clonal analysis of myogenesis. Science, **140**:1273–1284. Reprinted in Bell (ed.), 1965. Molecular and Cellular Aspects of Development. Harper & Row, Publishers, Incorporated, New York. Pp. 116–134.

————, 1964. The embryological origin of muscle. Sci. Am., **211** (August): 61–67.

Hay, Elizabeth D., 1965. Metabolic patterns in limb development and regeneration. In DeHaan and Ursprung (eds.), Organogenesis. Holt, Rinehart and Winston, Inc., New York. Pp. 315–336.

Milaire, J., 1965. Aspects of limb morphogenesis in mammals. In DeHaan and Ursprung (eds.), Organogenesis. Holt, Rinehart and Winston, Inc., New York. Pp. 283–300.

Skeleton

Goodrich, E. S., 1930. Studies on the Structure and Development of Vertebrates. The Macmillan Company, London. Reprinted by Dover Publications, Inc., New York. Chaps. 1, 3, 4, and 6.

Hamilton, W. J., J. D. Boyd, and H. W. Mossman, 1962. Human Embryology. The Williams & Wilkins Company, Baltimore. 3d ed. Chap. 13, Skeletal system.

Romer, A. S., 1962. The Vertebrate Body. W. B. Saunders Company, Philadelphia. 3d ed. Chaps. 7 and 8.

Moffett, B. C., Jr., 1965. The morphogenesis of joints. In DeHaan and Ursprung (eds.), Organogenesis. Holt, Rinehart and Winston, Inc., New York. Pp. 301–313.

Musculature

Romer, A. S., 1962. The Vertebrate Body. W. B. Saunders Company, Philadelphia. 3d ed. Chap. 9, The muscular system.

Gilbert, P. W., 1957. The origin and development of the human extrinsic ocular muscles. Carnegie Contrib. Embryol., **36**:61–78. 14 plates.

chapter **17**

THE
CIRCULATORY
SYSTEM

It is remarkable how small some animals are which possess blood-vascular systems and how early in development hearts begin to beat. Tiny crustacean larvae and molluscan embryos —mere specks to the naked eye—have beating hearts, and the human embryo begins to circulate blood when it is one-eighth of an inch or so in length.

The earliest function of the circulatory system appears to have been to distribute the products of digestion. At any rate, the blood and blood vessels arise in the splanchnopleure of the yolk-laden midgut (amphibians) or yolk sac (birds). Even in mammals, despite the almost complete absence of yolk, it is in the yolk sac that these elements of the circulatory system first appear. An early respiratory function is also indicated, for the first blood cells contain hemoglobin and serve in the transport of gases.

The circulatory system, like other systems of the embryo, passes through two phases in its development: a prefunctional phase, and a functional phase. During the prefunctional phase, the heart and major blood vessels self-differentiate independently of each other. Obviously, the heart must be formed before it can beat, and the major blood vessels must be present before there can be a circulation.

But this independent phase of self-differentiation does not continue. If the heart rudiment is experimentally removed, blood vessels nevertheless form; but in the absence of the heart and of a circulation, they soon degenerate. If a major blood vessel is occluded, the bloodstream will adopt a detour through a neighboring bed of capillaries, or an adjacent minor blood vessel will take on the structure of the vessel which was destroyed. Phenomena of this sort were observed long ago and given the name "functional adaptation."

The Origin of Blood Vessels

The blood vessels and the blood originate early in development from mesenchymal cells known as *angioblasts*. They appear first in the splanchnopleure of the midgut (amphibians) or yolk sac (birds and mammals) as clusters and cords which bear the name of blood islands. The outer cells of the islands flatten and become the endothelial linings of the vitelline blood vessels. The inner cells of the blood islands round up and become the first blood corpuscles of the embryo (Fig. 17–1, *A* and *B*).

Other blood vessels originate in a similar manner; that is, they form *in situ* from the mesenchyme of the embryo. Knots of angioblasts form in the mesenchyme, hollow out, link together, and become endothelial tubes. After the main channels have been established and circulation has begun, the *in situ* differentiation of blood vessels ceases. Further proliferation of blood vessels is by capillary "budding" (Fig. 17–1, *C–E*). Sprouts grow out from nearby vessels into the surrounding tissues. The sprouts branch, fuse with each other, and hollow out to form capillary loops through which blood at once commences to flow. Possibly a lowering of the oxygen tension or an increase in carbon dioxide content in the unvascularized tissue stimulate and guide the outgrowth of the sprouts. The final result is a mesh of capillaries.

The first blood vessels in any region of the embryo are capillaries. Differentiation of arteries and veins comes later and is apparently controlled by the volume and steadiness of the flow of the blood. Preferred channels through the capillary net enlarge and become surrounded by mesenchyme cells, which then transform into smooth-muscle and connective-tissue cells. Whether a blood vessel becomes

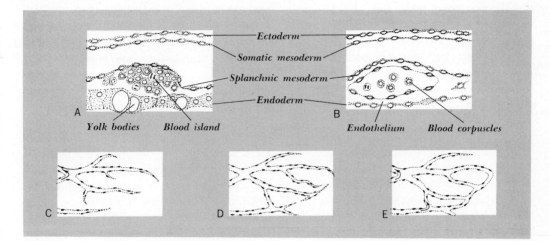

FIGURE 17-1 The origin of blood vessels. A and B. Sections through blood islands of chick embryos. C to E. Sequence showing the expansion of the capillary net. *(From L. B. Arey, 1965, Developmental Anatomy, W. B. Saunders Company, Philadelphia, 7th ed., Fig. 304.)*

an artery, with its thicker wall, or a vein has been shown to depend on the character of the blood flow within it.

The Origin of Blood Cells

The production of blood cells is a process known as hemopoiesis. It begins in the blood islands of the yolk sac; but from time to time the site of origin of blood shifts. The old sites usually cease to function. Hemopoiesis takes place when undifferentiated mesenchyme cells round up and become free, basic-staining cells known as *hemocytoblasts*. From this common source a considerable variety of blood cells are formed. The first are primitive red cells (erythroblasts) derived from the blood islands and containing hemoglobin. They serve in the transport of blood gases. Various sorts of white corpuscles differentiate later, apparently in response to the influences of different sur-

roundings. Thus, when cells of the yolk sac (which would have become red cells) are transplanted to the chorioallantoic membrane of a chick, they transform mainly into white corpuscles.

In the human embryo, the seat of blood production soon shifts from the yolk sac to the body mesenchyme. Next it centers in the liver, and still later in the thymus, spleen, and lymph nodes. Finally, the place of origin of the red corpuscles (erythrocytes) and most of the white corpuscles (leucocytes proper) is the bone marrow. Lymphocytes continue to be derived from the lymphoid tissues. The details are complex and may be found in textbooks of histology.

It will be noted that blood arises in the various regions of the embryo which are distant from the dorsal axial organs. Explants of the posterior and lateral blastoderm of a chick produce blood cells even when they do not produce other recognizable tissues. It would seem that the neuralizing factor (pp. 113–114) antagonizes the production of blood cells, while the mesodermalizing factor favors the process. Note, furthermore, that the bone marrow, which is one of the final hemopoietic tissues of mammals, has a mesodermalizing effect on differentiation.

Hemoglobin

The cells which are in the sequence leading to the production of hemoglobin are known as *erythroid cells*. The series begins with hemocytoblasts (stem cells), devoid of any trace of hemoglobin, and progresses through several stages of erythroblasts. It ends with erythrocytes (red corpuscles) which are 95 percent hemoglobin. The development of hemoglobin is a true epigenesis, the coming-into-being of a chemical which did not at first exist.

The development of erythroid cells is not limited to the embryo but goes on continuously in the hemopoietic centers of the body. At first the erythroblasts are but little different from generalized mesenchyme cells, but they become committed; that is, they acquire the capacity when isolated to self-differentiate into red blood corpuscles. Two or three cell divisions take place, but for the most part the genes are engaged in producing ribosomes and mRNA. Mitochondria also increase in number, and their enzymes are active. Then, in the case of mammals, a remarkable change takes place: the erythroid cells lose their nuclei, possibly by extruding them. The synthesis of hemoglobin, which has already begun, then becomes increasingly active.

The fact that hemoglobin is produced almost entirely *after* the cells have lost their nuclei means that transcription (the transfer of information from DNA to mRNA) precedes translation (the transfer of information from mRNA to newly synthesized protein) in time. When mRNA is produced, it is "masked." After the nuclei have been lost, the mRNA becomes unmasked, joins with the ribosomes to make polyribosomes, and at once begins the synthesis of polypeptide chains of hemoglobin.

Today the chemistry of hemoglobin is better understood than the chemistry of most proteins. Each molecule consists of four polypeptide chains which enclose a heme group. The four chains are not identical. In fetal hemoglobin two of the chains are known as alpha chains, and two are gamma chains. Beginning about the time of birth, beta chains are substituted for the gamma chains. This change of composition of the hemoglobin means that the mRNA which accomplishes the synthesis of the polypeptide chains is different. And the difference in the mRNA signifies that different genes have been called into action. The mechanism by which this is accomplished is not known. One suspects, however, that the change in the composition of the hemoglobin is adaptive. The oxygen tensions within the embryo during development are not the same as the oxygen tensions in the body following birth. Can it be that the changing oxygen tensions somehow influence the genes, possibly through some hormone mechanism? A hormone, erythropoietin, regulates the production of red cells in the adult body.

THE HEART

The Origin of the Heart

The early development of the amphibian heart was described in Chapter 5 and that of the chick in Chapter 8. The story of the mammal heart is similar. Typically, the heart forms ventral to the embryonic pharynx when folds of splanchnic mesoderm push medially and enfold the epithelial tubes of the primitive ventral blood vessels (Figs. 8-6 and 8-8). The enfolding mesoderm becomes the *epimyocardium;* the endothelia of the paired ventral blood vessels unite in the midline and become the *endocardium,* or lining of the heart. Between the two layers, there is for a time a sort of jelly.

Thus, the heart develops in the ventral mesentery of the pharynx. That part of the mesentery which is dorsal to the heart, namely, between the heart and the underside of the pharynx, is the *dorsal mesocardium* (Fig. 15-3, *A*). That part of the mesentery which is ventral to the heart, between the heart and the ventral body wall, is the *ventral mesocardium.* The latter immediately breaks through. In fact, mammals can hardly be said to have a ventral mesocardium at any time. Soon, the dorsal

mesocardium also disappears, so that the heart lies free in the pericardial cavity except, of course, at its two ends where the great blood vessels enter and leave.

In man and other primates, the splanchnic mesoderm surrounds the future endocardium even before there is a head fold or foregut. The heart forms anterior to the future head. Then, when the head, foregut, and subcephalic pocket form, the heart region swings posteriorly to its typical position beneath the pharynx.

As the mammalian heart grows in length, it coils within the pericardial cavity in the same way that the hearts of lower vertebrates coil. First, there is the C-shaped ventricular loop to the right (Fig. 17-2, showing a chick heart). Then the loop swings posteriorly so that the heart becomes a closely appressed S-shaped curve, with the truncus ventral to the atrium and the ventricular loop beneath the sinus venosus. As the atrium enlarges, it develops

FIGURE 17–2 Ventral view of the chick heart. A. C-shaped heart of a 33-hour chick. B. S-shaped heart of a 48-hour chick. C. Spiral heart of a 72-hour chick. The heart is represented as though transparent. (*Adapted from B. M. Patten, 1964, Foundations of Embryology, McGraw-Hill Book Company, New York, 2d ed., Figs. 13–17 and 13–18.*)

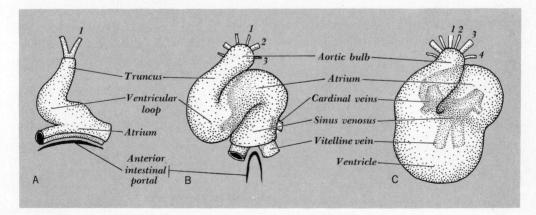

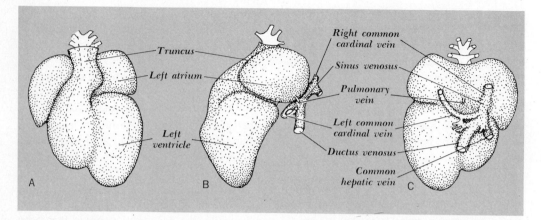

FIGURE 17–3   The heart of a 6-mm pig embryo. *A*. Ventral view. *B*. View from left. *C*. Dorsal view.

lateral lobes, the "auricles," which bulge ventrally and partially enfold the truncus (Fig. 17–3). At this stage the mammalian heart closely resembles the four chambers of a fish heart: sinus venosus, atrium, ventricle, and truncus.

At an early stage of development, the sinus venosus widens until it comes into contact with the lateral body walls. Fusion immediately takes place, forming the lateral mesocardia (Fig. 8–25, *A*). Across the lateral mesocardia as "bridges," the common cardinal veins carry blood from the body wall to the heart. Thus the sinus venosus receives four vessels: the *right* and *left vitelline veins* from the splanchnopleure, and the *right* and *left common cardinal veins* from the somatopleure (Fig. 17–2, *C*).

Why does the heart coil and assume its typical form? Is the process a matter of "mechanical expediency," as has been suggested, as though a limitation of space forced the heart to assume on S-shape? Does the presence of the sinus venosus on the dorsal side and the truncus on the ventral side pinch the atrium

and cause it to bulge laterally and produce sacculations?

Experiments have shown that neither of the above-mentioned explanations is the case. Bacon, working with amphibian embryos, found that when explants of prospective heart tissue are removed from a midgastrula, they develop their characteristic S-curves even though they are unconfined within constraining walls. This does not mean, however, that the surrounding tissues have no influence. In birds and mammals, a heart will not develop if the surrounding tissues are absent.

When and how does the heart beat originate? We noted in Chapter 8 that the twitching of the chick heart begins at about the 9-somite stage when the ventricle just begins to form. It increases in tempo and regularity when the atrium is added and again when the sinus venosus has formed and has begun its function as the "pacemaker" of the heart. The same principles apply to the developing mammal embryo.

The Partitioning of the Heart

The heart becomes partitioned into a right side and a left side in all air-breathing vertebrates. This takes place rather early in development, long before the lungs have begun to function at the time of birth. No such division occurs in the hearts of vertebrates which breathe in water. Their hearts pump just one kind of blood, namely, "venous blood" (blood low in oxygen); aeration takes place in the gills, and only "arterial blood" (blood rich in oxygen) is distributed to the body. The advent of lungs necessitated a radical change in the arrangement of the circulation. The heart now received two sorts of blood: venous blood from the body, and arterial blood from the lungs. It became important that the two streams be kept separate so that the venous blood could be sent to the lungs for aeration and only the arterial blood distributed to the body.

The partitioning of the circulation was not accomplished easily in the process of evolution. In fact, it is still incomplete in amphibians and reptiles. In these, the atrium is divided into right and left chambers, but the ventricle remains more or less undivided. Venous blood returning from the body enters the sinus venosus and is discharged into the right atrium only. Blood from the lungs comes by way of the pulmonary veins (which are new formations) and goes directly into the left atrium. A functional separation of the two bloodstreams is maintained as the blood passes through the undivided ventricle. The venous blood, which enters from the right atrium, is directed through the ventricle toward the sixth aortic arches and goes mainly to the lungs. The arterial blood from the left atrium is directed toward the third and fourth aortic arches and goes to the rest of the body.

Bird and mammal embryos make preparation for the ultimate separation of the bloodstreams very early in development, but it is not until the time of birth that the actual separation takes place. The switchover, when it does occur, is truly a remarkable phenomenon. (It is described in the last paragraphs of this chapter.)

Figures 17–4 and 17–5 depict in a somewhat diagrammatic manner the partitioning of the mammalian heart. The following take place:

1 The sinus venosus shifts to the right and is gradually absorbed into the wall of the right atrium. For a time it is bounded by the valves of the sinus venosus (Fig. 17–4), but ultimately the valves are absorbed.

2 Dorsal and ventral "endocardial cushions" develop in the canal between the atrium and ventricle (AV canal). These grow and come into contact with each other. They fuse and thus divide the original single AV canal into right and left AV canals.

3 A median partition, the septum primum (septum I), grows from the dorsal, anterior, and ventral walls of the atrium toward the endocardial cushions (Fig. 17–4). It divides the atrium into right and left chambers. However, the partition is at no time complete, for just before the septum reaches and fuses with the AV cushions (closing "interatrial foramen I") a new hole develops through it, known as interatrial foramen II (Fig. 17–5). This feature is important, for if the septum were to become complete, the left atrium would receive no blood other than the small amount which returns from the unexpanded lungs. The left atrium then would have almost no work to do, and the left ventricle would be underdeveloped. As it is, about one-half the blood flows through the interatrial foramen II and enters the left chambers of the heart.

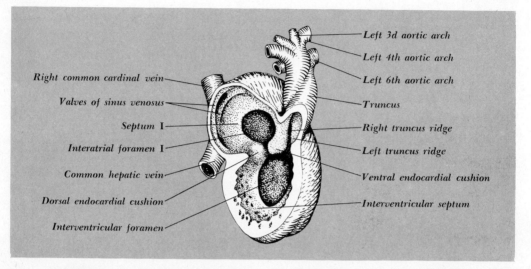

Right common cardinal vein
Valves of sinus venosus
Septum I
Interatrial foramen I
Common hepatic vein
Dorsal endocardial cushion
Interventricular foramen

Left 3d aortic arch
Left 4th aortic arch
Left 6th aortic arch
Truncus
Right truncus ridge
Left truncus ridge
Ventral endocardial cushion
Interventricular septum

FIGURE 17–4   Heart of a 6-mm pig embryo (or older) viewed obliquely from the ventral right. The lateral wall has been removed to show the cavities of the right atrium, right ventricle, and right (or pulmonary) trunk. The left chambers are partly visible through the open interatrial and interventricular foramina (somewhat schematic).

Soon a second partition, the septum secundum (septum II), appears to the right of septum primum and grows backward and dorsally in a plane parallel to septum primum (Fig. 17–5). Unlike the first septum, the second does not reach the endocardial cushions. Instead, an opening remains, known as the *foramen ovale,* through which blood continues to flow until the time of birth. Now, septum I overlaps septum II in such a manner that it serves as a valve for the foramen ovale. When the blood pressure in the right atrium is higher than in the left atrium, as is the case in the relaxed embryonic heart, the valve-like septum I is blown away from the foramen ovale, and blood flows freely through the holes from right to left. But when the atria contract (atrial systole), the increase in pressure in the left atrium presses septum I against the foramen ovale and closes it. The result is that the blood of the left atrium is forced into the left ventricle and then is pumped by the left ventricle into the aorta. We shall presently describe how, at the time of birth, the increase in pressure of the blood in the left atrium causes the foramen ovale to close and to remain closed.

It is of importance that the foramen ovale and interatrial foramen II are so situated that the blood which flows through them is mainly oxygen-rich blood from the posterior vena cava. Oxygen-poor blood from the anterior vena cava goes through the right AV canal into the right ventricle.

4 An *interventricular septum* gradually grows from the tip of the ventricular loop toward the AV cushions. It ultimately fuses with the cushions and by so doing divides the ventricle into right and left chambers. However, the interventricular foramen remains open for some time and presumably functions, as it does in adult amphibians and reptiles, by assuring that the blood pressures in the right and left ventricles will be equal (Fig. 17–5).

5 Meanwhile, longitudinal *endocardial thickenings,* similar histologically to the AV cushions, begin to form in the wall of the truncus. These are the *truncus ridges* (Fig. 17–4). They appear first at the base of the ventral aorta and then grow toward the ventricle, pursuing a spiral course. Ultimately they reach the ventricle and fuse with the interventricular septum and AV cushions. They also fuse with each other so that the truncus is divided into two spiral channels: the *pulmonary trunk* and the *aortic trunk* (Fig. 17–5). Note that the blood of

the right ventricle enters the pulmonary trunk and flows forward in a clockwise spiral. First it passes to the ventral side of the aortic trunk, then to its left, and finally to its dorsal side where it enters the sixth (pulmonary) aortic arches. Similarly, the blood of the left ventricle enters the aortic trunk, spirals to the dorsal side of the pulmonary trunk, then to its right, then ventral to it, and enters the third and fourth aortic arches.

The endocardial cushions of the AV canals hollow out on their ventricular side and give rise to the cusps of the AV valves: the *tricuspid valve* on the right and the *bicuspid* or *mitral valve* on the left. Similarly, the proximal ends of the endocardial thickenings of the truncus become excavated and give rise to the cusps of the semilunar valves: the *pulmonary valve* on the right and the *aortic valve* on the left.

FIGURE 17–5 Heart of a more advanced pig embryo than Fig. 17–4. The left atrioventricular canal can be seen through the incompletely closed interventricular foramen.

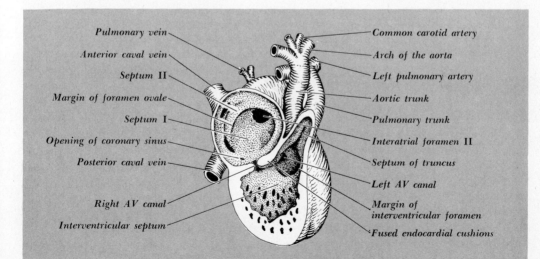

Pulmonary vein

Anterior caval vein

Septum II

Margin of foramen ovale

Septum I

Opening of coronary sinus

Posterior caval vein

Right AV canal

Interventricular septum

Common carotid artery

Arch of the aorta

Left pulmonary artery

Aortic trunk

Pulmonary trunk

Interatrial foramen II

Septum of truncus

Left AV canal

Margin of interventricular foramen

Fused endocardial cushions

THE ARTERIES

The Fate of the Aortic Arches

The early development of the mammalian arteries repeats, in most respects, the story which we told concerning the arteries of the chick (pp. 204–206). The ventral aortas become foreshortened and unite to form what has been called the "aortic sac" (Fig. 17–8). From the sac, the third, fourth, and sixth aortic arches radiate outward and, after bending upward through their respective branchial arches, join the dorsal aortas.

The mammalian aortic arches, like those of the chick, arise in sequence, beginning with the most anterior and ending with the sixth (or pulmonary) arches. Arches 1 and 2 disappear early; arches 3, 4, and 6 remain; the fifth arches are rudimentary. In the 6-mm pig, the third, fourth, and sixth arches carry blood, the sixth arches having just begun to function (Fig. 17–6). Soon after the 10-mm stage, the dorsal aortas between the third and fourth aortic arches (the so-called carotid ducts) regress (Fig. 17–7). Thus, the third arches become the roots of the *internal carotid arteries,* and the fourth arches become the roots of the dorsal aortas (Fig. 17–8).

The *external carotid arteries* replace the ventral aortas from the third arches forward. The segments of the ventral aortas between the third and fourth arches become the *common carotid arteries.* They are at first very short,

FIGURE 17–6   Arteries of a 6-mm pig embryo as seen from the left side.

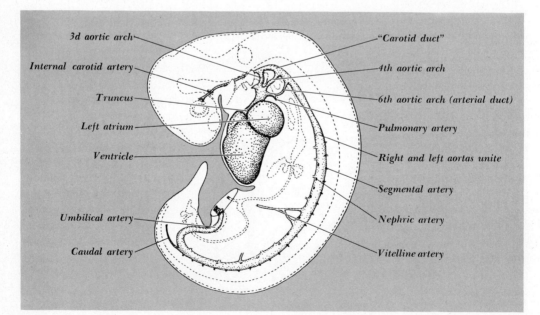

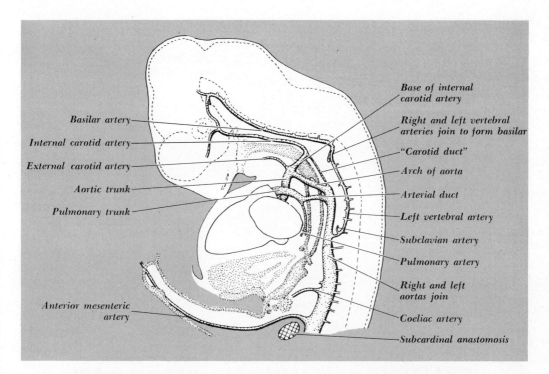

FIGURE 17-7 Arteries of the head and anterior trunk of a 12-mm pig embryo. *(Adapted from a reconstruction by F. T. Lewis, 1902, Amer. J. Anat., 2:211-226, Plate I.)*

but later when the neck forms and the heart retreats into the thorax, the common carotid arteries become greatly stretched out.

In mammals, the left fourth aortic arch persists as the permanent arch of the aorta. The right fourth arch remains as the root of the right subclavian artery. The ventral aorta proximal to the fourth arches is known as the *innominate artery*. In man, its branches are the right subclavian and the right common carotid arteries. The left subclavian enters the arch of the aorta separately. In the cat, the left common carotid is a branch of the innominate. Birds differ from mammals in that in them the

right fourth aortic arch becomes the permanent arch of the aorta (Fig. 17-9, *A*). The left fourth in this case is suppressed.

The sixth aortic arches sprout branches into the thick mesenchyme which surrounds the posterior part of the pharynx, and within which the lung buds ramify. These become the *pulmonary arteries* (Fig. 17-7). The connections of the sixth arches with the dorsal aortas are known as arterial ducts. In mammals, the right arterial duct disappears, and the left one remains open until the time of birth. This left duct is the definitive *ductus arteriosus*. At birth it gradually closes and becomes the *ligamen-*

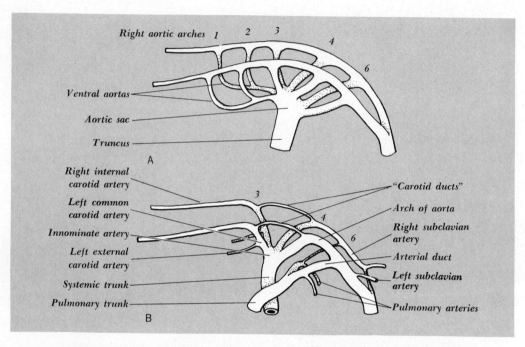

FIGURE 17–8 The varied fates of the aortic arches in a mammal embryo (diagrammatic). *A.* The complete set of arches (except for rudimentary arch 5). Actually all the arches are not functional at one time. *B.* The arches of a 10- or 12-mm pig embryo.

tum arteriosum, connecting the pulmonary trunk and dorsal aorta. In birds, the *right* arterial duct persists until hatching.

Experiments on the Aortic Arches

What decides that one aortic arch remains open while another arch closes? Are the determining factors within the arch itself (intrinsic), or do the surroundings control the closure (extrinsic factors)? To what extent do hemodynamic principles of blood pressure and volume of blood flow play determining roles? Specifically, why do arches 1, 2, and 5 disappear? Why, in mammals, does the fourth arch remain as the arch of the aorta on the left side, while

in birds the right arch persists? Why does the ductus arteriosus close at birth? These are intriguing questions which are subject to experimental analysis.

Some answers have been obtained. Rychter, by gently compressing the branchial arches of the chick with delicate silver clips (Fig. 17–9), found that when the right fourth arch (the definitive aortic arch of the bird) is constricted, the right third arch usually takes over its function (Fig. 17–9, *B*). In this case, the aorta between the third and fourth arches (the carotid duct) remains open. Sometimes, however, the left fourth arch functions vicariously (Fig. 17–9, *B*), and occasionally the left third, or any two of these, or even all three may remain in operation. The conclusion would seem to be that as long as the flow of blood in an arch continues, the arch remains open. The dorsal aortas between the third and fourth arches (the carotid ducts) normally regress, but they do not do so until arch 4 has begun to carry more blood than arch 3. Then the flow of blood in the carotid ducts stagnates, and the ducts close. However, when the fourth arch is occluded, the carotid duct continues to carry blood and remains open.

Consider the ductus arteriosus of the mammal. As long as the lungs remain unexpanded, the pressure in the pulmonary trunk is greater than the pressure in the aorta. Blood therefore continues to flow through the ductus, and the ductus remains open. But when, at birth, the lungs expand with air, the capillary bed of the lungs opens so that the blood pressure in the pulmonary trunk falls. As a result, the flow of blood through the arterial duct ceases or even reverses. It is then that the duct closes. Actually, explanations such as these barely touch the problems of why the blood vessels behave as they do.

Branches of the Dorsal Aorta

In all vertebrates, the paired dorsal aortas unite posterior to the pharynx and become a single median artery. The resulting central location of the aorta permits it to send branch arteries directly to the entire cross section of the embryo (Figs. 11–14, *A*, and 11–15, *A*):

FIGURE 17–9   Rychter's experiment on the aortic arches of the chick embryo. *A.* The arches of a 13-day chick embryo as seen from the ventral side. *B.* The result of having earlier (at 4 days) closed the right 4th arch with a silver clip. The right 3d and/or left 4th arch now functions in its stead. *C.* The result of closing the right 3d arch with a clip. The right 4th arch has taken over its role. *(Adapted from Z. Rychter, 1962, Advan. Morphogenesis, 2:333–371.)*

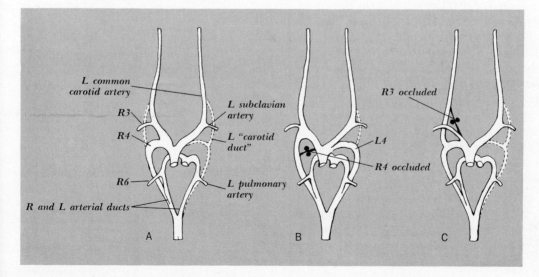

(1) dorsolateral branches to the back, body wall, and limbs; (2) lateral branches to the intermediate mesoderm; and (3) ventral branches to the gut.

1 The paired dorsolateral branches, known as *segmental arteries* (intersegmental arteries), grow into the septa between adjacent somites. Here each divides into a central ramus, which supplies the spinal cord and the deep muscles of the back, and a lateral ramus, which goes to the lateral and ventral body wall.

The seventh cervical arteries of the segmental series are of special interest, for they become the *subclavian arteries* to the anterior limb buds. They also have branches, the *vertebral arteries,* which go forward at the sides of the spinal cord until they join the median *basilar artery* beneath the hindbrain (Fig. 17–7). The basilar artery, in turn, joins the so-called arterial circle (of Willis) that forms a ring around the pituitary body. The arterial circle receives its main supply of blood from the internal carotid arteries and sends branches (cerebral arteries) to the forebrain.

2 Paired lateral branches from the dorsal aorta supply the excretory and reproductive structures formed from the mesomere. We shall call them *nephric arteries.* When the mesonephros regresses, certain nephric arteries remain and become the adrenal arteries to the adrenal glands, the renal arteries to the metanephroi, and the spermatic or ovarian arteries to the gonads.

3 The ventral branches of the dorsal aorta are derived from the original vitelline arteries which supply the yolk sac (Fig. 17–6). At first they are paired right and left, but they unite in the midline and become arteries which supply the gut (Fig. 17–7): (1) the *coeliac artery* to the stomach, spleen, liver, and first part of the duo-denum and pancreas; (2) the *anterior mesenteric artery* (original vitelline arteries) to the small intestine and the beginning of the large intestine; and (3) the *posterior mesenteric artery* to the remainder of the large intestine and rectum.

A fourth pair of ventral arteries become the *umbilical arteries* which supply the cloaca and its derivatives, including the urinary bladder, the allantois, and hence the placenta (Fig. 17–10). Although the umbilical arteries are at first splanchnic branches of the aorta, they soon adopt shortcuts from the dorsal aorta by way of the segmental arteries in the region of the hindlimb bud. These segmental arteries are known as the *iliac arteries*. After birth, the portion of each umbilical artery which is distal to the bladder closes, and their remnants (together with the remnant of the allantoic stalk) become a cord in the ventral ligament of the bladder. Proximal to the bladder, the umbilical arteries remain as the *hypogastric arteries* (internal iliac arteries).

THE VEINS

The story of the veins is even more complicated than that of the arteries. It involves the opening up of new channels and the abandonment of old ones. It also involves changes in terminology from the names applied to embryonic blood vessels to those used in adult anatomy.

The Common Hepatic and Hepatic Portal Veins

Four veins discharge into the sinus venosus of the young embryo. These are the right and left common cardinal veins, and the right and left vitelline veins (Fig. 17–2). The common

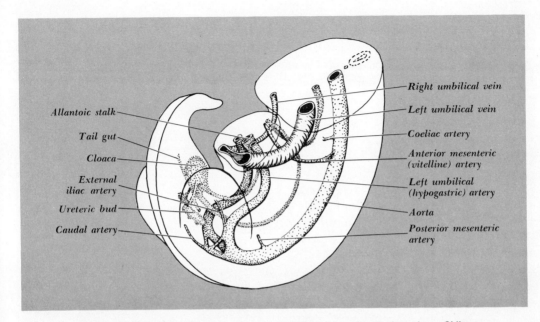

FIGURE 17–10 Arteries of the posterior trunk and tail of a 10-mm pig embryo. Oblique view from the ventral left.

cardinal veins come from the body wall (somat-opleure) via the lateral mesocardia. The vitelline veins come from the wall of the gut (splanchnopleure) and flow through the liver on their way to the heart. Between the liver and the heart, the vitelline veins unite to form the *common hepatic vein*—"common" because it receives tributaries from the liver known as *hepatic veins* (Fig. 17–12). Later, the common hepatic vein becomes the proximal segment of the posterior vena cava.

Within the liver, the vitelline veins are invaded by endodermal cords from the hepatic diverticulum and are broken into sinusoids (Fig. 14–9). In birds a central passageway remains, for a time, known as the "ductus venosus." It transports vitelline blood from the yolk

sac directly to the heart. It is not the same as the ductus venosus of mammals, which is a new-formed passage through the liver which carries umbilical blood.

Posterior to the liver, the vitelline veins become the *hepatic portal vein*. A portal vein, it will be recalled, is one which subdivides into a bed of capillaries, or "sinusoids," on its way to the heart. The pathway of the hepatic portal vein is that of the original vitelline veins modified by the development of certain anastomoses (Fig. 17–12). As a result, the blood seemingly takes a spiral course around the intestine. Ultimately, it is the intestine which spirals, and the route of the hepatic portal vein is the more direct route. Let us trace the flow of blood (Fig. 17–12). (1) The blood leaves

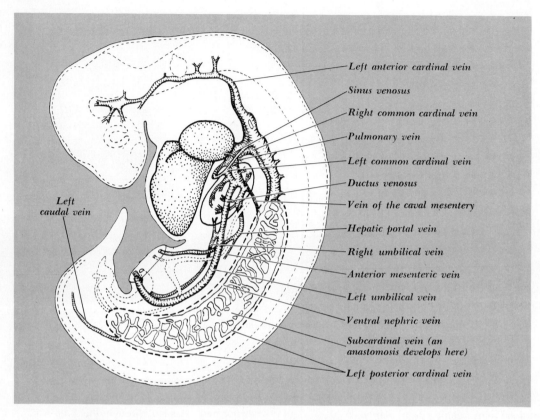

Left anterior cardinal vein

Sinus venosus

Right common cardinal vein

Pulmonary vein

Left common cardinal vein

Ductus venosus

Vein of the caval mesentery

Hepatic portal vein

Right umbilical vein

Anterior mesenteric vein

Left umbilical vein

Ventral nephric vein

Subcardinal vein (an
anastomosis develops here)

Left posterior cardinal vein

Left
caudal vein

**FIGURE 17–11 Veins of a 6-mm pig embryo from the left (diagrammatic). Note that
the left umbilical still retains an outlet into the left common cardinal vein, although
the expanding liver has tapped it.**

the yolk sac by the right and left vitelline veins.
When it reaches the level of the posterior end
of the duodenal loop, the blood of the right
vitelline vein crosses over by an anastomosis
ventral to the gut and joins the left vitelline
vein. By so doing it avoids taking a circuitous
route around the outer border (right side) of
the duodenal loop. (2) The combined bloodstream then traverses a second anastomosis,
this time dorsal to the gut, and joins the right
vitelline vein. It then enters the right lobe of

the liver. (3) A third anastomosis is located
within liver tissue. By it a part of the blood of
the hepatic portal vein returns, by a channel
ventral to the gut, to the left side of the liver.
Thus the blood is distributed to the sinusoids
of both lobes of the liver.

As the vitelline blood crosses from left to
right dorsal to the duodenum (second anastomosis), it receives tributary veins from the dorsal mesentery of the gut: First it receives the
anterior and *posterior mesenteric veins* from

the small and large intestines (Fig. 17–11). Then it receives the *splenic vein* from the stomach and spleen.

The Anterior Vena Cava

Three systems of veins are derived from the cardinal veins of the embryo, namely, the *anterior vena cava* (precaval vein), the *posterior vena cava* (postcaval vein), and the *umbilical vein* (Fig. 8–28). All three systems are at first tributaries of the common cardinal veins; but the latter two systems of veins, namely, the posterior vena cava and the umbilical vein, are tapped by the liver as it expands within the ventral mesentery. As a result they acquire new and direct passages through the liver to

the heart. The anterior vena cava, on the other hand, retains much of the original pathway of the anterior cardinal vein.

Within the head, the anterior cardinal veins shift closer to the brain and become in part the dural sinuses within the skull. Posterior to the head, the anterior cardinal veins, now known as the *internal jugular veins,* are joined by the *subclavian veins* from the anterior limb buds (Fig. 17–15). The *external jugular veins* are superficial tributaries of the subclavian veins and drain blood from the sides of the head and neck.

In many mammals, including man and the cat, an oblique anastomosis develops between the left anterior cardinal vein, where the left subclavian vein enters it, and the right anterior

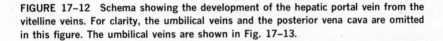

FIGURE 17–12 Schema showing the development of the hepatic portal vein from the vitelline veins. For clarity, the umbilical veins and the posterior vena cava are omitted in this figure. The umbilical veins are shown in Fig. 17–13.

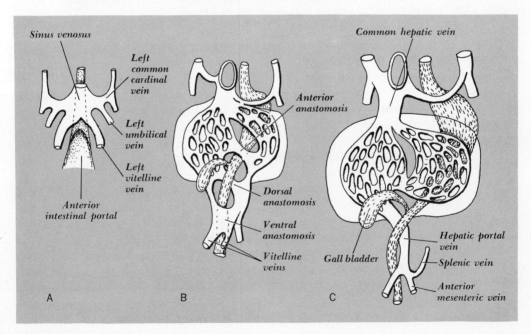

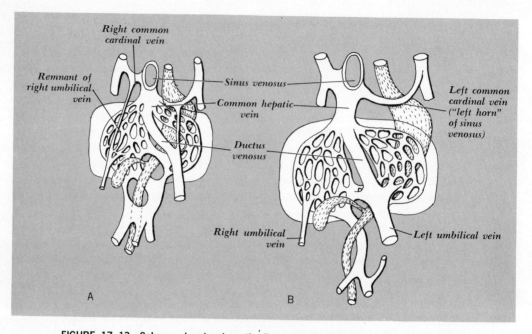

FIGURE 17–13 Schema showing how the liver taps the right and left umbilical veins and develops the ductus venosus to carry left umbilical blood.

cardinal vein (Fig. 17–14). Blood from the left side finds its way across this anastomosis and abandons its original route to the heart. The anastomosis is known as the *left innominate vein* (Fig. 17–15). The *right innominate vein,* in contrast, is a short segment of the right anterior cardinal vein.

When the sinus venosus shifts to the right side of the heart and is absorbed into the wall of the right atrium, its "left horn" becomes drawn out transversely across the dorsal side of the heart. As a result of the closure of the left anterior and posterior cardinal veins, it no longer receives blood from the body wall. Instead, it remains as the *coronary sinus* into which veins of the heart wall pour their blood.

The right anterior cardinal vein proximal to the junction of the two innominate veins, to-gether with the right common cardinal vein, constitute the *anterior vena cava* (precaval vein). Its principal tributary is the right posterior cardinal vein, now known as the *azygos vein.* The azygos vein drains the dorsal wall of the thorax (Fig. 17–15).

The Posterior Vena Cava

The story of the posterior vena cava (postcaval vein) is even more complicated than that of the anterior vena cava. It will be recalled that the posterior cardinal veins course forward in the body wall dorsolateral to the tubules of the mesonephros (Fig. 8–26). In fact, the mesonephric tubules invade the posterior cardinal veins and break them up into a net of sinusoids. However, several more or less continuous venous channels remain within the mesoneph-ros (Figs. 11–15 and 17–11): (1) the posterior cardinal veins themselves, located dorsolater-ally; (2) ventral nephric veins; (3) median chan-nels, known as the *subcardinal veins;* and finally (4) dorsomedial channels, which are rather late in making their appearance and which are known as *supracardinal veins.* These latter veins may be thought of as the result of

a shift of the posterior cardinal veins nearer to the midline.

Two important anastomoses take place: The first or *subcardinal anastomosis* is an opening between the right and left subcardinal veins ventral to the aorta where the two mesonephroi are in contact with each other (Figs. 11–15, 17–11, and 17–14). The second or *caval an-astomosis* arises within the caval mesentery

FIGURE 17–14 The development of the anterior and posterior vena cavas. An early stage showing the intercardinal, subcardinal, and iliac anastomoses. Note also the vein of the caval mesentery. Segments which have been obliterated are shown in solid black.

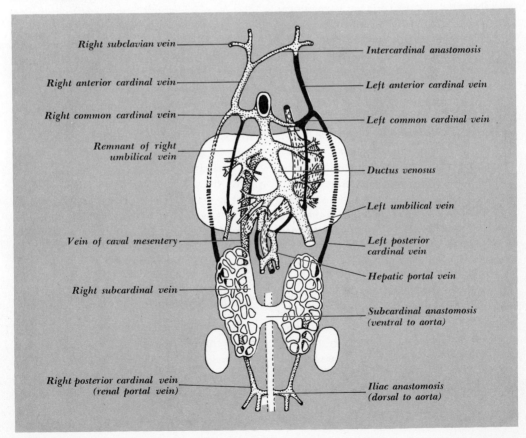

between the right subcardinal vein and the hepatic veins of the right lobe of the liver. In the 6-mm pig only a few capillaries cross over this mesentery from the mesonephros to the liver (Figs. 11–14, *A,* and 17–11), but the connection enlarges rapidly until it becomes an open channel which ultimately carries the entire blood flow from the posterior parts of the body. We shall call this the *vein of the caval mesentery* (Fig. 17–14).

In vertebrates which have a functional mesonephros, that is, in the lower vertebrates and in the embryos of the higher vertebrates, the blood of the posterior parts of the body journeys forward in the right and left posterior cardinal veins (or in the supracardinal veins which replace them) until it reaches the mesonephroi. It then filters between the mesonephric tubules and collects in the subcardinal veins. The blood of the left subcardinal vein flows across the subcardinal anastomosis into the right subcardinal. The combined blood then proceeds forward by way of the vein of the caval mesentery (caval anastomosis) through the liver to the heart. Because the blood flow is interrupted by the tubules of the mesonephroi, the cardinal veins and their tributaries posterior to the mesonephroi are at this stage known as the *renal portal system* (Fig. 17–14). In the development of the higher vertebrates, however, the mesonephroi regress and are replaced by permanent kidneys (metanephroi) which, unlike the mesonephroi, do not interrupt the course of the blood (Fig. 17–15). We therefore no longer use the term "renal portal vein," but refer to the same vessels as the postrenal segments of the posterior vena cava. At first they are two: right and left. Their tributaries are the *adrenal veins,* the *renal veins* (from the metanephroi), and the *spermatic* or *ovarian veins,* as the case may be.

Usually an anastomosis (iliac anastomosis) develops between the right and left posterior cardinal veins in the region of the hindlimb buds, dorsal to the aorta (Fig. 17–14). Across it, the blood of the entire left hind quarter of the body flows to the right and enters the right posterior vena cava. The left posterior vena cava survives only as the outlet of the left adrenal, renal, and spermatic (or ovarian) veins (Fig. 17–15).

These embryonic relations are of especial interest to anatomists, for they explain peculiarities which are encountered in the laboratory: (1) They explain the otherwise strange fact that the veins are ventral to the arteries in the kidney region, whereas in the pelvic region they are dorsal. (2) They explain the asymmetry of the veins which come from the adrenal bodies, kidneys, and gonads (Fig. 17–15). (3) They also account for numerous anomalies of the veins of the abdomen. In the comparative anatomy laboratory, cats are sometimes encountered which have two venae cavae posterior to the kidneys. Occasionally the left postrenal vena cava persists instead of the right one. Sometimes the posterior vena cava is lateral to the ureter; usually it is medial. Rarely is there a cat in which the right vena cava continues forward to the heart by way of the azygos vein. This is a persistence of the posterior cardinal vein.

The Umbilical Vein

At first the umbilical veins are somatic veins which course forward in the ventrolateral body wall and enter the common cardinal veins just before they cross the lateral mesocardia to the sinus venosus of the heart (Fig. 8–21). They are homologues of the lateral abdominal veins of the shark and the ventral abdominal veins of the frog. In bird and mammal embryos they acquire connections posteriorly with the allantois by way of its broad ventral mesentery (Figs. 8–28 and 17–11).

As the liver expands laterally within the septum transversum, it taps the umbilical veins and captures their flow of blood (Figs. 17–13, A, and 17–14). The right umbilical vein gradually regresses and finally disappears. The left umbilical vein develops a diagonal channel across the substance of the liver from the left side to the common hepatic vein on the right side. This is the mammalian *ductus venosus* (Figs. 17–13 and 17–14). At birth the left umbilical vein and ductus venosus close, and all that remains are cords in the margin of the falciform ligament and in the liver (Fig. 17–15).

FIGURE 17–15 The anterior and posterior vena cavas at a later stage than that illustrated in Fig. 17–14, namely at the stage of the unborn fetus. The umbilical vein and the ductus venosus, which close at birth, are shown in solid black.

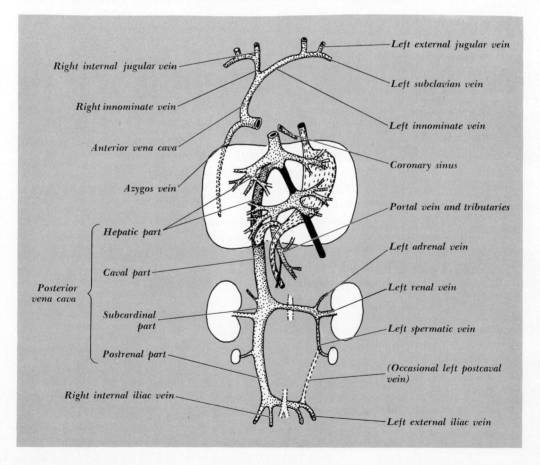

Pulmonary Veins

The pulmonary veins are new formations which are formed when endothelial cells of the left atrium invade the dense mesenchymal tissue which invests the lung buds (Figs. 11–13, *B*, 17–3, and 17–11). At first there is one evagination, but it bifurcates and makes connections with the capillary net around the developing lungs. Later the proximal segment of the single pulmonary vein is absorbed into the wall of the left atrium so that two pulmonary veins (four in man) enter the left atrium separately.

THE LYMPHATIC SYSTEM

The blood which enters the capillaries from the arteries leaves by the veins, except for a small amount of liquid which filters through the capillary walls into the intercellular spaces. The force which drives it through is capillary blood pressure. If the liquid were not continually carried away, the intercellular spaces would swell, and the tissues would become turgid (edema). The lymphatic system is developed to drain the intercellular spaces and return the intercellular fluid (lymph) to the heart.

Lymph vessels are delicate and difficult to study. They form long after the circulation of the blood has been established. They probably originate as clefts in the mesenchyme which become enclosed in an endothelium of flattened mesenchyme cells. The vesicles thus formed elongate, branch, and unite to form nets of lymph capillaries in almost every organ of the body. Large lymph vessels are produced by channeling within the capillary net. They serve to transport the lymph toward the heart, and finally to discharge it into veins near the

heart. Thus in man the *thoracic duct* carries lymph from the lower body forward through the thorax and empties it near the junction of the left internal jugular and subclavian veins.

THE FETAL CIRCULATION AND THE CHANGES AT BIRTH

The Fetal Circulation

The most oxygenated blood of the mammalian embryo is that which returns to the embryo via the umbilical vein (Fig. 17–16). However, as it enters the liver it is diluted by venous blood from the posterior vena cava and hepatic veins. It therefore enters the right atrium of the heart as partially mixed blood. Venous blood (blood poor in oxygen) from the anterior vena cava also enters the right atrium.

The blood of the right atrium leaves by two routes: (1) Approximately half of the blood goes through the foramen ovale into the left atrium and from thence, via the bicuspid valve, to the left ventricle of the heart. By injecting radio-opaque substance into the bloodstream and then observing the flow of blood with the help of X-rays, it has been demonstrated that most of this blood is mixed blood from the liver and posterior vena cava. The left ventricle pumps the blood into the aortic trunk, which distributes it to the heart wall (via the coronary arteries), the head (via the common carotid arteries), and the arms (via the subclavian arteries). (2) The other half of the blood which enters the right atrium goes by way of the tricuspid valve into the right ventricle (Fig. 17–16). Radio-opaque experiments have shown that this is mainly venous blood from the anterior vena cava which has returned from the head and the forward parts of the body.

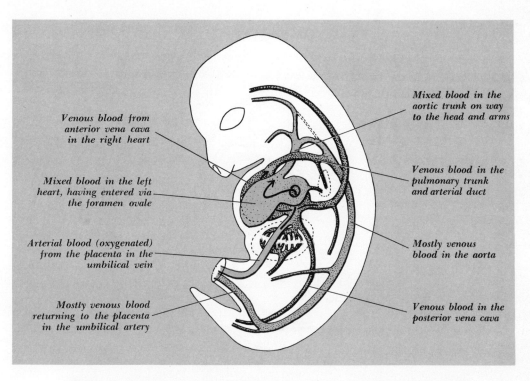

Venous blood from
anterior vena cava
in the right heart

Mixed blood in the left
heart, having entered via
the foramen ovale

Arterial blood (oxygenated)
from the placenta in the
umbilical vein

Mostly venous blood
returning to the placenta
in the umbilical artery

Mixed blood in the
aortic trunk on way
to the head and arms

Venous blood in the
pulmonary trunk
and arterial duct

Mostly venous
blood in the aorta

Venous blood in the
posterior vena cava

FIGURE 17–16 Diagram of the fetal circulation. The vessel which carries arterial blood from the placenta (namely, the umbilical vein) is not stippled. Those vessels which carry mixed arterial and venous blood to the head and arms are lightly stippled. Those which carry venous blood are heavily stippled.

The right ventricle pumps it into the pulmonary trunk, and then most of it goes through the arterial duct into the dorsal aorta and is distributed to the lower thoracic and abdominal regions of the body. Much of it goes by way of the umbilical arteries to the placenta. Very little goes through the pulmonary arteries to the yet unexpanded lungs.

Changes in the Circulation at the Time of Birth

Radical changes take place in the circulation at the time of birth. When the diaphragm and external chest muscles contract, the lowered pressure within the thoracic cavity draws air into the lungs. For the same reason the pulmonary capillaries are dilated, and blood is drawn into the lungs. The flow of blood through the lungs increases from 3 to 10 times. The result is that the pressure of blood in the pulmonary trunk decreases, while that in the pulmonary vein—and hence in the left atrium—increases (Fig. 17–17).

The consequence of the fall of pressure in the pulmonary trunk is that blood ceases to flow from the pulmonary trunk through the ar-

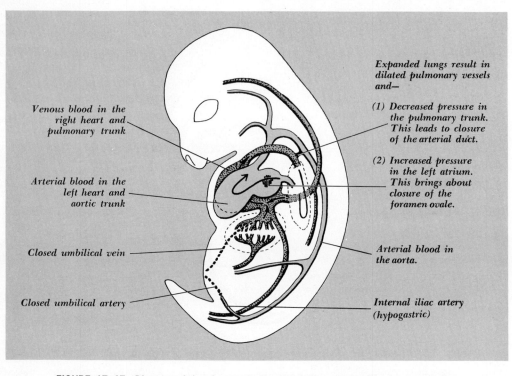

Venous blood in the
right heart and
pulmonary trunk

Arterial blood in the
left heart and
aortic trunk

Closed umbilical vein

Closed umbilical artery

Expanded lungs result in
dilated pulmonary vessels
and—

(1) Decreased pressure in
the pulmonary trunk.
This leads to closure
of the arterial duct.

(2) Increased pressure
in the left atrium.
This brings about
closure of the
foramen ovale.

Arterial blood in
the aorta.

Internal iliac artery
(hypogastric)

FIGURE 17–17   Diagram of the changes in the circulation which take place at the time
of birth.

terial duct into the aorta. It may even flow in the opposite direction. This may be advantageous, for it assures that there will be an adequate supply of blood going to the lungs to meet respiratory needs.

The rise in pressure in the left atrium forces the valve-like septum primum against the foramen ovale in the septum secundum, thereby stopping the passage of blood. Actual fusion of the two septa follows gradually.

In the so-called blue baby the foramen ovale fails to close, with the result that venous blood continues to enter the left atrium (from the right atrium) and mingle with an inadequate supply of arterial blood from the lungs. In Bla-

lock's famous operation on blue babies, the ductus arteriosus is reopened in order to increase the pressure in the left atrium (by backflow from the aorta) and thus cause the foramen ovale to close.

The closure of the ductus arteriosus takes place gradually. Ultimately only a cord remains which connects the pulmonary trunk with the aorta; this is known as *the arterial ligament.*

At birth, the walls of the umbilical vessels contract. Shortly after birth, their lumina become obliterated. The proximal parts of the umbilical arteries remain, however, as the *internal iliac* or *hypogastric arteries* which supply the bladder and pelvic regions. Their distal parts, the parts between the bladder and the umbilicus, become cords in the margin of the ventral ligament of the bladder. As previously stated, the closed umbilical vein becomes a cord in the posterior margin of the falciform ligament.

THE THYMUS GLAND AND IMMUNOLOGICAL RESPONSE

It is now appropriate to discuss the development of immunological reactions. Warm-blooded animals, that is, birds and mammals, respond to foreign proteins (antigens) by producing immune substances (antibodies) which destroy the toxic properties of the invader. In a somewhat similar manner, bird and mammal tissues recognize and reject grafts of tissues from an organism with a different genotype (heredity) than their own. But this property is not present in the embryo. It arises during the course of development, and it arises earlier in some animals than in others. Thus a sheep embryo of six weeks is able to provide antibodies against viruses. At 12 weeks it responds to and rejects grafts of tissues from other individuals of its own (homoplastic grafts) or another species (heteroplastic grafts). In other animals, these same capacities arise later in development. A newborn chick or mouse, for example, does not give immune reactions; and a rat two weeks old will accept grafts of foreign tissue. Antiviral activity cannot be detected in chicks until three or four weeks after hatching. Any natural immunity which may be present is probably a passive immunity that results from antibodies being supplied by the mother through the yolk and albumen of the egg.

The development of immunological responses is correlated with the development of the thymus gland (Fig. 14-7). Apparently this gland functions in the young animal by producing lymphocytes and, seemingly, a hormone. After it has performed these services, it regresses. (In addition to the thymus gland, chicks have an organ of similar histological structure and function which is attached to the cloaca. This is known as the *bursa of Fabricius* and, like the thymus, is active only in young animals.)

The thymus of a newborn mouse possesses an outer layer or cortex which is packed with lymphocytes of a distinctive sort. These are small, possess large nuclei, and have very little cytoplasm. In fact, the ratio of DNA to total protoplasm is greater in the juvenile thymus than in any other tissue of the body. From the thymus, the small lymphocytes are carried by the blood to lymphoid tissues throughout the body, notably to the spleen, lymph nodes, and to Peyer's patches in the walls of the intestine. Supposedly the lymphocytes from the thymus "seed" these organs and, by multiplying, accumulate in them as nodular masses surrounding "germinal centers." If the thymus of a newborn mouse is removed, the spleen and lymph nodes remain underdeveloped. Relatively few lymphocytes are produced. Otherwise, the young mouse develops quite normally for the first three or four months. Then deterioration sets in, and in a short time the mouse is dead. However, if the removal of the thymus gland is delayed until the mouse is several weeks old, the operation has little or no effect. Presumably the seeding of lymphoid tissue has already been accomplished.

Seemingly the thymus gland has also a chemical or hormonal function. If the thymus of a newborn mouse is removed and a few days later another thymus is implanted in the mouse's abdominal cavity—but this time the thymus is enclosed in a "diffusion chamber" with pores so small that cells cannot escape— the lymph nodes and spleen react by producing lymphocytes of their own. This influence which emanates from the implanted thymus presumably is of a hormonal nature.

The fact that embryos do not produce antibodies has made it possible to cultivate viruses, bacteria, and foreign tissues, including malignant cells, in incubated chick eggs. The material to be cultured is injected into the yolk sac of 7-day embryos, and 4 days later it is "harvested" from the embryonic and extra-embryonic body fluids.

How are immune reactions brought about? It is probable that at first the lymphocytes in the lymphoid tissues are "uncommitted"; that is, they are not yet ready to produce particular antibodies. But when foreign antigens are presented, some of the lymphocytes respond by transforming into plasma cells which produce specific antibodies. However, this reaction takes time. For example, a skin graft to an adult mouse is retained for 10 to 12 days before it is rejected. The now "committed" plasma cells then continue to be a self-replen-

ishing source of the specific antibody. When a second time the body is exposed to the same antigen, an immunological response takes place promptly and more effectively. The graft is rejected.

How do the lymphocytes distinguish between the body's own antigens and the antigens brought in by the invader? How do they recognize "self and nonself"? The answer is partly genetic. An animal accepts tissue of its own genotype but rejects tissue of another genotype. However, not all the genes are involved. It seems that there are a few genes which produce "isoantigens," that is, substances which are alike in every cell of the animal's body. At the crucial time in development when immunological capacities are being acquired, the body's lymphocytes learn not to respond to the body's own isoantigens. They "learn by experience." If foreign antigens are presented to an animal (as by a graft of foreign tissue) before this crucial time, the animal tolerates the graft. In addition, it continues to accept grafts of tissues of the same donor as though they were its own. If the foreign antigens are presented after the crucial time, they are rejected.

SELECTED READINGS

General References

Arey, L. B., 1965. Developmental Anatomy. W. B. Saunders Company, Philadelphia. 7th ed. Chap. 19, The vascular system. Chap. 20, The heart and circulation changes.

Patten, B. M., 1964. Foundations of Embryology. McGraw-Hill Book Company, New York. 2d ed. Chap. 23, The development of the circulatory system.

Goodrich, E. S., 1930. Studies on the Structure and Development of Vertebrates. The Macmillan Company, London. Reprinted by Dover Publications, Inc., New York. Chap. 10, Vascular system and heart.

Copenhaver, W. M., 1955. Heart, blood vessels, blood, and entodermal organs. In Willier, Weiss, and Hamburger (eds.), Analysis of Development. W. B. Saunders Company, Philadelphia. Pp. 440–461.

Hemoglobin

Ebert, J. D., 1965. Interacting Systems in Development. Holt, Rinehart and Winston, Inc., New York. Pp. 116–117 and 134–135.

Baglioni, C., and C. E. Sparks, 1963. A study of hemoglobin differentiation in Rana catesbiana. Develop. Biol., **8**:272–285. Reprinted in Bell (ed.), 1965. Molecular and Cellular Aspects of Development. Harper & Row, Publishers, Incorporated, New York. Pp. 236–245.

Perutz, M. F., 1964. The hemoglobin molecule. Sci. Am., **211**(November): 64–76.

Zucherkandl, E., 1965. The evolution of hemoglobin. Sci. Am., **212**(May):110–118.

Marks, P. A., and J. S. Kovach, 1966. Development of mammalian erythroid cells. In Moscona and Monroy (eds.), Current Topics in Developmental Biology. Academic Press Inc., New York. Vol. 1, pp. 213–252.

The Heart

Patten, B. M., and T. C. Kramer, 1933. The initiation of contraction in the embryonic chick heart. Am. J. Anat., **53**:349–375.

Goss, C. M., 1942. The physiology of the embryonic mammalian heart before circulation. Am. J. Physiol., **137**:146–159.

Ebert, J. D., 1959. The first heart beats. Sci. Am., **200** (March):87–96.

Harari, I., 1962. Heart cells in vitro. Sci. Am., **206** (May):141–152.

DeHaan, R. L., 1964. Cell interactions and oriented movements during development. J. Exptl. Zool., **157**:127–138.

————,1965. Morphogenesis of the vertebrate heart. In DeHaan and Ursprung (eds.), Organogenesis. Holt, Rinehart and Winston, Inc., New York. Pp. 377–419.

Arteries and Veins

Evans, H. M., 1909. On the development of the aortae, cardinal and umbilical veins, and other blood-vessels of vertebrate embryos from capillaries. Anat. Record, **3**:498–518.

Sabin, Florence R., 1922. Direct growth of veins by sprouting. Carnegie Contrib. Embryol., **14**:1–10.

Butler, E. G., 1927. The relative role played by the embryonic veins in the development of the mammalian vena cava posterior. Am. J. Anat., **39**:267–353.

Rychter, Z., 1962. Experimental morphology of the aortic arches and the heart loop in chick embryos. Advan. Morphogenesis, **2**:333–371.

The Fetal Circulation and the Changes at Birth

Blalock, A., C. R. Hanlon, and H. W. Scott, 1949. The surgical treatment of congenital cyanotic heart disease. Sci. Monthly, **69**:360–367.

Barclay, A. E., K. J. Franklin, and M. M. L. Prichard, 1944. The Foetal Circulation and Cardiovascular System, and the Changes That They Undergo at Birth. Blackwell Scientific Publications, Ltd., Oxford.

Slaughter, F. G., 1950. Heart surgery. Sci. Am., **182** (January):14–17.

The Thymus Gland and the Development of Immunity

Burnet, Sir M., 1961. The mechanism of immunity. Sci. Am., **204** (January):58–67.

————, 1962. The thymus gland. Sci. Am., **207** (November):50–57.

Auerbach, R., 1961. Genetic control of thymus lymphoid differentiation. Proc. Natl. Acad. Sci., **47**:1175–1181. Reprinted in Bell (ed.), 1965. Molecular and Cellular Aspects of Development. Harper & Row, Publishers, Incorporated, New York. Pp. 230–235.

————, 1965. Experimental analysis of lymphoid differentiation in the mammalian thymus and spleen. In DeHaan and Ursprung (eds.), Organogenesis. Holt, Rinehart and Winston, Inc., New York. Pp. 539–557.

Ebert, J. D., 1965. Interacting Systems in Development. Holt, Rinehart and Winston, Inc., New York. Chapter 12, Developmental aspects of immunity.

chapter **18**

THE
UROGENITAL
SYSTEM

The excretory and reproductive systems are related to each other in their evolutionary origin and in their embryonic development. They continue to be associated in the mature organism. Why is this so?

The explanation is an evolutionary one. An animal composed of ectoderm and endoderm only, such as *Hydra,* has no need of a multi-cellular excretory mechanism, for each tissue can excrete directly to the outside or into a cavity which is open to the outside. Similarly the gonads discharge to the outside, and there is no need of reproductive ducts. But animals with an organized intermediate layer, the mesoderm, have an internal medium, or tissue fluid, in which metabolic wastes accumulate and within which germ cells form so that ducts are needed to carry these products away.

It is probable that the first true body cavities (coelomic cavities) were formed within gonads. If so, it is not surprising to find that even in the highest animals the gonads are formed from the lining of the body cavity, namely, the coelomic epithelium. In cyclostomes, both eggs and sperms collect in the body cavity and find their way out by ducts; but in the rest of the vertebrates, only the eggs are discharged into the coelom, and the sperm make use of excretory tubules.

The nephridia of annelids and the pronephric tubules of vertebrates have openings (nephrostomes) from the body cavity. (Nephridia and pronephric tubules arise in different ways and are hence examples of convergence in the process of evolution.) The mechanism of these tubules can be surmised. Probably the ancestors of vertebrates lived in freshwater, and water entered their bodies by osmosis. Water, wastes, and nutrients absorbed from the alimentary tract accumulated in the coelomic fluid. The problem therefore was to eliminate excess water and wastes without loss of nutrients and valuable salts. This was solved by a dual process: The nephrostomes possessed cilia which propelled the fluid of the body cavity into the excretory tubules. In the tubules absorption occurred, useful salts and nutrients were taken back into the bloodstream, while excess water and waste remained in the tubules and were discharged to the outside.

EXCRETORY ORGANS

The excretory organs and the ducts of the reproductive system develop from the intermediate mesoderm (mesomeres), namely, the longitudinal strands of mesoderm which lie between the axial mesoderm and the lateral mesoderm (Fig. 8-3, *E*). Much as the axial mesoderm becomes segmented into a series of somites, beginning in the hindbrain region and progressing posteriorly, so the intermediate mesoderm segments into *nephrotomes.* However, the nephrotomes are clearly formed and distinct only in several anterior segments. Farther back, the intermediate mesoderm continues to be unsegmented *nephrogenic tissue* (Fig. 8-17).

It has been found convenient in embryology to refer to a sequence of three "kidneys" which form in succession, one after the other. These are the pronephros, the mesonephros, and the metanephros (Fig. 18-1). This distinction between the kidneys is not clear, however, in fish and amphibians.

Pronephros

The pronephroi (plural for pronephros) are the first excretory organs to form (Fig. 18-2, *A*). They arise well forward near the anterior end.

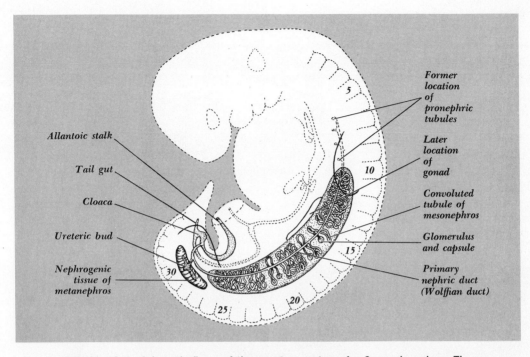

Allantoic stalk

Tail gut

Cloaca

Ureteric bud

Nephrogenic
tissue of
metanephros

Former
location
of
pronephric
tubules

Later
location
of
gonad

Convoluted
tubule of
mesonephros

Glomerulus
and capsule

Primary
nephric duct
(Wolffian duct)

FIGURE 18–1 Schematic figure of the excretory system of a 6-mm pig embryo. The pronephric tubules are represented from an earlier stage, and the gonads and metanephros are introduced from a later stage of development.

One tubule forms from the somatic layer of each nephrotome as a sort of bud which sprouts outwardly. At its proximal end, it retains an opening into the body cavity, known as a *nephrostome*. At its distal or free end it joins with the tubule just posterior to it. Thus a duct is formed, the *primary nephric duct* (archinephric duct). After several pronephric tubules have thus joined, the duct grows backward in the space between the axial and lateral mesoderms until it reaches the posterior limit of the lateral mesoderm. Here it comes into contact with the endoderm of the cloaca and fuses with it. The primary nephric ducts are thus, at first, pronephric ducts. Later, they are

adopted by the mesonephros, and finally they become the sperm ducts of the adult male.

Near each open nephrostome, or just inside the opening, there is a tuft of capillaries which is fed by a short nephric artery directly from the aorta (Figs. 6–7, *C*, and 18–3, *C*). If the tuft protrudes into the body cavity, it is known as an *external glomerulus* or *glomus* if several such tufts are fused together. But if the tuft is enclosed within the nephrostome, it is called an *internal glomerulus*. Presumably these tufts facilitate the diffusion of substances from the blood into the body cavity.

The glomeruli, nephrostomes, and tubules arise from different embryonic sources, yet

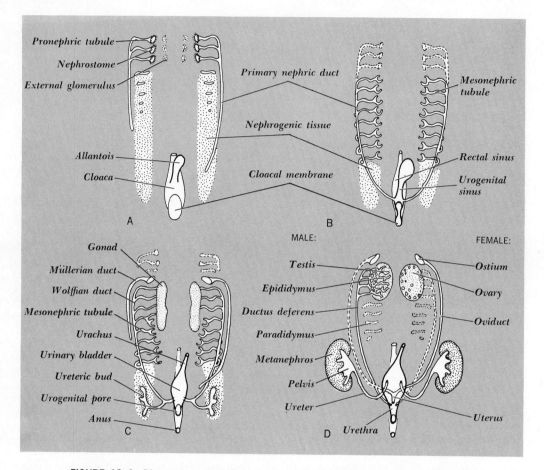

FIGURE 18–2 Diagrams representing the development of the urogenital system at successive stages in development. *A.* Stage of the pronephros and the elongating primary nephric duct. *B.* Stage of the mesonephros and the division of the cloaca. *C.* Stage of the indifferent gonad and reproductive ducts; the early metanephros. *D.* The differentiation of the male and female reproductive organs.

they integrate together and form unified structures. Transplantation experiments on amphibian embryos have shown that there is mutual interaction between these primordia, for if one of the primordia is absent, the others develop imperfectly.

Experiments on amphibians have also shown that although the primary nephric ducts grow autonomously from in front backward, yet they are guided and find their way in relation to adjacent parts. If an obstruction is placed in the path of a duct, the duct either fails to grow backward or it detours and then regains its proper position. A primary nephric duct will unite with any piece of cloacal tissue which is planted in its path.

Pronephroi are present and function in the embryonic and larval stages of fish and amphibians. They develop but apparently do not function in the embryos of reptiles, birds, and mammals. In all vertebrates, however, they give origin to the primary nephric ducts.

Mesonephros

The right and left mesonephroi are the second kidneys to form in higher vertebrates (Fig. 18-2, *B*). In fish and amphibians, the second kidneys are comparable to combined mesonephroi and metanephroi and have been termed opisthonephroi. Each forms in the mid-body region from intermediate mesoderm. At first, one tubule is formed per segment, but the tubules multiply by a sort of budding until there may be as many as eight tubules opposite each somite. All evidence of segmentation in the mesonephroi is then lost. At first, each tubule is a ball of cells which elongates and becomes S-shaped. Finally, it becomes much convoluted (Fig. 18-4).

A few of the first tubules to form have nephrostomes with internal glomeruli at their median ends. Most of them, however, begin blindly by wrapping around and nearly enclosing a glomerulus and forming a capsule,

Bowman's capsule (Fig. 18-4, *B*). A capsule and a glomerulus together constitute a *renal corpuscle*. A corpuscle and the convoluted tubule which leads from it make an excretory unit known as a *nephron*.

As a rule, each mesonephric tubule grows laterally and joins the primary nephric duct. Thus the duct, which was first the duct of the pronephros, becomes the mesonephric or *Wolffian duct*. In some of the lower vertebrates, however, the more posterior mesonephric tubules join together and form a duct of their own, namely, a secondary nephric duct or ureter, which grows backward and joins the cloaca. By reason of this fact, these more posterior tubules are comparable to those of the metanephros of the higher vertebrates rather than to those of a mesonephros.

Each mesonephros is a fairly compact structure which forms a ridge along the dorsal side of the body cavity, parallel to the mesentery of the gut, and extending the full length of the peritoneal cavity (Fig. 11-15). It is large in pig embryos but small in those of man, as we shall presently note.

It is quite clear that the glomeruli of renal corpuscles are organs of filtration through which an exudate (filtrate) of the blood is driven by arterial blood pressure. It also seems evi-

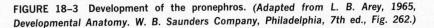

FIGURE 18-3 Development of the pronephros. (Adapted from L. B. Arey, 1965, Developmental Anatomy. W. B. Saunders Company, Philadelphia, 7th ed., Fig. 262.)

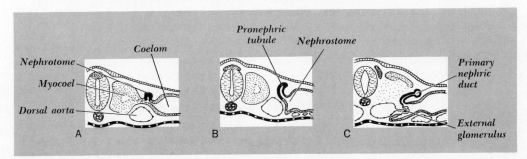

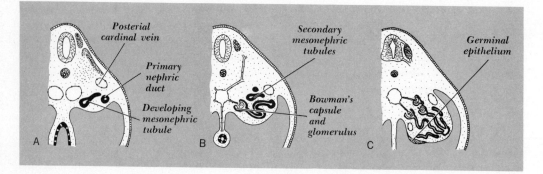

Posterial
cardinal vein

Primary
nephric
duct

Developing
mesonephric
tubule

Secondary
mesonephric
tubules

Bowman's
capsule
and
glomerulus

Germinal
epithelium

A   B   C

FIGURE 18–4   Development of the mesonephros. (Adapted from several sources.)

dent that the convoluted tubules are organs of reabsorption by which useful nutrients and salts are returned to the bloodstream. This return, in the case of the mesonephros, is facilitated by the close relation of the tubules to the flow of blood in the posterior cardinal veins. The latter become broken into braided streams of capillaries when they are invaded by the tubules.

The coelomic epithelium, which covers the anterior medial surface of each mesonephros, thickens and gives rise to a *germinal epithelium* from which a gonad is formed (Fig. 18–4, *C*).

Mesonephroi (or opisthonephroi) develop in all vertebrates. They are the functioning adult kidneys of fish and amphibians. They are present and functional in the embryos of most reptiles, birds, and mammals. However, in many rodents and in carnivores and primates, including man, they are not well developed. In these mammals, the excretory function is performed largely by the placenta.

Metanephros

The metanephroi are the third and definitive kidneys of reptiles, birds, and mammals. They

develop far posteriorly in the region of the cloaca and hindlimb buds (Fig. 18–2, *C* and *D*). Each begins when a bud (metanephric or *ureteric bud*) grows dorsally from near the posterior end of a primary nephric duct (Wolffian duct) and pushes into adjacent nephrogenic tissue. The bud expands, branches within the nephrogenic tissue, and gives rise to the pelvis and collecting tubules of an adult kidney (Fig. 18–5). Its stem becomes a secondary nephric duct or *ureter*.

The nephrogenic tissue reacts to the metanephric bud in the same way that, in the mesonephric region, it reacted to the primary nephric duct; that is, it differentiates nephrons. As in the former case, the primordium of each nephron is at first a small ball of cells which then elongates and becomes S-shaped. Its outer end enfolds a glomerulus and forms a capsule (*Bowman's capsule*) about it. Its inner end unites with the tip of a collecting tubule. The intervening duct elongates until it has formed a much-coiled *convoluted tubule* (Fig. 18–5, *E*). The adult human kidney is said to have about one million such nephrons.

A metanephros does not interrupt the course of a blood vessel as does a mesonephros. Instead, it receives an abundant supply of blood by way of its own renal artery and vein (Fig. 17–15). It is said that in man more than one-fifth of the blood which leaves the heart flows through the kidneys.

Experiments on chick development have demonstrated that when a metanephric bud is prevented from forming, induction does not take place, and the nephrogenic tissue does not form nephrons. Something of this nature probably explains why occasionally a person lacks a kidney on one side.

Metanephroi are the functional kidneys of adult reptiles, birds, and mammals. They are presumably homologous with the posterior part of the kidneys (opisthonephroi) of those fish and amphibians which have secondary nephric ducts (ureters).

In the later development of mammals, the mesonephroi and metanephroi shift positions with respect to each other. The metanephroi move anteriorly along the dorsal side of the body cavity, possibly as a result of the straightening out of the body. The mesonephroi glide posteriorly, passing ventral to the metanephroi, until they come to lie posterior to them (Figs. 18–7 to 18–10).

Urinary Bladder and Urethra

The posterior organ of the embryonic alimentary tract is the cloaca (Fig. 18–6). Into it empty, not only the intestine, but also the primary and, later, secondary nephric ducts. Thus, in most vertebrates, the cloaca serves

FIGURE 18–5 Development of the metanephros or permanent kidney. The convoluted tubules, in *E*, are shown at three stages in their development.

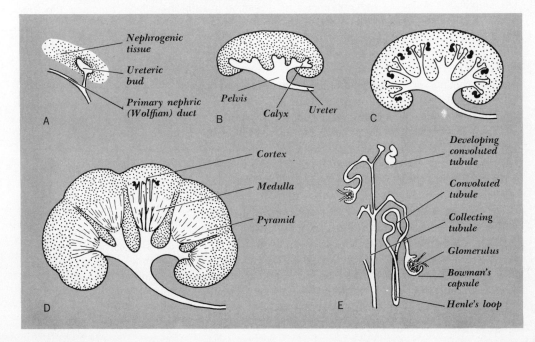

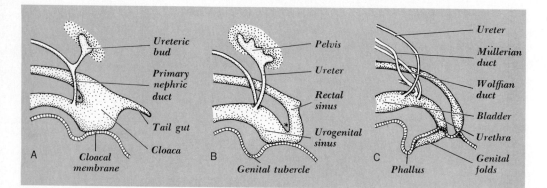

FIGURE 18–6 Development of the cloaca and associated ducts. The asterisks mark the septum which, growing backward in a frontal plane, divides the cloaca into the rectal sinus and urogenital sinus.

as the common passage to the outside for fecal, urinary, and reproductive materials. It disappears in higher fish, and in mammals it becomes divided into rectal and urogenital sinuses.

In amphibians following metamorphosis, a saclike downpocketing from the floor of the cloaca becomes the urinary bladder. In reptiles, birds, and mammals, this downpocketed sac is precociously developed. It expands far beyond the confines of the embryo proper and is the *allantois* (Fig. 8–11). In birds and most reptiles, only a narrow stalk of the allantois is present in the body of the embryo; but in mammals, the proximal portion of the allantoic stalk enlarges and contributes to the formation of the urinary bladder. The distal part of the stalk remains narrow and is known as the *urachus* (Fig. 18–2, *C*).

In placental mammals, the cloaca promptly divides in a frontal plane into a dorsal portion, the *rectum,* and a ventral part, the *urogenital sinus* (Fig. 18–6). The partitioning begins at the forward end where the allantoic stalk and

intestine meet. It progresses posteriorly until it reaches the cloacal membrane (anal plate). Thus the cloacal membrane becomes divided into an anal plate proper which breaks through and becomes the definitive *anus* and a urogenital membrane which becomes the *urogenital aperture* (Fig. 18–2, *C* and *D*). The anterior part of the urogenital sinus (together with some of the adjacent allantoic stalk) expands and forms the *urinary bladder.* The posterior part of the urogenital sinus becomes the *urethra.*

At first, the primary nephric ducts (Wolffian ducts) enter at the sides of the cloaca near its anterior end where the allantoic stalk is attached. After the partitioning of the cloaca has taken place, the ducts enter that part of the urogenital sinus which becomes the bladder (Fig. 18–6, *A*).

In mammals, a remarkable shifting of the openings of the ducts takes place (Fig. 18–6, *C*). The primary nephric ducts are absorbed into the walls of the urogenital sinus to such an extent that the primary ducts (Wolffian ducts) and secondary ducts (ureters) acquire separate entrances. The secondary ducts continue to open into the sides of the bladder, but the openings of the primary ducts glide dorsally and posteriorly until they open onto the dorsal side of the urethra. Later, new ducts, known as the Müllerian ducts, are formed parallel to the Wolffian ducts. We will describe these in connection with the reproductive system.

Experiments on the Embryonic Kidney

Experiments on amphibian gastrulas by means of vital stains show that the prospective kidney-forming material is located in the ventrolateral region of the marginal zone, i.e., the zone of the future chorda-mesoderm (Fig. 5–7). At this early stage, no cells are as yet committed to become excretory structures. When a fragment of the future ectoderm or mesoderm of a gastrula is transplanted to the nephric region of an older embryo, it shares in the production of nephric tubules. When dorsal lip material of an amphibian gastrula or cells from the region of the anterior end of the primitive streak of a head-process chick are explanted, they self-differentiate into a variety of ectodermal and mesodermal tissues, including nephric tubules; but this does not mean that the fates of individual cells has already been determined.

Pronephric mesoderm reaches its permanent location in the embryo during the formative movements of gastrulation and neurulation. At the same time, processes take place by which the fates of the cells are largely determined. For example, if the pronephric primordium of a late neurula of an amphibian is removed and reimplanted upside down, the tubules which develop are upside down. If a portion of the pronephric primordium is removed, the pronephros which forms is imperfect. The mesonephric region becomes determined somewhat later, namely, at about the tail-bud stage.

There is experimental evidence that the primary nephric duct originates in the pronephric region and then grows backward in the manner which has already been described. For instance, if the region of the pronephros is vitally stained with Nile blue sulfate, the duct which grows backward is stained even though the nephrogenous tissue through which it passes is unstained. However, the capacity to form a primary nephric duct is not confined to the pronephric nephrotomes, for if these are removed, a duct may still be formed by the nephrotomes next in sequence posteriorly.

Mesonephric tubules develop in the nephrogenous mesenchyme of the trunk region of the embryo in response to the inductive influence of the backward growing primary nephric duct. In the same manner, the convoluted tubules of the metanephros form as a response of posterior nephrogenous material to the invading ureteric bud. The interaction between these two tissues is mutual. On the one hand, the nephrogenic mesenchyme is induced by the epithelium of the ureteric bud to form convoluted tubules and glomeruli. On the other hand, the ureteric bud is stimulated by the nephric mesenchyme to branch and form collecting tubules.

Grobstein has analyzed the interaction between the nephric epithelium and nephric mesenchyme in the development of the laboratory mouse. First he placed metanephric primordia of 11-day mice briefly in calcium-free solutions containing a small amount of trypsin. The treatment enabled him to separate the epithelium from the surrounding mesenchyme.

Next he cultivated the two tissues on plasma clots separately. He found, as he had found in his previous experiments on the salivary gland, that tissues thus isolated do not continue to differentiate. But when he placed the epithelium and mesenchyme together on the same plasma clot, they joined, interacted, and in due course produced the structures of a kidney.

Are these capacities to act and react specific? Will nephric epithelium respond only to nephric mesenchyme, or will it also respond to other mesenchymes? Grobstein has performed experiments which have answered these questions. He finds that nephric epithelium will not respond to foreign mesenchyme such as salivary mesenchyme. But his results are different in the case of the mesenchyme. Nephric mesenchyme reacts to salivary epithelium by forming tubules. Strangely enough, it responds also to neural tissue from the dorsal side of the spinal cord! It does not respond to tissue from the ventral side of the cord nor to any other neural tissues which he tested.

What is the nature of the inductive agent which is transmitted from the epithelium to the mesenchyme and calls forth its differentiation? To seek an answer to this question, Grobstein placed the reacting tissues, mesenchyme and dorsal neural tissue, on opposite sides of a thin porous membrane, namely, a millipore filter. He found that induction can take place through a filter only 20 μ thick, with pores only 0.4 μ in diameter. The tubules differentiate even though there is no cytoplasmic contact. This and other experiments have led him to conclude that the inducing agent is a mobile macromolecule normally present in the immediate vicinity of cells but separable from them.

REPRODUCTIVE ORGANS

The reproductive system develops relatively late in the history of the embryo. In fact, its full development cannot be said to have been completed until after puberty. Moreover, the reproductive system remains remarkably labile and subject in varying degrees to the influence of the sex hormones.

The Indifferent Stage

At first male and female embryos cannot be distinguished (Fig. 18-7). This stage is known as the indifferent stage of development. The gonads may become either ovaries or testes. If the gonad is to be an ovary, an outer layer of cells, the cortex, progresses, and an inner core, the medulla, retrogresses. If it is to be a testis, the opposite takes place.

An indifferent embryo also has two sets of primordia for the reproductive ducts. If the embryo is to be a male, the primordia of the male ducts develop and those of the female ducts retrogress (Fig. 18-9). If it is to be a female, the opposite occurs (Fig. 18-10). An indifferent embryo possesses parts which are capable of becoming molded into the external genital organs of either a male or a female. Thus at an early stage an embryo is potentially both male and female; and an important problem of embryology is to discover how and why it becomes one or the other.

Gonads

The germ glands, or *gonads,* first appear as thickenings of the coelomic epithelium on the medial side of each mesonephros. These are the *germinal epithelia* (Fig. 18-4, *C*). They give rise to the outer layer, or *cortex,* of the

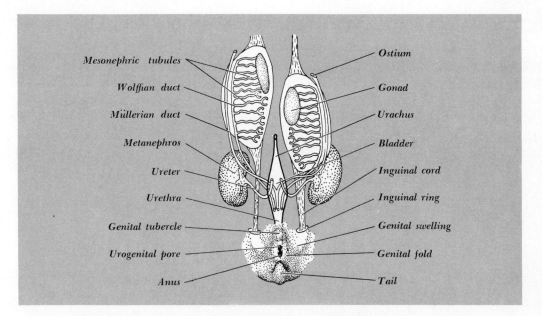

Mesonephric tubules

Wolffian duct

Müllerian duct

Metanephros

Ureter

Urethra

Genital tubercle

Urogenital pore

Anus

Ostium

Gonad

Urachus

Bladder

Inguinal cord

Inguinal ring

Genital swelling

Genital fold

Tail

FIGURE 18–7 Diagram of the indifferent stage in the development of the urogenital system. In ventral view. The right side (embryo's left) represents a somewhat later stage of development than the left. The mesonephros has begun to glide posteriorly across the ventral surface of the metanephros.

gonad. Soon each bulges into the body cavity as a longitudinal ridge, the *genital ridge* (Fig. 18–8, *A* and *B*). The cores of the ridges are mesenchyme. Then cordlike clumps of cells, *medullary cords,* from the germinal epithelium push into the mesenchymal stroma, and together the cords and stroma become the *medulla* of the gonad. Some of the cells of the germinal epithelium (cortex) and the medullary cords are larger and rounder; they also stain differently than the other cells. These are the so-called *primordial germ cells.*

There has been much dispute as to the origin of the primordial germ cells. The bulk of the evidence indicates that they are not derived from the initial germinal epithelium but migrate

into it by way of blood or blood vessels from some other part of the embryo. A crescent-shaped area of extra-embryonic blastoderm in front of the head of a presomite chick contains cells which have been interpreted as primordial germ cells. If this area is destroyed by cautery or surgery without damage to the rest of the embryo, a genital ridge may form which lacks germ cells, or it may not form at all.

In tailed amphibians, cells which appear to be primordial germ cells have been recognized at the late gastrula or early neurula stage. They are located in the prospective lateral mesoderm (hypomere) lateral and ventral to the blastopore. If this area is removed, the genital ridges are smaller than normal and lack

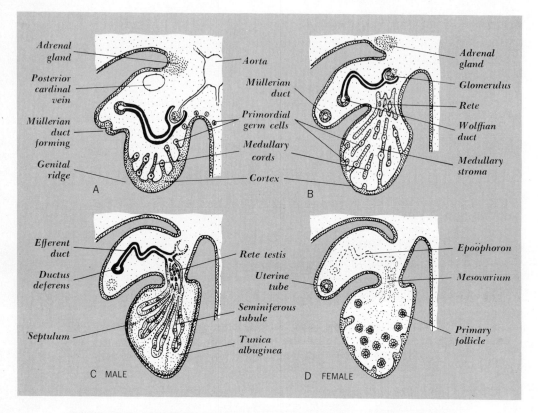

FIGURE 18–8 Stages in the development of the gonads. *A* to *C*. Stages in the formation of the testis. *A, B,* and *D*. Stages in the development of the ovary. In *C* and *D*
the parts which are regressing are indicated by broken lines. *(Adapted in part from
Arey.)*

germ cells. Moreover, if the mesoderm of this
area is transplanted to a late gastrula of another species (heteroplastic transplantation),
the genital ridge which then develops in the
host may have primordial germ cells of the
donor species. This definitely seems to show
that the cells of the genital ridge and the primordial germ cells are independent in origin.

There is, however, some uncertainty as to
whether the primordial germ cells give rise to
the functional germ cells of the mature testis

or ovary. That they play a role in favoring the
formation of the genital ridge is established,
but most of them regress. The cells which
finally ripen into germ cells may possibly originate from the germinal epithelium. This is
unlikely in view of the probable differentiation
of body tissues.

Formation of the testis   In the male, the medullary cords push into the mesenchymal core
from the germinal epithelium and become

seminiferous tubules (Fig. 18–8, *A* to *C*). Their walls consist of immature germ cells (spermatogonia) and accompanying nurse cells (cells of Sertoli). Ultimately, the spermatogonia transform into mature sperm. The nurse cells, as their name implies, serve to nourish and sustain bundles of sperm cells.

The germinal epithelium of the developing testis becomes a thin membrane like the rest of the peritoneum (Fig. 18–8, *C*). Between this membrane and the tubules, a connective tissue capsule forms, known as the *tunica albuginea*. The tubules acquire connections with a net of cords, the *rete testis,* at the base of the testis. The tubules and rete are at first solid, but at about the time of birth they hollow out. The rete then acquires connections with some of the adjacent mesonephric tubules. The latter are now known as efferent ductules (ductuli efferentes).

The rest of the mesenchyme of the medulla forms a connective-tissue framework within the testis. This includes septa (septula) between groups of seminiferous tubules. Some of the cells become interstitial cells which supposedly secrete the male sex hormones.

Formation of the ovary   In the female, the cells of the first set of sex cords regress. Later a second set forms and breaks into clusters of one or more immature germ cells (oogonia) surrounded by a layer of follicle cells (Fig. 18–8, *D*). (See Chapter 9, page 212 and Fig. 9–1.) The follicle cells multiply until the follicle wall is several cells thick. The follicle also becomes invested by a connective tissue capsule derived from the mesenchymal stroma, known as a theca.

At maturity, the oocytes of birds grow tremendously by the accumulation of layers of yolk (pp. 160–161). In mammals, the oocytes also grow, although in this case little or no yolk is laid down. At ovulation in both birds and mammals the follicle bursts, and the oocyte and accompanying follicle cells are swept into the body cavity.

There are several hundred thousand oogonia in a human ovary, yet only about four hundred mature during a woman's lifetime. The remainder degenerate and disappear. The same principle applies to other mammals. Indeed, there is some dispute as to whether any of the original oogonia remain to become mature ova.

Reproductive Ducts

The sperm ducts of the male are derived from the primary nephric ducts, known also as the Wolffian ducts (Fig. 18–9). When, at maturity, the ripe sperm cells leave the seminiferous tubules, they pass through the rete testis and then enter the adjacent efferent ductules (once mesonephric tubules) and pass down the Wolffian duct (originally the primary nephric duct). From this time on, the Wolffian duct is known as the *ductus deferens* (vas deferens).

In the males of some of the lower vertebrates, each Wolffian duct serves both as an excretory and as a reproductive duct. But more frequently a secondary nephric duct, or ureter, is present to take over at least a part of the function of carrying urine. In postnatal reptiles, birds, and mammals, the metanephros and ureter perform the entire excretory function.

The efferent ductules and adjacent portion of the ductus deferens become highly convoluted and constitute an appendage to the testis, known as the *epididymis*. The rest of the mesonephric tubules, since they no longer have a function, regress and become a vestige known as the *paradidymis* (Fig. 18–9).

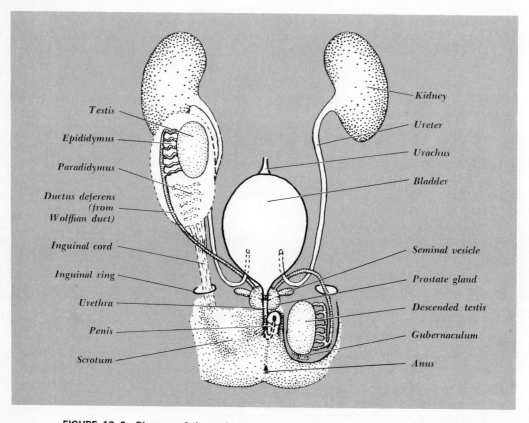

FIGURE 18–9 Diagram of the male reproductive organs. On the right the left testis is shown as having descended into the scrotal sac.

The oviducts of the female are new formations known as the *Müllerian ducts* (Figs. 18–6, *C*, and 18–10). In some lower vertebrates (shark), they arise by the longitudinal splitting of the Wolffian ducts. In most vertebrates, however, they appear late in development as the result of a longitudinal infolding of the epithelium of the mesonephros lateral to the Wolffian duct (Fig. 18–8, *A, B, D*). In any case, each Müllerian duct is open at its anterior end into the body cavity. The eggs which burst from the ovary at the time of ovulation are swept into this *ostium* of the oviduct. The borders of the ostium usually more or less enfold the ovary.

Posterior to the mesonephroi, the Müllerian ducts swing ventrally beneath the Wolffian ducts and grow posteriorly to join the urogenital sinus on its dorsal side (Fig. 18–2, *D*). In placental mammals, the posterior ends of the Müllerian ducts fuse together to form a single median duct (Fig. 18–7).

The derivatives of the Müllerian ducts are as follows (Fig. 18–10): The anterior portions become the oviducts proper, known in mammals as the *uterine* (fallopian) *tubes*. Farther back, the Müllerian ducts give rise to the *uterus*. In monotremes and some marsupials, this region of the Müllerian ducts remains separate so that there are two uteri, a condition known as *uterus duplex* (Fig. 18–11). In most mammals, however, the ducts are partially united in the midline so that either the uterus has two horns (*bicornuate uterus*) or is single but divided internally (*bipartite uterus*). In primates, the uterine portions of the Müllerian ducts are completely united, producing a *uterus simplex*. The posterior parts of the Mül-

lerian ducts, along with some contribution from the cloaca, unite to produce a single *vagina*—single at least in placental mammals.

External Genitalia

External genitalia are developed in those vertebrates in which fertilization is internal. They serve as a means for the transfer of sperm from the male to the female. The external genitals of the male and female are derived from the same embryonic source, and at the indifferent stage they cannot be distinguished. In mammals, the primordia include

FIGURE 18–10 Diagram of the female reproductive organs. On the right side the left ovary is shown as having partially descended.

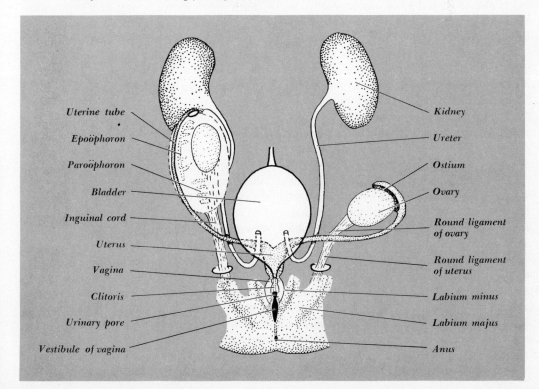

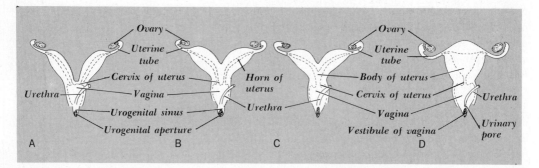

FIGURE 18–11 Types of mammal uteri. A. Duplex uterus as found in many marsupials. B. Bipartite uterus as found in rodents. C. Bicornuate uterus as found in carnivores
and ungulates. D. Simplex uterus as found in primates. (Adapted from several sources.)

a *genital tubercle* (phallus) on the ventral side
of the body just anterior to the urogenital opening, a pair of *genital folds,* one on each side
of the opening, and a pair of *genital swellings*
lateral to the folds (Fig. 18–7).

In the male the phallus grows large and elongates to form the *penis* (Fig. 18–9). A groove
on its ventral surface is bounded by the genital
folds. These close in beneath and form a tube,
thereby extending the urethra to the tip of the
penis. The genital folds terminate in a *prepuce,*
or foreskin, at the end of the penis. The genital
swellings become the *scrotal sacs.*

In the female, the phallus is smaller and is
known as the *clitoris* (Fig. 18–10). The genital
folds do not enfold and extend the urethra as
they do in the male. Instead, the urogenital
sinus remains open and receives the urinary
pore and the vestibule of the vagina. The genital folds and genital swellings become the
labia minora and *labia majora,* respectively.

Both the penis and the clitoris develop
cavernous bodies derived from mesenchyme
and capable of being distended by blood. Both
also end in sensitive tips, the *glans penis* and
the *glans clitoris.*

Descent of the Testes and Ovaries

Each mesonephros is part of a longitudinal
fold, the *nephric ridge,* which extends the full
length of the roof of the peritoneal cavity
lateral to the mesentery of the gut. Anteriorly
the ridges begin in the region of the pleural
canals and share in the formation of the pleuro-
peritoneal membranes (Fig. 15–6). They end
posteriorly by swinging to the midline and fus-
ing together beneath the gut (rectum).

Now, there are cords or fibers in the nephric
ridges which do not swing to the midline but
which instead continue backward along the
dorsal wall of the body cavity until they reach
the region of the genital swellings (future
scrotal sacs or labia majora). Here they anchor
in the posterior wall of the body cavity. They
are the so-called *inguinal cords* (Fig. 18–7).
As the embryo grows, they seem to shorten;
but mostly they fail to grow in length as the
embryo grows. At any rate, they drag the
mesonephroi posteriorly across the ventral
surface of the metanephroi until the mesonephroi come to lie posterior to the metanephroi (Figs. 18–9 and 18–10).

After a time, the developing muscular walls of the abdomen surround the posterior ends of the inguinal cords and produce pockets (vaginal processes) of peritoneum at the bottoms of which the cords are anchored. The openings through the muscle layers are known as *inguinal rings*. The passages lined by peritoneum are termed *inguinal canals*. In the male, the pockets penetrate all layers of the abdominal wall and invade the scrotal sac. In the female they are less well developed.

About the time of birth, the inguinal cords contract. In the male, each cord, now known as a *gubernaculum testis,* drags the testis and the accompanying epididymus from the abdominal cavity through the inguinal rings and canal into the scrotal sac (Fig. 18–9). As it descends, the testis drags behind it the ductus deferens, the spermatic artery and vein, and the accompanying nerves. Enclosed in folds of peritoneal origin, these structures constitute the *spermatic cord*.

In some mammals, the inguinal canals remain open so that the testes descend into the scrotal sac at the beginning of each breeding season and then return to the abdominal cavity at other times of the year. But in most mammals the inguinal canal closes after the descent of the testis. A failure to close is known as congenital hernia. A later rupture of the inguinal rings is termed secondary hernia.

A somewhat comparable, although more limited, descent of the ovaries takes place in the female (Fig. 18–10). The ovaries move downward, but normally they do not leave the abdominal cavity. Instead, the ovaries, uterine tubes, and uterus become intraperitoneal; that is, they become enfolded by peritoneum and acquire mesenteries by which they are suspended from the dorsal body wall. The mesenteries are known as the mesovarium, meso-salpinx, and broad ligament of the uterus, respectively. In the female, the inguinal cords remain as the round ligaments of the ovary and uterus. They extend from the ovary to the base of the uterine tube, and then to the inguinal ring which serves to anchor the uterus posteriorly and ventrally.

THE DETERMINATION OF SEX

What determines the sex of the new individual? A remote answer to this question is that the chromosomes of the zygotic nucleus decide whether the new individual is to be male or female. In the fruit fly, the males have one X chromosome and one Y chromosome; the females have two X chromosomes. Each X chromosome has a predominantly female-producing tendency. The rest of the chromosomes have an overall male-producing tendency. Now, the female tendency of one X chromosome is not sufficient to counteract the male tendencies of the other chromosomes. Hence, the result is a male. The female tendency of two X chromosomes, however, is stronger than the male tendencies of the rest of the chromosomes, so that the result is a female. For a long time this was thought to be the case in man also, but now it has been demonstrated that the dominant male-producing potency is in the Y chromosome. Hence an individual with the sex chromosome formula XO (one X chromosome, no Y chromosome) is a female, not a male, as in the fruit fly. This is an example of the risk of transferring conclusions from one species to another. (In birds, the relationship is the reverse: Embryos with two sex chromosomes in their nuclei become males, while those with one sex chromosome become females.)

These explanations are purely genetic. They do not give any real insight into the embryological mechanisms by which sex is controlled. One might be led to expect that the chromosomes of the primordial germ cells would decide whether the cells are to become sperms or eggs and that the germ cells in turn would determine the differentiation of the gonads in which they develop. But this is not the case. The somatic tissues of a gonad determine the differentiation of the germ cells. It is not vice versa. The mechanism is as follows: (We are now concerned with the determination of the primary sex character of the new individual, that is, as to whether the gonad is a future testis producing sperm cells or an ovary producing eggs. We are not yet considering the control of the secondary sex characteristics of the reproductive ducts and external genitalia.)

Each gonad at the indifferent stage possesses two "structural elements," a cortex and a medulla. Between the two a sort of antagonism exists, such that when one gains ascendency, the other regresses. In mammals, any factor, whether genetic or environmental, which favors the development of the medulla results in the gonad becoming a testis, and the individual normally develops as a male. Any factor which does not favor the medulla permits the cortex to develop and become an ovary. The individual is then a female as though by default. In birds it seems to be the opposite. In them the cortex is predominant, and any influence which favors the cortex results in the production of ovarian tissue. Any influence which does not favor the cortex permits the medulla to develop and become a testis.

Under normal circumstances the genes differentially favor the development of either the medulla or the cortex, and thus they are the critical factors which determine sex.

However, there are other factors besides the genes which may affect the development of the cortex versus the medulla. It has been found that raising the temperature of a developing frog's egg inhibits the cortex and favors the medulla. Indeed, in some frogs a rise in temperature may cause the ovaries of adult frogs to gradually transform into testes. Also, it has been found that lowering the temperature has the opposite effect: It favors the cortex and produces more females. Age may also have an effect. In the European frog, *Rana temporaria,* the cortex regresses late in life, with the result that the oldest frogs are all males. Half of them are genetic males, and half are genetic females turned neomales.

Sex hormones also may have an influence on the embryonic gonad. In amphibian eggs immersed in a solution containing male sex hormones, the cortex regresses whether the embryo is genetically a male or female. The opposite can occur if female sex hormones are used. No clearly comparable effects have been obtained in birds and mammals. But the facts suggest the likelihood that the antagonisms between the cortex and medulla are themselves hormonal in nature.

The control by the gonads of the secondary sex characteristics has been extensively studied and is better understood. Details will be found in textbooks of endocrinology. Briefly stated, sex hormones are produced by the interstitial cells of the ovary and testis. These not only control the development of the reproductive ducts and external genitalia, but they are required to maintain them in a functional state. It is an impressive fact that the secondary sex characteristics remain plastic throughout life. When gonads are removed, the body returns to a more juvenile state. Under treatment with appropriate hormones, the secondary sexual characters of an individual may be reversed.

In mammals, the male sex hormone produced by the testis (testosterone) is prepotent over the female hormones. It stimulates the development of the derivatives of the Wolffian ducts and the regression of those of the Müllerian ducts. It has comparable effects on the external genital organs. Many years ago Lillie described how in cattle, when twin calves of unlike sex develop side by side and their placental circulations become intermingled, the sex hormones of the male calf overpower the sex hormones of the female. The male calf develops normally, but the female calf becomes a false hermaphrodite of a type known as a freemartin. Fortunately such a situation is not known to occur in the case of human fraternal twins.

In birds, the female sex hormones of the ovary (estrogens) suppress the development of masculine characteristics.

SELECTED READINGS

Excretory Organs

Arey, L. B., 1965. Developmental Anatomy. W. B. Saunders Company, Philadelphia. 7th ed. Chap. 17, The urinary system.

Kerr, J. G., 1919. Textbook of Embryology. Macmillan & Co., Ltd., London. Vol. 2, Vertebrata with the Exception of Mammalia.

Goodrich, E. S., 1930. Studies on the Structure and Development of Vertebrates. The Macmillan Company, London. Reprinted by Dover Publications, Inc., New York. Chap. 13, Excretory organs and genital ducts.

Romer, A. S., 1962. The Vertebrate Body. W. B. Saunders Company, Philadelphia. 3d ed. Pp. 365–386.

Fraser, Elizabeth A., 1950. The development of the vertebrate excretory system. Biol. Rev., 25:159–187.

Burns, R. K., 1955. Vertebrate organogenesis: Urogenital system. In Willier, Weiss, and Hamburger (eds.), Analysis of Development. W. B. Saunders Company, Philadelphia. Pp. 462–469.

Torrey, T. W., 1965. Morphogenesis of the vertebrate kidney. In DeHaan and Ursprung (eds.), Organogenesis. Holt, Rinehart and Winston, Inc., New York. Pp. 559–579.

Experiments on Embryonic Kidneys

Grobstein, C., 1955. Inductive interactions in the development of the mouse metanephros. J. Exptl. Zool., 130:319–339.

———, and A. J. Dalton, 1957. Kidney tubule induction in mouse metanephric mesenchyme without cytoplasmic contact. J. Exptl. Zool., 135:57–74.

Reproductive Organs

Arey, L. B., 1965. Developmental Anatomy. W. B. Saunders Company, Philadelphia. 7th ed. Chap. 18, The genital system.

Hamilton, W. J., J. D. Boyd, and H. W. Mossman, 1962. Human Embryology. The Williams & Wilkins Company, Baltimore. 3d ed. Chap. 11, Urogenital system.

Burns, R. K., 1955. Vertebrate organogenesis: Urogenital system. In Willier, Weiss, and Hamburger (eds.), Analysis of Development. W. B. Saunders Company, Philadelphia. Pp. 470–491.

Balinsky, B. I., 1965. An Introduction to Embryology. W. B. Saunders Company, Philadelphia. 2d ed. Pp. 447–466.

The Determination of Sex

Gowen, J. W., 1961. Cytologic and genetic basis of sex. In Young and Corner (eds.), Sex and Internal Secretions. The Williams & Wilkins Company, Baltimore. 3d ed. Pp. 3–75.

Burns, R. K., 1961. Role of hormones in the differentiation of sex. In Young and Corner (eds.), Sex and Internal Secretions. The Williams & Wilkins Company, Baltimore. 3d ed. Pp. 76–158.

Gallien, L. G., 1965. Genetic control of sexual differentiation in vertebrates. In DeHaan and Ursprung (eds.), Organogenesis. Holt, Rinehart and Winston, Inc., New York. Pp. 583–610.

Jost, A., 1965. Gonadal hormones in the sex differentiation of the mammalian fetus. In DeHaan and Ursprung (eds.), Organogenesis. Holt, Rinehart and Winston, Inc., New York. Pp. 611–628. (Also see other papers in this same volume by Price and Ortiz, Chieffi, Wells, Hamilton and Teng, and Charniaux-Cotton.)

EPILOGUE

Those who read this Survey of Embryology will no doubt be impressed with the extent and variety of what is known concerning development. Actually, we understand very little. We see germ cells maturing, sperm cells swimming and being received by eggs, blastomeres dividing, and processes of folding, molding, and differentiation taking place. But how little we know of what goes on behind the scenes! Our knowledge consists more in our increased ability to ask questions than in our ability to answer them.

Today we are excited—and rightly so—by what we have learned concerning genes: their method of replication and the manner in which they determine individual and species characteristics. We are impressed also with the astounding stability of heredity, which has been reproducing the same fundamental biochemical mechanisms for a billion years and the same basic pattern of vertebrate structure for a half-billion years or more. But we still have not answered the most basic question of all: What "turns genes on and off" during development? What tells each gene when and where to act? It seems clear that the "physical basis of heredity" is the genes plus the epigenetic factors which control them.

Present-day embryologists are concerned with cells and the specific macromolecules which largely compose them. But we must not overlook the fact that cells and macromolecules are in the service of the organism as a whole. It is the whole organism which endures in nature. Cells are the units in most of the ordinary processes of living and are therefore of primary interest to biochemists, physiologists, and those who speculate concerning the origin and nature of life. Cells are also the units in the transmission of the genes by the processes of mitosis, meiosis, and fertilization. As such, they are of basic importance in the field of genetics. But organisms as wholes, not individual cells, are of overriding significance to morphologists, systematists, ecologists, and students of the history of life on the earth. The organism gives the cells and the macromolecules their meaning and their reason for being.

At the molecular level, structure arises when molecules fit themselves together as they do when forming a crystal. The form of a virus particle is probably the result of the self-assembly of prefabricated macromolecules, as Edgar and others have shown. The same is possibly true of cell organelles such as cellular membranes and the apparatus of cilia. But organisms belong to a higher and different level of organization, their morphogenesis is by differentiation rather than by integration.

The central problem of development is differentiation. Somewhat arbitrarily we may divide differentiation into *intracellular differentiation* such as we find in protozoa, in the ontogeny of sperm cells, and even in uncleaved eggs, and *intercellular differentiation* as seen in the development of many-celled plants and animals. Perhaps it would be better if we referred to the former as noncellular differentiation and to the latter as supracellular differentiation. Note, however, that in both cases it is regions, either of parts of cells or of groups of cells, not usually single cells, which undergo differentiation. In some cases, for example, in the tunicate *Styela,* differentiations take place in the uncleaved egg which do not occur until the many-celled blastula stage in vertebrates.

The transcendence of the organism over its cells is seen in the behavior of embryonic cells in tissue culture. When cells of unlike origin are dissociated and intermingled, they temporarily adhere; but then they segregate and take positions with respect to each other which are comparable to the positions they occupy in an embryo. In doing this they are acting as though they were still parts of an organic whole. We do not easily discern the organismic influences which are at work when development is proceeding smoothly; but when we interfere with the normal course, the influences at work stand out more clearly. We may be certain that epigenetic factors continue to be present, holding differentiated cells in their places, maintaining their character, and preventing them from wandering astray.

The organismic view is the opposite of integration. It is epigenesis. An organism is not fabricated by the assembling of already specialized parts. A phenomenon of this kind would defy analysis, for it would credit parts with capacities without explaining how they obtained these capacities. Rather, the organism precedes its parts. It is in relation to the organism that the parts acquire their properties and become different.

Apparently development is always forward. As such, it is an historic process. At least this seems to be true with respect to the cytoplasm. Regions which at first are competent to form any part of the whole become more and more restricted as to their fates. This should not be thought of as a negative process, that is, as a progressive loss of hypothetic "potencies." It is a positive process in which new enzyme systems are built up and new capacities are acquired. Sometimes it may appear that development turns backward—dedifferentiation has been described. But it usually turns out that old cells are resorbed and new cells take their place. Whether or not nuclei as well as cytoplasm undergo irreversible changes during development is still a debatable question.

Some degree of plasticity certainly remains throughout life. Although an adult mammal cannot regenerate a lost appendage in the manner of a larval salamander, yet it can heal a wound. Developmental processes never wholly cease as long as life endures.

SOURCES OF EMBRYOLOGICAL INFORMATION

The following books and journals will be of help to the student who seeks further information concerning development. The books which provide useful bibliographies are indicated.

Books which Combine Descriptive and Experimental Embryology

Balinsky, B. I., 1965. An Introduction to Embryology. W. B. Saunders Company, Philadelphia. 2d ed. A stimulating text which combines descriptive embryology with numerous insights into the results of experiment. Contains chapters on metamorphosis, regeneration, and asexual reproduction. Includes a valuable bibliography of nearly 800 titles.

Barth, L. G., 1953. Embryology. Holt, Rinehart and Winston, Inc., and Mentzer, Bush and Company, New York. Rev. ed. Based on lectures on embryology, with some emphasis on experimental analysis of development. No bibliography.

Witschi, E., 1956. Development of Vertebrates. W. B. Saunders Company, Philadelphia. A critical treatment which follows the story of the development of amphibian, chick, and mammal to late stages. Contains a useful bibliography.

Waddington, C. H., 1956. Principles of Embryology. The Macmillan Company, New York. An interpretation of the processes of development of invertebrates as well as vertebrates.

Accounts of Descriptive Vertebrate Embryology

Huettner, A. F., 1949. Fundamentals of Comparative Embryology of the Vertebrates. The Macmillan Company, New York. Rev. ed. A basic descriptive text which takes up in sequence the early development of the frog, chick, and mammal. Notable for its original illustrations.

McEwen, R. S., 1957. Vertebrate Embryology. Holt, Rinehart and Winston, Inc., New York. 4th ed. Mainly a descriptive text. Amphioxus, frog, fish, chick, and mammal are treated in succession. Useful bibliographies.

Nelsen, O. E., 1953. Comparative Embryology of Vertebrates. McGraw-Hill Book Company, New York. A topical treatment illustrated with abundant figures. The comparative approach is maintained throughout. The bibliographies include the older literature as well as the new.

Patten, B. M., 1964. Foundations of Embryology. McGraw-Hill Book Company, New York. 2d ed. A standard text, emphasizing the chick and the mammal (pig and man). Excellently illustrated. The useful bibliography is organized topically.

Rugh, R. 1964. Vertebrate Embryology: The Dynamics of Development. Harcourt, Brace, & World, Inc., New York. Nine chapters, including descriptive accounts of the development of the frog, chick, mouse, pig, and man. A glossary of nearly 1500 embryological terms.

Shumway, W., and F. B. Adamstone, 1954. Introduction to Vertebrate Embryology. John Wiley & Sons, Inc., New York. 5th ed. A brief but clear statement of the more essential facts of vertebrate development. Appendices include an introduction to embryological technique and a glossary.

Accounts Dealing with Special Groups

Rugh, R., 1951. The Frog: Its Reproduction and Development. McGraw-Hill Book Company, New York. An orderly account of the development of the frog, intended as a background for further studies in embryology and as a basis for work in experimental embryology. Contains a glossary of embryological terms.

Patten, B. M., 1951. Embryology of the Chick. McGraw-Hill Book Company, New York. 4th ed. A concise account, addressed to students beginning the study of embryology. Clear figures. Bibliography arranged by topics.

Hamilton, H. L., 1952. Lillie's Development of the Chick. Holt, Rinehart and Winston, Inc., New York. 3d ed. The classic text on the chick embryo. The 3d edition includes plates illustrating normal stages in the development of the chick. A bibliography of select references on chick development.

Romanoff, A., 1959. The Avian Embryo. The Macmillan Company, New York. A comprehensive reference to the development of bird embryos.

Patten, B. M., 1948. Embryology of the Pig. McGraw-Hill Book Company, New York. 3d ed. A clear statement of mammalian development for the student. The pig is chosen because of its availability for laboratory study. A useful bibliography.

Human Embryologies

Arey, L. B., 1965. Developmental Anatomy. W. B. Saunders Company, Philadelphia. 7th ed. A valuable reference. Long the standard textbook of human embryology. An excellent laboratory manual for the study of the chick and pig is included as the final division of the book.

Hamilton, W. J., J. D. Boyd, and H. W. Mossman, 1962. Human Embryology, The Williams & Wilkins Company, Baltimore. 3d ed. A detailed and well-illustrated account of human development. The last chapter concerns comparative embryology.

Patten, B. M., 1953. Human Embryology. McGraw-Hill Book Company, New York. 2d ed. Clearly written and well illustrated.

Cytologies and Histologies

Knowledge of the physics, chemistry, and structure of living cells is needed if one is to understand the spirit and concerns of present day embryologists. It is not always clear, however, as to how much of this information has to do with development.

Brachet, J. (ed.), 1959–1961. The Cell: Biochemistry, Physiology, Morphology. Academic Press Inc., New York.
Vol. 1. Methods and problems of cell biology.
Vol. 2. Cells and their component parts.
Vol. 3. Mitosis and meiosis.
Vols. 4 and 5. Specialized cells.

DeRobertis, E. D. P., W. W. Nowinski, and F. A. Saez, 1965. Cell Biology, W. B. Saunders Company, Philadelphia. 4th ed.

Giesy, Arthur, 1962. Cell Physiology. W. B. Saunders Company, Philadelphia. 2d ed.

Bloom, W., and D. W. Fawcett, 1962. A Textbook of Histology. W. B. Saunders Company, Philadelphia. 8th ed.

Greep, R. O. (ed.), 1966. Histology. McGraw-Hill Book Company, New York. 2d ed. The contributions of 22 authors cover the structure, physiology, and development of cells and tissues.

Porter, K., and Mary Bonneville, 1964. Introduction to the Fine Structure of Cells and Tissues. 2d ed. Lea & Febiger, Philadelphia. An atlas of electron micrographs.

Fawcett, D. W., 1966. The Cell: Its Organelles and its Inclusions. W. B. Saunders Company, Philadelphia. A splendid atlas of electron micrographs together with descriptive legends.

Monographs on Experimental Embryology

The embryological texts of Balinsky, Barth, Waddington, and Witschi, which are listed above, incorporate a great deal of experimental material along with the descriptive. Arey has a brief chapter devoted to experimental embryology. The following books are concerned with the analysis of development. They are listed chronologically, since each volume represents an epoch in the history of the subject.

Wilson, E. B., 1925. The Cell in Development and Heredity. The Macmillan Company, New York. 3d ed. The classic monograph in its field. A masterful account of earlier work on the cell.

Morgan, T. H., 1927. Experimental Embryology. Columbia University Press, New York. Summarizes and interprets the earlier experiments on eggs and embryos. Much of the work was done on invertebrate materials.

Huxley, J. S., and G. R. deBeer, 1934. The Elements of Experimental Embryology. Cambridge University Press, New York. A clear presentation and broad interpretation of the processes of development.

Spemann, H., 1938. Embryonic Development and Induction. Yale University Press, New Haven, Conn. Based on lectures at Yale University which summarized the analysis of development by means of surgical techniques.

Weiss, Paul, 1939. Principles of Development. Holt, Rinehart and Winston, Inc., New York. An attempt to organize and interpret the results of experiments on embryos. The last section of the book deals with neurogenesis.

Child, C. M., 1941. Patterns and Problems of Development. The University of Chicago Press, Chicago. A comprehensive restatement of the concept of gradients as applied to the problems of development and regeneration.

Brachet, J., 1950. Chemical Embryology. Interscience Publishers, New York. A classic in its field. Vast progress has been made since this was written, but this has not destroyed its usefulness.

Willier, B. H., P. A. Weiss, and V. Hamburger (eds.), 1955. Analysis of Development. W. B. Saunders Company, Philadelphia. A collection of authoritative articles by 28 leaders in embryological research.

McElroy, W. D., and B. Glass (eds.), 1958. A Symposium on the Chemical Basis of Development. The Johns Hopkins Press, Baltimore. Contributions by 43 experimental embryologists on the basic problems of cellular and molecular embryology.

Brachet, J., 1960. The Biochemistry of Development. Pergamon Press Inc., New York. Summarizes ten years of progress in the biochemistry of development.

DeHaan, R. L., and H. Ursprung (eds.), 1965. Organogenesis. Holt, Rinehart and Winston, Inc., New York. Thirty-one reviews of experimental work on vertebrate tissues and organ systems by experts in their several fields. Thorough and up-to-date bibliographies.

Weber, R. (ed.), 1965. The Biochemistry of Animal Development. Academic Press Inc., New York. Vol. I. (Vol. II to follow.)

Bell, E. (ed.), 1965. Molecular and Cellular Aspects of Development. Harper & Row, Publishers, Incorporated, New York. Contains the texts of 45 important papers of recent date, together with editorial introductions.

Flickinger, R. A., 1966. Developmental Biology. Wm. C. Brown Company, Dubuque, Iowa. Each of its sixteen chapters is a reprint of a current paper concerned with development.

Supplementary Texts in Paperback Edition

Several publishing houses are now providing up-to-date reading in the form of paperbound supplementary textbooks. These are written by experts in their several fields and are a welcome addition to the literature available to the beginning student.

Foundations of Modern Biology Series. Prentice-Hall, Inc. Englewood Cliffs, N.J.

White, E. H., 1965. Chemical Background for the Biological Sciences.

Swanson, C. P., 1965. The Cell. 2d ed.

McElroy, W. D., 1965. Cell Physiology and Biochemistry.

Sussman, M., 1964. Growth and Development.

Hartman, P. E., and S. R. Suskind, 1965. Gene Action. (In Genetics Series.)

Modern Biology Series. Holt, Rinehart and Winston, Inc., New York.

Loewy, A., and P. Siekevitz, 1963. Cell Structure and Function.

Levine, R. P., 1962. Genetics.

Ebert, J. D., 1965. Interacting Systems in Development.

Principles of Biology Series. Addison-Wesley Publishing Company, Inc., Reading, Mass.

Barth, Lucena J., 1964. Development: Selected Topics.

Selected Topics in Modern Biology. Reinhold Publishing Corporation, New York.

Spratt, N. J., Jr., 1964. Introduction to Cell Differentiation.

Many valuable articles have appeared from time to time in Scientific American.

Journals which Publish Original Papers in Embryology

Numerous journals publish original papers reporting embryological research. Among them are the following. The serious student will do well to thumb through current numbers to see what is being done in the laboratories.

Archiv für Entwicklungs-mechanik. The pioneer journal, established by Wilhelm Roux in 1888.

Biological Bulletin. Woods Hole Marine Biological Association.

Developmental Biology. Academic Press Inc., New York. Although recently established, this has become a principal outlet for embryological research.

Journal of Embryology and Experimental Morphology. Clarenden Press, Oxford. Contains many of the papers of British and French embryologists.

Journal of Experimental Zoology. Wistar Institute, Philadelphia. Long the favored medium for the publication of experimental work by American biologists. The Wistar Institute also publishes the Journal of Morphology, American Journal of Anatomy, Anatomical Record, and the Journal of Comparative Neurology,

all of which have reports on embryological studies.

The following cytological journals have many embryological papers:

Experimental Cell Research. Academic Press Inc., New York.

Journal of Cell Biology. Rockefeller University Press, New York. Formerly the Journal of Biophysics, Biochemistry, and Cytology.

Journal of Molecular Biology. Academic Press, London.

Symposia and Reviews of Developmental Biology

The progress now being made in experimental embryology and other aspects of developmental biology is rapid, and the literature has become voluminous. Current work is well summarized in the proceedings of societies of biologists and in volumes of reviews.

Symposia of the Society for the Study of Development and Growth, now known as the Society for Developmental Biology:

Rudnick, Dorothea (ed.), 1956. Cellular Mechanisms in Differentiation and Growth. (14th symposium.) Princeton University Press, Princeton, N.J.

————, 1957. Rhythmic and Synthetic Processes of Growth. (15th symposium.) Princeton University Press, Princeton, N.J.

————, 1959. Developmental Cytology. (16th symposium.) The Ronald Press Company, New York.

————, 1959. Cell, Organism, and Milieu. (17th symposium.) The Ronald Press Company, New York.

————, 1960. Developing Cell Systems and their Control. (18th symposium.) The Ronald Press Company, New York.

————, 1961. Molecular and Cellular Structure. (19th symposium.) The Ronald Press Company, New York.

———, 1962. Regeneration. (20th symposium.) The Ronald Press Company, New York.

Locke, M. (ed.), 1963. Cytodifferentiation and Macromolecular Synthesis. (21st symposium.) Academic Press Inc., New York.

———, 1964. Cellular Membranes in Development. (22d symposium.) Academic Press Inc., New York.

———, 1965. The Role of Chromosomes in Development. (23d symposium.) Academic Press Inc., New York.

———, 1966. Reproduction: Molecular, Subcellular, and Cellular. (24th symposium.) Academic Press Inc., New York.

———, 1967. Major Problems in Developmental Biology. (25th symposium.) Academic Press Inc., New York.

———, 1968. Control Mechanisms in Developmental Processes. (26th symposium.) **Academic Press Inc., New York.**

Reviews of progress in developmental biology:
Abercrombie, M., and J. Brachet (eds.), 1961–**1967. Advances in Morphogenesis. Academic Press Inc., New York. Vols. 1–6. A volume of reviews is published each year.**

Moscona, A. A., and A. Monroy (eds.), 1966–**1967. Current Topics in Developmental Biology. Academic Press Inc., New York. Vols. 1–2. A volume will appear each year interpreting molecular, cellular, and genetic advances in the field of development.**

Bourne, G. H., and J. F. Danielli (eds.), 1952–1966. International Reviews of Cytology. Academic Press Inc., New York. Vols. 1–20. One or two volumes appear each year. Many of the articles are embryological.

INDEX

Page numbers in *italics* refer to illustrations.